Joseph Fourier

The Analytical Theory of Heat

Joseph Fourier

The Analytical Theory of Heat

ISBN/EAN: 9783743308831

Manufactured in Europe, USA, Canada, Australia, Japa

Cover: Foto ©berggeist007 / pixelio.de

Manufactured and distributed by brebook publishing software (www.brebook.com)

Joseph Fourier

The Analytical Theory of Heat

THE
ANALYTICAL THEORY OF HEAT

BY

JOSEPH FOURIER.

TRANSLATED, WITH NOTES,

BY

ALEXANDER FREEMAN, M.A.,
FELLOW OF ST JOHN'S COLLEGE, CAMBRIDGE.

EDITED FOR THE SYNDICS OF THE UNIVERSITY PRESS.

PREFACE.

In preparing this version in English of Fourier's celebrated treatise on Heat, the translator has followed faithfully the French original. He has, however, appended brief foot-notes, in which will be found references to other writings of Fourier and modern authors on the subject: these are distinguished by the initials A. F. The notes marked R. L. E. are taken from pencil memoranda on the margin of a copy of the work that formerly belonged to the late Robert Leslie Ellis, Fellow of Trinity College, and is now in the possession of St John's College. It was the translator's hope to have been able to prefix to this treatise a Memoir of Fourier's life with some account of his writings; unforeseen circumstances have however prevented its completion in time to appear with the present work.

TABLE

OF

CONTENTS OF THE WORK[1].

CHAPTER I.

Introduction.

SECTION I.

STATEMENT OF THE OBJECT OF THE WORK.

[1] Each paragraph of the Table indicates the matter treated of in the articles indicated at the left of that paragraph. The first of these articles begins at the page marked on the right.

SECTION II.

General Notions and preliminary Definitions.

SECTION III.

Principle of the Communication of Heat.

SECTION IV.

Of the Uniform and Linear Movement of Heat.

SECTION V.

Law of the Permanent Temperatures in a Prism of Small Thickness.

SECTION VI.

The Heating of Closed Spaces.

SECTION VII.

OF THE UNIFORM MOVEMENT OF HEAT IN THREE DIMENSIONS.

SECTION VIII.

MEASURE OF THE MOVEMENT OF HEAT AT A GIVEN POINT OF A GIVEN SOLID.

CHAPTER II.

Equation of the Movement of Heat.

SECTION I.

EQUATION OF THE VARIED MOVEMENT OF HEAT IN A RING.

SECTION II.

Equation of the Varied Movement of Heat in a Solid Sphere.

SECTION III.

Equation of the Varied Movement of Heat in a Solid Cylinder.

SECTION IV.

Equations of the Varied Movement of Heat in a Solid Prism of Infinite Length.

SECTION V.

Equations of the Varied Movement of Heat in a Solid Cube.

SECTION VI.

General Equation of the Propagation of Heat in the Interior of Solids.

SECTION VII.

General Equation Relative to the Surface.

SECTION VIII.

Application of the General Equations.

SECTION IX.

General Remarks.

CHAPTER III.

Propagation of Heat in an infinite rectangular solid.

SECTION I.

Statement of the Problem.

SECTION II.

First Example of the use of Trigonometric Series in the Theory of Heat.

SECTION III.

Remarks on these Series.

SECTION IV.

General Solution.

SECTION V.

FINITE EXPRESSION OF THE RESULT OF THE SOLUTION.

SECTION VI.

DEVELOPMENT OF AN ARBITRARY FUNCTION IN TRIGONOMETRIC SERIES.

SECTION VII.

Application to the actual Problem.

CHAPTER IV.

Of the linear and varied Movement of Heat in a ring.

SECTION I.

GENERAL SOLUTION OF THE PROBLEM.

SECTION II.

OF THE COMMUNICATION OF HEAT BETWEEN SEPARATE MASSES.

CHAPTER V.

Of the Propagation of Heat in a solid sphere.

SECTION I.

GENERAL SOLUTION.

SECTION II.

DIFFERENT REMARKS ON THIS SOLUTION.

CHAPTER VI.

Of the Movement of Heat in a solid cylinder.

CHAPTER VII.

Propagation of Heat in a rectangular prism.

CHAPTER VIII.

Of the Movement of Heat in a solid cube.

CHAPTER IX.

Of the Diffusion of Heat.

SECTION I.

OF THE FREE MOVEMENT OF HEAT IN AN INFINITE LINE.

SECTION II.

OF THE FREE MOVEMENT OF HEAT IN AN INFINITE SOLID.

SECTION III.

THE HIGHEST TEMPERATURES IN AN INFINITE SOLID.

SECTION IV.

Comparison of the Integrals.

ERRATA.

Page 9, line 28, *for* III. *read* IV.

Pages 54, 55, for k read K.

Page 189, line 2, The equation should be denoted (A).

Page 205, last line but one, for x read X.

Page 298, line 18, for $\frac{du}{dr}$ read $\frac{du}{dx}$.

Page 299, line 16, for *of* read *in*.

" " last line, *read*

$$\int_0^\pi du\, \phi\,(t \sin u) = \pi\phi + tS_1\phi' + \frac{t^2}{\lfloor 2} S_2\phi'' + \&c.$$

Page 300, line 3, *for* A_2, A_4, A_6, *read* πA_2, πA_4, πA_6.

Page 407, line 12, for $d\phi$ read dp.

Page 432, line 13, read $(x - a)$.

Corrections to the Edition of *Fourier's Analytical Theory of Heat*, by A. Freeman, M.A., Cambridge, 1878.

PAGE	LINE	ERROR	CORRECTION	
9	28	III.	IV.	
19	10	*et passim.* Conductibility	Conductivity	
—	14	ratio of their capacities	inverse ratio of their capacities	
26	25	solids and liquids increase in volume	*add*, in most cases	
27	27	dissolve	melt	
28	2	oecupy	occupy	
54 55		*through* § 71, k	K	
57	4	right	*add*, or left (pro re nata)	
58		§§ 73—80, k	K	
66	6	as	*omit*, as	
—	11	$m-n=a$	$m=a$	
67	31	ω	π	
72	4	$j\left(1+\frac{He}{K}\right)$	$j\left(1+\frac{He}{K}\right)+1.$	
—	13	*in denominator*	*add*, 1	
—	14	j	$j+1$	
90	16	v^2	v_2	
146	9	$\cos(2m-3x)$	$\cos(2m-3)x$	
—	20	$-\frac{1}{2^2m^3}$	$+\frac{1}{2^3m^3}$	
152	17	$+$	$-$	
—	26	$\pm K$	$+K$	
—	27	$+\frac{K}{2^2m^2}$	$-\frac{K}{2^2m^2}$	*if K is defined as on line 14*
—	29	$+\frac{K}{2^3m^3}$	$-\frac{K}{2^3m^3}$	
156	11	$e^{-5x}\cos 5y$	$\frac{1}{5}e^{-5x}\cos 5y$	
162	21	$\phi(x, y)$	$f(x, y)$	
164	22	1	0	

PAGE	LINE	ERROR	CORRECTION
169	14	B	$-B$
—	14	D	$-D$
172	2—4	the numerals	should be squared
—	10	6^2	5^2
—	26	A_2	A_1
174	30	$1^2 2^2 3^2 4^2 5^2$	1, 2, 3, 4, 5
180	last	*remove* (A)	*to end of line* 11
181	23	216	215.
182	9	$-\frac{1}{2}$	$-\frac{1}{1^2}$
184	18	$\frac{s}{n}$	s
189	2	$\frac{1}{2}\pi$	$\frac{1}{4}\pi$
—	—	*denote the equation by*	(A), *for sake of note* p. 191
194	18	$\frac{1}{2}\pi$	π
195	12	$\frac{2}{i^3}$ when i is even	$(-1)^n\frac{2}{i^3}$ when i is of form $2n+1$
—	13	$-\frac{\pi}{i^2}$	$+\frac{\pi}{i^2}$
—	14	*within the brackets*	$\frac{\cos x}{1^3}-\frac{\cos 3x}{3^3}+\frac{\cos 5x}{5^3}-\frac{\cos 7x}{7^3}+$ &c.
—	note	$\left(\frac{\pi}{2}\right)^2-x$	$\left(\frac{\pi}{2}\right)^2-x^2$
205	17	x	X
210	5	$2r$	r
211	15	*in value of* $2F(y, p)$	*insert* + before $e^{-(y-p\sqrt{-1})}$
212	15	proportianal	proportional
216	6, 7	F	f
—	16	$e^{-\frac{t}{r^2}}$	$e^{-k\frac{t}{r^2}}$
218	28	heat	temperature
—	29	*within the brackets*	*the signs of all terms* +
—	—	*instead of* M	*its value*, $\frac{b}{\pi}(1-e^{-2\pi})$
220	31	2π	$2\pi r$
—	36	$2\pi rM$	$2\pi rMCDS$
221		*through* § 245, a_0	b_0
226	17	$\alpha+(\alpha-\beta)\frac{\omega}{m}$	$\alpha-(\alpha-\beta)\frac{\omega}{m}$
227	23	$-\frac{1}{2}$, *twice*	$+\frac{1}{2}$, *twice*
228	13—17	k	1
229	18	$\beta+(a-\beta)\frac{\omega}{m}$	$\beta+(\alpha-\beta)\frac{\omega}{m}$
232	5	B	$-B$
—	6	$a_1 \sin mu$, and $\frac{a_1}{\sin u}\sin(m-1)u$	$\frac{a_1}{\sin u}\sin mu$, and $\frac{a_0}{\sin u}\sin(m-1)u$

CORRECTIONS.

PAGE	LINE	ERROR	CORRECTION
232	16	$q+2=\cos u$	$q+2=2\cos u$
239	2	a	α
270	20	$oi\omega i$	$ol\omega l$
271	3	$l-hX$	$1-hX$
284	4	0·006500	0·006502
286	14	$vd\left(\frac{4\pi x^3}{3}\right)$	$CDvd\left(\frac{4\pi x^3}{3}\right)$
—	last	$2n_i X -$ &c.	$n_i X -$ &c.
295	1	$\frac{d^{i+1}y}{d\theta^{i+2}}$	$\frac{d^{i+1}y}{d\theta^{i+1}}$
—	3	$\frac{d^{i+1}y}{d\theta^{i+1}}$	$\frac{d^{i+2}y}{d\theta^{i+2}}$
300	3	$A_2,\ A_4,\ A_6$	$\pi A_2,\ \pi A_4,\ \pi A_6$
304	3	$a_2 u_1$	$a_1 u_1$
—	18	$\frac{d^2u}{dx}$	$\frac{d^2u}{dx^2}$
—	22	$\int u \frac{d^2\sigma}{dx} dx$	$\int u \frac{d^2\sigma}{dx^2} dx$
307	18	$\int x\psi(\mu)\psi(\nu)\,dx$	$\int x\psi(\mu x)\psi(\nu x)\,dx$
309	6	$\frac{h^2X}{2^2\theta}$	$\frac{h^2X^2}{2^2\theta_1}$
311	13	$\sqrt{n^2}+p^2$	$\sqrt{n^2+p^2}$
—	15	$\frac{du}{dz}$	$\frac{dv}{dz}$
—	16	*the order of the equations*	*should be interchanged*
313	27	$n\tan\nu l$	$n\tan nl$
324	29, 30	*comma after bracket*	*dele comma*
325	25, 26	*do.*	*do.*
326	8	$YZ=\frac{dX}{dt}$	$YZ\frac{dX}{dt}$
—	13	$\frac{k}{K}$, *thrice*	$\frac{h}{K}$, *thrice*
335	27	after	at
336	6	$k\frac{d^2u}{dz^2}$	$k\frac{d^2u}{dx^2}$
—	8	$q_1,\ q_2,\ q_3$	$a_1,\ a_2,\ a_3$
—	9	$a_1,\ a_2,\ a_3$	$q_1,\ q_2,\ q_3$
337	20	$\cos qx$	$\cos q_j x$
339	26	Q	$\frac{\pi}{2}Q$
—	27	Q	$\frac{\pi}{2}Q$
341	20	$\dot{u}_1$	u
—	23	$a_1\sin\frac{\pi}{X}$	$a_1\sin\frac{\pi x}{X}$
345	2	$\cos qa$	$\sin qa$
355	15	n	n_1

PAGE	LINE	ERROR	CORRECTION
355	16	*a bracket*	*is missing*
356	5	*sign of last term*	*should be* +
—	12	*sign of first term*	*should be* +
—	25	c^{-q}	e^{-q^2}
359	5	$e^{-\frac{HLt}{CDS}}$	$ue^{-\frac{HLt}{CDS}}$
360	23	0·00	1·00
362	18	e^{-kt}	e^{-ht}
372	1	$\sqrt{\pi}$ *in the denominator*	*should be in the numerator*
392	2	$\frac{f}{2}$	$\frac{f}{\sqrt{2}}$
396	3	3 *in numerator*	$3^{\frac{3}{2}}$
407	12	$d\phi$	dp
—	28	equation	integration
432	13	$(x-a)$	$(x-\alpha)$

The Editor takes this opportunity of expressing his thanks to ROBERT E. BAYNES, Esq. and to WALTER G. WOOLCOMBE, Esq. for the majority of these corrections.

ADDENDUM. An article "*On the linear motion of heat, Part II.*", written by Sir WM THOMSON under the signature N.N., will be found in the *Cambridge Mathematical Journal*, Vol. III. pp. 206—211, and in Vol. I. of the Author's collected writings. It examines the conditions, subject to which an arbitrary distribution of heat in an infinite solid, bounded by a plane, may be supposed to have resulted, by conduction, in course of time, from some previous distribution. [A. F.]

MURSTON RECTORY, SITTINGBOURNE, KENT.
June 21*st*, 1888.

PRELIMINARY DISCOURSE.

PRIMARY causes are unknown to us; but are subject to simple and constant laws, which may be discovered by observation, the study of them being the object of natural philosophy.

Heat, like gravity, penetrates every substance of the universe, its rays occupy all parts of space. The object of our work is to set forth the mathematical laws which this element obeys. The theory of heat will hereafter form one of the most important branches of general physics.

The knowledge of rational mechanics, which the most ancient nations had been able to acquire, has not come down to us, and the history of this science, if we except the first theorems in harmony, is not traced up beyond the discoveries of Archimedes. This great geometer explained the mathematical principles of the equilibrium of solids and fluids. About eighteen centuries elapsed before Galileo, the originator of dynamical theories, discovered the laws of motion of heavy bodies. Within this new science Newton comprised the whole system of the universe. The successors of these philosophers have extended these theories, and given them an admirable perfection: they have taught us that the most diverse phenomena are subject to a small number of fundamental laws which are reproduced in all the acts of nature. It is recognised that the same principles regulate all the movements of the stars, their form, the inequalities of their courses, the equilibrium and the oscillations of the seas, the harmonic vibrations of air and sonorous bodies, the transmission of light, capillary actions, the undulations of fluids, in fine the most complex effects of all the natural forces, and thus has the thought of Newton been confirmed: *quod tam paucis tam multa præstet geometria gloriatur*[1].

But whatever may be the range of mechanical theories, they do not apply to the effects of heat. These make up a special order of phenomena, which cannot be explained by the principles of motion and equilibrium. We have for a long time been in possession of ingenious instruments adapted to measure many of these effects; valuable observations have been collected; but in this manner partial results only have become known, and not the mathematical demonstration of the laws which include them all.

I have deduced these laws from prolonged study and attentive comparison of the facts known up to this time: all these facts I have observed afresh in the course of several years with the most exact instruments that have hitherto been used.

To found the theory, it was in the first place necessary to distinguish and define with precision the elementary properties which determine the action of heat. I then perceived that all the phenomena which depend on this action resolve themselves into a very small number of general and simple facts; whereby every physical problem of this kind is brought back to an investigation of mathematical analysis. From these general facts I have concluded that to determine numerically the most varied movements of heat, it is sufficient to submit each substance to three fundamental observations. Different bodies in fact do not possess in the same degree the power to *contain* heat, *to receive or transmit it across their surfaces*, nor to *conduct* it through the interior of their masses. These are the three specific qualities which our theory clearly distinguishes and shews how to measure.

It is easy to judge how much these researches concern the physical sciences and civil economy, and what may be their influence on the progress of the arts which require the employment and distribution of heat. They have also a necessary connection with the system of the world, and their relations become known when we consider the grand phenomena which take place near the surface of the terrestrial globe.

[1] *Philosophiæ naturalis principia mathematica. Auctoris præfatio ad lectorem.* Ac gloriatur geometria quod tam paucis principiis aliunde petitis tam multa præstet. [A. F.]

In fact, the radiation of the sun in which this planet is incessantly plunged, penetrates the air, the earth, and the waters; its elements are divided, change in direction every way, and, penetrating the mass of the globe, would raise its mean temperature more and more, if the heat acquired were not exactly balanced by that which escapes in rays from all points of the surface and expands through the sky.

Different climates, unequally exposed to the action of solar heat, have, after an immense time, acquired the temperatures proper to their situation. This effect is modified by several accessory causes, such as elevation, the form of the ground, the neighbourhood and extent of continents and seas, the state of the surface, the direction of the winds.

The succession of day and night, the alternations of the seasons occasion in the solid earth periodic variations, which are repeated every day or every year: but these changes become less and less sensible as the point at which they are measured recedes from the surface. No diurnal variation can be detected at the depth of about three metres [ten feet]; and the annual variations cease to be appreciable at a depth much less than sixty metres. The temperature at great depths is then sensibly fixed at a given place: but it is not the same at all points of the same meridian; in general it rises as the equator is approached.

The heat which the sun has communicated to the terrestrial globe, and which has produced the diversity of climates, is now subject to a movement which has become uniform. It advances within the interior of the mass which it penetrates throughout, and at the same time recedes from the plane of the equator, and proceeds to lose itself across the polar regions.

In the higher regions of the atmosphere the air is very rare and transparent, and retains but a minute part of the heat of the solar rays: this is the cause of the excessive cold of elevated places. The lower layers, denser and more heated by the land and water, expand and rise up: they are cooled by the very fact of expansion. The great movements of the air, such as the trade winds which blow between the tropics, are not determined by the attractive forces of the moon and sun. The action of these celestial bodies produces scarcely perceptible oscillations in a fluid so rare and at so great a distance. It is the changes of temperature which periodically displace every part of the atmosphere.

The waters of the ocean are differently exposed at their surface to the rays of the sun, and the bottom of the basin which contains them is heated very unequally from the poles to the equator. These two causes, ever present, and combined with gravity and the centrifugal force, keep up vast movements in the interior of the seas. They displace and mingle all the parts, and produce those general and regular currents which navigators have noticed.

Radiant heat which escapes from the surface of all bodies, and traverses elastic media, or spaces void of air, has special laws, and occurs with widely varied phenomena. The physical explanation of many of these facts is already known; the mathematical theory which I have formed gives an exact measure of them. It consists, in a manner, in a new catoptrics which has its own theorems, and serves to determine by analysis all the effects of heat direct or reflected.

The enumeration of the chief objects of the theory sufficiently shews the nature of the questions which I have proposed to myself. What are the elementary properties which it is requisite to observe in each substance, and what are the experiments most suitable to determine them exactly? If the distribution of heat in solid matter is regulated by constant laws, what is the mathematical expression of those laws, and by what analysis may we derive from this expression the complete solution of the principal problems? Why do terrestrial temperatures cease to be variable at a depth so small with respect to the radius of the earth? Every inequality in the movement of this planet necessarily occasioning an oscillation of the solar heat beneath the surface, what relation is there between the duration of its period, and the depth at which the temperatures become constant?

What time must have elapsed before the climates could acquire the different temperatures which they now maintain; and what are the different causes which can now vary their mean heat? Why do not the annual changes alone in the distance of the sun from the earth, produce at the surface of the earth very considerable changes in the temperatures?

From what characteristic can we ascertain that the earth has not entirely lost its original heat; and what are the exact laws of the loss?

If, as several observations indicate, this fundamental heat is not wholly dissipated, it must be immense at great depths, and nevertheless it has no sensible influence at the present time on the mean temperature of the climates. The effects which are observed in them are due to the action of the solar rays. But independently of these two sources of heat, the one fundamental and primitive, proper to the terrestrial globe, the other due to the presence of the sun, is there not a more universal cause, which determines *the temperature of the heavens,* in that part of space which the solar system now occupies? Since the observed facts necessitate this cause, what are the consequences of an exact theory in this entirely new question; how shall we be able to determine that constant value of *the temperature of space,* and deduce from it the temperature which belongs to each planet?

To these questions must be added others which depend on the properties of radiant heat. The physical cause of the reflection of cold, that is to say the reflection of a lesser degree of heat, is very distinctly known; but what is the mathematical expression of this effect?

On what general principles do the atmospheric temperatures depend, whether the thermometer which measures them receives the solar rays directly, on a surface metallic or unpolished, or whether this instrument remains exposed, during the night, under a sky free from clouds, to contact with the air, to radiation from terrestrial bodies, and to that from the most distant and coldest parts of the atmosphere?

The intensity of the rays which escape from a point on the surface of any heated body varying with their inclination according to a law which experiments have indicated, is there not a necessary mathematical relation between this law and the general fact of the equilibrium of heat; and what is the physical cause of this inequality in intensity?

Lastly, when heat penetrates fluid masses, and determines in them internal movements by continual changes of the temperature and density of each molecule, can we still express, by differential equations, the laws of such a compound effect; and what is the resulting change in the general equations of hydrodynamics?

Such are the chief problems which I have solved, and which have never yet been submitted to calculation. If we consider further the manifold relations of this mathematical theory to civil uses and the technical arts, we shall recognize completely the extent of its applications. It is evident that it includes an entire series of distinct phenomena, and that the study of it cannot be omitted without losing a notable part of the science of nature.

The principles of the theory are derived, as are those of rational mechanics, from a very small number of primary facts, the causes of which are not considered by geometers, but which they admit as the results of common observations confirmed by all experiment.

The differential equations of the propagation of heat express the most general conditions, and reduce the physical questions to problems of pure analysis, and this is the proper object of theory. They are not less rigorously established than the general equations of equilibrium and motion. In order to make this comparison more perceptible, we have always preferred demonstrations analogous to those of the theorems which serve as the foundation of statics and dynamics. These equations still exist, but receive a different form, when they express the distribution of luminous heat in transparent bodies, or the movements which the changes of temperature and density occasion in the interior of fluids. The coefficients which they contain are subject to variations whose exact measure is not yet known; but in all the natural problems which it most concerns us to consider, the limits of temperature differ so little that we may omit the variations of these coefficients.

The equations of the movement of heat, like those which express the vibrations of sonorous bodies, or the ultimate oscillations of liquids, belong to one of the most recently discovered branches of analysis, which it is very important to perfect. After having established these differential equations their integrals must be obtained; this process consists in passing from a common expression to a particular solution subject to all the given conditions. This difficult investigation requires a special analysis founded on new theorems, whose object we could not in this place make known. The method which is derived from them leaves nothing vague and indeterminate in the solutions, it leads them up to the final numerical applications, a necessary condition of every investigation, without which we should only arrive at useless transformations.

The same theorems which have made known to us the equations of the movement of heat, apply directly to certain problems of general analysis and dynamics whose solution has for a long time been desired.

Profound study of nature is the most fertile source of mathematical discoveries. Not only has this study, in offering a determinate object to investigation, the advantage of excluding vague questions and calculations without issue; it is besides a sure method of forming analysis itself, and of discovering the elements which it concerns us to know, and which natural science ought always to preserve: these are the fundamental elements which are reproduced in all natural effects.

We see, for example, that the same expression whose abstract properties geometers had considered, and which in this respect belongs to general analysis, represents as well the motion of light in the atmosphere, as it determines the laws of diffusion of heat in solid matter, and enters into all the chief problems of the theory of probability.

The analytical equations, unknown to the ancient geometers, which Descartes was the first to introduce into the study of curves and surfaces, are not restricted to the properties of figures, and to those properties which are the object of rational mechanics; they extend to all general phenomena. There cannot be a language more universal and more simple, more free from errors and from obscurities, that is to say more worthy to express the invariable relations of natural things.

Considered from this point of view, mathematical analysis is as extensive as nature itself; it defines all perceptible relations, measures times, spaces, forces, temperatures; this difficult science is formed slowly, but it preserves every principle which it has once acquired; it grows and strengthens itself incessantly in the midst of the many variations and errors of the human mind.

Its chief attribute is clearness; it has no marks to express confused notions. It brings together phenomena the most diverse, and discovers the hidden analogies which unite them. If matter escapes us, as that of air and light, by its extreme tenuity, if bodies are placed far from us in the immensity of space, if man wishes to know the aspect of the heavens at successive epochs separated by a great number of centuries, if the actions of gravity and of heat are exerted in the interior of the earth at depths which will be always inaccessible, mathematical analysis can yet lay hold of the laws of these phenomena. It makes them present and measurable, and seems to be a faculty of the human mind destined to supplement the shortness of life and the imperfection of the senses; and what is still more remarkable, it follows the same course in the study of all phenomena; it interprets them by the same language, as if to attest the unity and simplicity of the plan of the universe, and to make still more evident that unchangeable order which presides over all natural causes.

The problems of the theory of heat present so many examples of the simple and constant dispositions which spring from the general laws of nature; and if the order which is established in these phenomena could be grasped by our senses, it would produce in us an impression comparable to the sensation of musical sound.

The forms of bodies are infinitely varied; the distribution of the heat which penetrates them seems to be arbitrary and confused; but all the inequalities are rapidly cancelled and disappear as time passes on. The progress of the phenomenon becomes more regular and simpler, remains finally subject to a definite law which is the same in all cases, and which bears no sensible impress of the initial arrangement.

All observation confirms these consequences. The analysis from which they are derived separates and expresses clearly, 1° the general conditions, that is to say those which spring from the natural properties of heat, 2° the effect, accidental but continued, of the form or state of the surfaces; 3° the effect, not permanent, of the primitive distribution.

In this work we have demonstrated all the principles of the theory of heat, and solved all the fundamental problems. They could have been explained more concisely by omitting the simpler problems, and presenting in the first instance the most general results; but we wished to shew the actual origin of the theory and its gradual progress. When this knowledge has been acquired and the principles thoroughly fixed, it is preferable to employ at once the most extended analytical methods, as we have done in the later investigations. This is also the course which we shall hereafter follow in the memoirs which will be added to this work, and which will form in some manner its complement[1]; and by this means we shall have reconciled, so far as it can depend on ourselves, the necessary development of principles with the precision which becomes the applications of analysis.

The subjects of these memoirs will be, the theory of radiant heat, the problem of the terrestrial temperatures, that of the temperature of dwellings, the comparison of theoretic results with those which we have observed in different experiments, lastly the demonstrations of the differential equations of the movement of heat in fluids.

The work which we now publish has been written a long time since; different circumstances have delayed and often interrupted the printing of it. In this interval, science has been enriched by important observations; the principles of our analysis, which had not at first been grasped, have become better known; the results which we had deduced from them have been discussed and confirmed. We ourselves have applied these principles to new problems, and have changed the form of some of the proofs. The delays of publication will have contributed to make the work clearer and more complete.

The subject of our first analytical investigations on the transfer of heat was its distribution amongst separated masses; these have been preserved in Chapter III., Section II. The problems relative to continuous bodies, which form the theory rightly so called, were solved many years afterwards; this theory was explained for the first time in a manuscript work forwarded to the Institute of France at the end of the year 1807, an extract from which was published in the *Bulletin des Sciences* (*Société Philomatique,* year 1808, page 112). We added to this memoir, and successively forwarded very extensive notes, concerning the convergence of series, the diffusion of heat in an infinite prism, its emission in spaces

[1] These memoirs were never collectively published as a sequel or complement to the *Théorie Analytique de la Chaleur.* But, as will be seen presently, the author had written most of them before the publication of that work in 1822. [A. F.]

void of air, the constructions suitable for exhibiting the chief theorems, and the analysis of the periodic movement at the surface of the earth. Our second memoir, on the propagation of heat, was deposited in the archives of the Institute, on the 28th of September, 1811. It was formed out of the preceding memoir and the notes already sent in; the geometrical constructions and those details of analysis which had no necessary relation to the physical problem were omitted, and to it was added the general equation which expresses the state of the surface. This second work was sent to press in the course of 1821, to be inserted in the collection of the Academy of Sciences. It is printed without any change or addition; the text agrees literally with the deposited manuscript, which forms part of the archives of the Institute[1].

In this memoir, and in the writings which preceded it, will be found a first explanation of applications which our actual work

[1] It appears as a memoir and supplement in volumes IV. and V. of the *Mémoires de l'Académie des Sciences*. For convenience of comparison with the table of contents of the *Analytical Theory of Heat*, we subjoin the titles and heads of the chapters of the printed memoir:

Théorie du mouvement de la chaleur dans les corps solides, par M. Fourier. [*Mémoires de l'Académie Royale des Sciences de l'Institut de France. Tome IV.* (for year 1819). *Paris* 1824.]

I. *Exposition.*
II. *Notions générales et définitions préliminaires.*
III. *Equations du mouvement de la chaleur.*
IV. *Du mouvement linéaire et varié de la chaleur dans une armille.*
V. *De la propagation de la chaleur dans une lame rectangulaire dont les températures sont constantes.*
VI. *De la communication de la chaleur entre des masses disjointes.*
VII. *Du mouvement varié de la chaleur dans une sphère solide.*
VIII. *Du mouvement varié de la chaleur dans un cylindre solide.*
IX. *De la propagation de la chaleur dans un prisme dont l'extrémité est assujettie à une température constante.*
X. *Du mouvement varié de la chaleur dans un solide de forme cubique.*
XI. *Du mouvement linéaire et varié de la chaleur dans les corps dont une dimension est infinie.*

Suite du mémoire intitulé: Théorie du mouvement de la chaleur dans les corps solides; par M. Fourier. [*Mémoires de l'Académie Royale des Sciences de l'Institut de France. Tome V.* (for year 1820). *Paris*, 1826.]

XII. *Des températures terrestres, et du mouvement de la chaleur dans l'intérieur d'une sphère solide, dont la surface est assujettie à des changemens périodiques de température.*
XIII. *Des lois mathématiques de l'équilibre de la chaleur rayonnante.*
XIV. *Comparaison des résultats de la théorie avec ceux de diverses expériences.*
[A. F.]

does not contain; they will be treated in the subsequent memoirs[1] at greater length, and, if it be in our power, with greater clearness. The results of our labours concerning the same problems are also indicated in several articles already published. The extract inserted in the *Annales de Chimie et de Physique* shews the aggregate of our researches (Vol. III. page 350, year 1816). We published in the *Annales* two separate notes, concerning radiant heat (Vol. IV. page 128, year 1817, and Vol. VI. page 259, year 1817).

Several other articles of the same collection present the most constant results of theory and observation; the utility and the extent of thermological knowledge could not be better appreciated than by the celebrated editors of the *Annales*[2].

In the *Bulletin des Sciences* (*Société philomatique* year 1818, page 1, and year 1820, page 60) will be found an extract from a memoir on the constant or variable temperature of dwellings, and an explanation of the chief consequences of our analysis of the terrestrial temperatures.

M. Alexandre de Humboldt, whose researches embrace all the great problems of natural philosophy, has considered the observations of the temperatures proper to the different climates from a novel and very important point of view (Memoir on Isothermal lines, *Société d'Arcueil,* Vol. III. page 462); (Memoir on the inferior limit of perpetual snow, *Annales de Chimie et de Physique,* Vol. V. page 102, year 1817).

As to the differential equations of the movement of heat in fluids[3] mention has been made of them in the annual history of the Academy of Sciences. The extract from our memoir shews clearly its object and principle. (*Analyse des travaux de l'Académie des Sciences,* by M. De Lambre, year 1820.)

The examination of the repulsive forces produced by heat, which determine the statical properties of gases, does not belong

[1] See note, page 9, and the notes, pages 11—13.

[2] Gay-Lussac and Arago. See note, p. 13.

[3] *Mémoires de l'Académie des Sciences, Tome XII., Paris,* 1833, contain on pp. 507—514, *Mémoire d'analyse sur le mouvement de la chaleur dans les fluides, par M. Fourier. Lu à l'Académie Royale des Sciences,* 4 Sep. 1820. It is followed on pp. 515—530 by *Extrait des notes manuscrites conservées par l'auteur.* The memoir is signed Jh. Fourier, Paris, 1 Sep. 1820, but was published after the death of the author. [A. F.]

to the analytical subject which we have considered. This question connected with the theory of radiant heat has just been discussed by the illustrious author of the *Mécanique céleste,* to whom all the chief branches of mathematical analysis owe important discoveries. (*Connaissance des Temps,* years 1824-5.)

The new theories explained in our work are united for ever to the mathematical sciences, and rest like them on invariable foundations; all the elements which they at present possess they will preserve, and will continually acquire greater extent. Instruments will be perfected and experiments multiplied. The analysis which we have formed will be deduced from more general, that is to say, more simple and more fertile methods common to many classes of phenomena. For all substances, solid or liquid, for vapours and permanent gases, determinations will be made of all the specific qualities relating to heat, and of the variations of the coefficients which express them[1]. At different stations on the earth observations will be made, of the temperatures of the ground at different depths, of the intensity of the solar heat and its effects, constant or variable, in the atmosphere, in the ocean and in lakes; and the constant temperature of the heavens proper to the planetary regions will become known[2]. The theory itself

[1] *Mémoires de l'Académie des Sciences, Tome VIII., Paris* 1829, contain on pp. 581—622, *Mémoire sur la Théorie Analytique de la Chaleur, par M. Fourier.* This was published whilst the author was Perpetual Secretary to the Academy. The first only of four parts of the memoir is printed. The contents of all are stated. I. Determines the temperature at any point of a prism whose terminal temperatures are functions of the time, the initial temperature at any point being a function of its distance from one end. II. Examines the chief consequences of the general solution, and applies it to two distinct cases, according as the temperatures of the ends of the heated prism are periodic or not. III. Is historical, enumerates the earlier experimental and analytical researches of other writers relative to the theory of heat; considers the nature of the transcendental equations appearing in the theory; remarks on the employment of arbitrary functions; replies to the objections of M. Poisson; adds some remarks on a problem of the motion of waves. IV. Extends the application of the theory of heat by taking account, in the analysis, of variations in the specific coefficients which measure the capacity of substances for heat, the permeability of solids, and the penetrability of their surfaces. [A. F.]

[2] *Mémoires de l'Académie des Sciences, Tome VII., Paris,* 1827, contain on pp. 569—604, *Mémoire sur les températures du globe terrestre et des espaces planétaires, par M. Fourier.* The memoir is entirely descriptive; it was read before the Academy, 20 and 29 Sep. 1824 (*Annales de Chimie et de Physique,* 1824, XXVII. p. 136). [A. F.]

will direct all these measures, and assign their precision. No considerable progress can hereafter be made which is not founded on experiments such as these; for mathematical analysis can deduce from general and simple phenomena the expression of the laws of nature; but the special application of these laws to very complex effects demands a long series of exact observations.

The complete list of the Articles on Heat, published by M. Fourier, in the *Annales de Chimie et de Physique, Series* 2, is as follows:

1816. III. pp. 350—375. *Théorie de la Chaleur* (*Extrait*). Description by the author of the 4to volume afterwards published in 1822 without the chapters on radiant heat, solar heat as it affects the earth, the comparison of analysis with experiment, and the history of the rise and progress of the theory of heat.

1817. IV. pp. 128—145. *Note sur la Chaleur rayonnante.* Mathematical sketch on the sine law of emission of heat from a surface. Proves the author's paradox on the hypothesis of equal intensity of emission in all directions.

1817. VI. pp. 259—303. *Questions sur la théorie physique de la chaleur rayonnante.* An elegant physical treatise on the discoveries of Newton, Pictet, Wells, Wollaston, Leslie and Prevost.

1820. XIII. pp. 418—438. *Sur le refroidissement séculaire de la terre* (*Extrait*). Sketch of a memoir, mathematical and descriptive, on the waste of the earth's initial heat.

1824. XXVII. pp. 136—167. *Remarques générales sur les températures du globe terrestre et des espaces planétaires.* This is the descriptive memoir referred to above, *Mém. Acad. d. Sc. Tome VII.*

1824. XXVII. pp. 236—281. *Résumé théorique des propriétés de la chaleur rayonnante.* Elementary analytical account of surface-emission and absorption based on the principle of equilibrium of temperature.

1825. XXVIII. pp. 337—365. *Remarques sur la théorie mathématique de la chaleur rayonnante.* Elementary analysis of emission, absorption and reflection by walls of enclosure uniformly heated. At p. 364, M. Fourier promises a *Théorie physique de la chaleur* to contain the applications of the *Théorie Analytique* omitted from the work published in 1822.

1828. XXXVII. pp. 291—315. *Recherches expérimentales sur la faculté conductrice des corps minces soumis à l'action de la chaleur, et description d'un nouveau thermomètre de contact.* A thermoscope of contact intended for lecture demonstrations is also described. M. Emile Verdet in his *Conférences de Physique, Paris*, 1872. Part I. p. 22, has stated the practical reasons against relying on the theoretical indications of the thermometer of contact. [A. F.]

Of the three notices of memoirs by M. Fourier, contained in the *Bulletin des Sciences par la Société Philomatique*, and quoted here at pages 9 and 11, the first was written by M. Poisson, the mathematical editor of the *Bulletin*, the other two by M. Fourier. [A. F.]

THEORY OF HEAT.

Et ignem regunt numeri.—PLATO[1].

CHAPTER I.

INTRODUCTION.

FIRST SECTION.

Statement of the Object of the Work.

1. THE effects of heat are subject to constant laws which cannot be discovered without the aid of mathematical analysis. The object of the theory which we are about to explain is to demonstrate these laws; it reduces all physical researches on the propagation of heat, to problems of the integral calculus whose elements are given by experiment. No subject has more extensive relations with the progress of industry and the natural sciences; for the action of heat is always present, it penetrates all bodies and spaces, it influences the processes of the arts, and occurs in all the phenomena of the universe.

When heat is unequally distributed among the different parts of a solid mass, it tends to attain equilibrium, and passes slowly from the parts which are more heated to those which are less; and at the same time it is dissipated at the surface, and lost in the medium or in the void. The tendency to uniform distribution and the spontaneous emission which acts at the surface of bodies, change continually the temperature at their different points. The problem of the propagation of heat consists in

[1] Cf. Plato, *Timæus*, 53, B.

ὅτε δ' ἐπεχειρεῖτο κοσμεῖσθαι τὸ πᾶν, πῦρ πρῶτον καὶ γῆν καὶ ἀέρα καὶ ὕδωρ......... διεσχηματίσατο [ὁ θεὸς] εἴδεσί τε καὶ ἀριθμοῖς. [A. F.]

determining what is the temperature at each point of a body at a given instant, supposing that the initial temperatures are known. The following examples will more clearly make known the nature of these problems.

2. If we expose to the continued and uniform action of a source of heat, the same part of a metallic ring, whose diameter is large, the molecules nearest to the source will be first heated, and, after a certain time, every point of the solid will have acquired very nearly the highest temperature which it can attain. This limit or greatest temperature is not the same at different points; it becomes less and less according as they become more distant from that point at which the source of heat is directly applied.

When the temperatures have become permanent, the source of heat supplies, at each instant, a quantity of heat which exactly compensates for that which is dissipated at all the points of the external surface of the ring.

If now the source be suppressed, heat will continue to be propagated in the interior of the solid, but that which is lost in the medium or the void, will no longer be compensated as formerly by the supply from the source, so that all the temperatures will vary and diminish incessantly until they have become equal to the temperatures of the surrounding medium.

3. Whilst the temperatures are permanent and the source remains, if at every point of the mean circumference of the ring an ordinate be raised perpendicular to the plane of the ring, whose length is proportional to the fixed temperature at that point, the curved line which passes through the ends of these ordinates will represent the permanent state of the temperatures, and it is very easy to determine by analysis the nature of this line. It is to be remarked that the thickness of the ring is supposed to be sufficiently small for the temperature to be sensibly equal at all points of the same section perpendicular to the mean circumference. When the source is removed, the line which bounds the ordinates proportional to the temperatures at the different points will change its form continually. The problem consists in expressing, by one equation, the variable

form of this curve, and in thus including in a single formula all the successive states of the solid.

4. Let z be the constant temperature at a point m of the mean circumference, x the distance of this point from the source, that is to say the length of the arc of the mean circumference, included between the point m and the point o which corresponds to the position of the source; z is the highest temperature which the point m can attain by virtue of the constant action of the source, and this permanent temperature z is a function $f(x)$ of the distance x. The first part of the problem consists in determining the function $f(x)$ which represents the permanent state of the solid.

Consider next the variable state which succeeds to the former state as soon as the source has been removed; denote by t the time which has passed since the suppression of the source, and by v the value of the temperature at the point m after the time t. The quantity v will be a certain function $F(x, t)$ of the distance x and the time t; the object of the problem is to discover this function $F(x, t)$, of which we only know as yet that the initial value is $f(x)$, so that we ought to have the equation $f(x) = F(x, o)$.

5. If we place a solid homogeneous mass, having the form of a sphere or cube, in a medium maintained at a constant temperature, and if it remains immersed for a very long time, it will acquire at all its points a temperature differing very little from that of the fluid. Suppose the mass to be withdrawn in order to transfer it to a cooler medium, heat will begin to be dissipated at its surface; the temperatures at different points of the mass will not be sensibly the same, and if we suppose it divided into an infinity of layers by surfaces parallel to its external surface, each of those layers will transmit, at each instant, a certain quantity of heat to the layer which surrounds it. If it be imagined that each molecule carries a separate thermometer, which indicates its temperature at every instant, the state of the solid will from time to time be represented by the variable system of all these thermometric heights. It is required to express the successive states by analytical formulæ, so that we

may know at any given instant the temperatures indicated by each thermometer, and compare the quantities of heat which flow during the same instant, between two adjacent layers, or into the surrounding medium.

6. If the mass is spherical, and we denote by x the distance of a point of this mass from the centre of the sphere, by t the time which has elapsed since the commencement of the cooling, and by v the variable temperature of the point m, it is easy to see that all points situated at the same distance x from the centre of the sphere have the same temperature v. This quantity v is a certain function $F(x, t)$ of the radius x and of the time t; it must be such that it becomes constant whatever be the value of x, when we suppose t to be nothing; for by hypothesis, the temperature at all points is the same at the moment of emersion. The problem consists in determining that function of x and t which expresses the value of v.

7. In the next place it is to be remarked, that during the cooling, a certain quantity of heat escapes, at each instant, through the external surface, and passes into the medium. The value of this quantity is not constant; it is greatest at the beginning of the cooling. If however we consider the variable state of the internal spherical surface whose radius is x, we easily see that there must be at each instant a certain quantity of heat which traverses that surface, and passes through that part of the mass which is more distant from the centre. This continuous flow of heat is variable like that through the external surface, and both are quantities comparable with each other; their ratios are numbers whose varying values are functions of the distance x, and of the time t which has elapsed. It is required to determine these functions.

8. If the mass, which has been heated by a long immersion in a medium, and whose rate of cooling we wish to calculate, is of cubical form, and if we determine the position of each point m by three rectangular co-ordinates x, y, z, taking for origin the centre of the cube, and for axes lines perpendicular to the faces, we see that the temperature v of the point m after the time t, is a function of the four variables x, y, z, and t. The quantities of heat

which flow out at each instant through the whole external surface of the solid, are variable and comparable with each other; their ratios are analytical functions depending on the time t, the expression of which must be assigned.

9. Let us examine also the case in which a rectangular prism of sufficiently great thickness and of infinite length, being submitted at its extremity to a constant temperature, whilst the air which surrounds it is maintained at a less temperature, has at last arrived at a fixed state which it is required to determine. All the points of the extreme section at the base of the prism have, by hypothesis, a common and permanent temperature. It is not the same with a section distant from the source of heat; each of the points of this rectangular surface parallel to the base has acquired a fixed temperature, but this is not the same at different points of the same section, and must be less at points nearer to the surface exposed to the air. We see also that, at each instant, there flows across a given section a certain quantity of heat, which always remains the same, since the state of the solid has become constant. The problem consists in determining the permanent temperature at any given point of the solid, and the whole quantity of heat which, in a definite time, flows across a section whose position is given.

10. Take as origin of co-ordinates x, y, z, the centre of the base of the prism, and as rectangular axes, the axis of the prism itself, and the two perpendiculars on the sides: the permanent temperature v of the point m, whose co-ordinates are x, y, z, is a function of three variables $F(x, y, z)$: it has by hypothesis a constant value, when we suppose x nothing, whatever be the values of y and z. Suppose we take for the unit of heat that quantity which in the unit of time would emerge from an area equal to a unit of surface, if the heated mass which that area bounds, and which is formed of the same substance as the prism, were continually maintained at the temperature of boiling water, and immersed in atmospheric air maintained at the temperature of melting ice.

We see that the quantity of heat which, in the permanent state of the rectangular prism, flows, during a unit of time, across a certain section perpendicular to the axis, has a determinate ratio to the quantity of heat taken as unit. This ratio is not the same for all sections: it is a function $\phi(x)$ of the distance x, at which the section is situated. It is required to find an analytical expression of the function $\phi(x)$.

11. The foregoing examples suffice to give an exact idea of the different problems which we have discussed.

The solution of these problems has made us understand that the effects of the propagation of heat depend in the case of every solid substance, on three elementary qualities, which are, its capacity for heat, its own conducibility, and the exterior conducibility.

It has been observed that if two bodies of the same volume and of different nature have equal temperatures, and if the same quantity of heat be added to them, the increments of temperature are not the same; the ratio of these increments is the ratio of their capacities for heat. In this manner, the first of the three specific elements which regulate the action of heat is exactly defined, and physicists have for a long time known several methods of determining its value. It is not the same with the two others; their effects have often been observed, but there is but one exact theory which can fairly distinguish, define, and measure them with precision.

The proper or interior conducibility of a body expresses the facility with which heat is propagated in passing from one internal molecule to another. The external or relative conducibility of a solid body depends on the facility with which heat penetrates the surface, and passes from this body into a given medium, or passes from the medium into the solid. The last property is modified by the more or less polished state of the surface; it varies also according to the medium in which the body is immersed; but the interior conducibility can change only with the nature of the solid.

These three elementary qualities are represented in our formulæ by constant numbers, and the theory itself indicates experiments suitable for measuring their values. As soon as they are determined, all the problems relating to the propagation of heat depend only on numerical analysis. The knowledge of these specific properties may be directly useful in several applications of the physical sciences; it is besides an element in the study and description of different substances. It is a very imperfect knowledge of bodies which ignores the relations which they have with one of the chief agents of nature. In general, there is no mathematical theory which has a closer relation than this with public economy, since it serves to give clearness and perfection to the practice of the numerous arts which are founded on the employment of heat.

12. The problem of the terrestrial temperatures presents one of the most beautiful applications of the theory of heat; the general idea to be formed of it is this. Different parts of the surface of the globe are unequally exposed to the influence of the solar rays; the intensity of their action depends on the latitude of the place; it changes also in the course of the day and in the course of the year, and is subject to other less perceptible inequalities. It is evident that, between the variable state of the surface and that of the internal temperatures, a necessary relation exists, which may be derived from theory. We know that, at a certain depth below the surface of the earth, the temperature at a given place experiences no annual variation: this permanent underground temperature becomes less and less according as the place is more and more distant from the equator. We may then leave out of consideration the exterior envelope, the thickness of which is incomparably small with respect to the earth's radius, and regard our planet as a nearly spherical mass, whose surface is subject to a temperature which remains constant at all points on a given parallel, but is not the same on another parallel. It follows from this that every internal molecule has also a fixed temperature determined by its position. The mathematical problem consists in discovering the fixed temperature at any given point, and the law which the solar heat follows whilst penetrating the interior of the earth.

This diversity of temperature interests us still more, if we consider the changes which succeed each other in the envelope itself on the surface of which we dwell. Those alternations of heat and cold which are reproduced every day and in the course of every year, have been up to the present time the object of repeated observations. These we can now submit to calculation, and from a common theory derive all the particular facts which experience has taught us. The problem is reducible to the hypothesis that every point of a vast sphere is affected by periodic temperatures; analysis then tells us according to what law the intensity of these variations decreases according as the depth increases, what is the amount of the annual or diurnal changes at a given depth, the epoch of the changes, and how the fixed value of the underground temperature is deduced from the variable temperatures observed at the surface.

13. The general equations of the propagation of heat are partial differential equations, and though their form is very simple the known methods[1] do not furnish any general mode of integrating them; we could not therefore deduce from them the values of the temperatures after a definite time. The numerical interpretation of the results of analysis is however necessary, and it is a degree of perfection which it would be very important to give to every application of analysis to the natural sciences. So long as it is not obtained, the solutions may be said to remain incomplete and useless, and the truth which it is proposed to discover is no less hidden in the formulæ of analysis than it was in the physical problem itself. We have applied ourselves with much care to this purpose, and we have been able to overcome the difficulty in all the problems of which we have treated, and which contain the chief elements of the theory of heat. There is not one of the problems whose solution does not provide convenient and exact means for discovering the numerical values of the temperatures acquired, or those of the quantities of heat which

[1] For the modern treatment of these equations consult

Partielle Differentialgleichungen, von B. Riemann, Braunschweig, 2nd Ed., 1876. The fourth section, *Bewegung der Wärme in festen Körpern.*

Cours de physique mathématique, par E. Matthieu, Paris, 1873. The parts relative to the differential equations of the theory of heat.

The Functions of Laplace, Lamé, and Bessel, by I. Todhunter, London, 1875. Chapters XXI. XXV.—XXIX. which give some of Lamé's methods.

Conférences de Physique, par E. Verdet, Paris, 1872 [*Œuvres*, Vol. IV. Part I.]. *Leçons sur la propagation de la chaleur par conductibilité.* These are followed by a very extensive bibliography of the whole subject of conduction of heat.

For an interesting sketch and application of Fourier's Theory see

Theory of Heat, by Prof. Maxwell, London, 1875 [4th Edition]. Chapter XVIII. On the diffusion of heat by conduction.

Natural Philosophy, by Sir W. Thomson and Prof. Tait, Vol. I. Oxford, 1867. Chapter VII. Appendix D, On the secular cooling of the earth. [A. F.]

have flowed through, when the values of the time and of the variable coordinates are known. Thus will be given not only the differential equations which the functions that express the values of the temperatures must satisfy; but the functions themselves will be given under a form which facilitates the numerical applications.

14. In order that these solutions might be general, and have an extent equal to that of the problem, it was requisite that they should accord with the initial state of the temperatures, which is arbitrary. The examination of this condition shews that we may develop in convergent series, or express by definite integrals, functions which are not subject to a constant law, and which represent the ordinates of irregular or discontinuous lines. This property throws a new light on the theory of partial differential equations, and extends the employment of arbitrary functions by submitting them to the ordinary processes of analysis.

15. It still remained to compare the facts with theory. With this view, varied and exact experiments were undertaken, whose results were in conformity with those of analysis, and gave them an authority which one would have been disposed to refuse to them in a new matter which seemed subject to so much uncertainty. These experiments confirm the principle from which we started, and which is adopted by all physicists in spite of the diversity of their hypotheses on the nature of heat.

16. Equilibrium of temperature is effected not only by way of contact, it is established also between bodies separated from each other, which are situated for a long time in the same region. This effect is independent of contact with a medium; we have observed it in spaces wholly void of air. To complete our theory it was necessary to examine the laws which radiant heat follows, on leaving the surface of a body. It results from the observations of many physicists and from our own experiments, that the intensities of the different rays, which escape in all directions from any point in the surface of a heated body, depend on the angles which their directions make with the surface at the same point. We have proved that the intensity of a ray diminishes as the ray

makes a smaller angle with the element of surface, and that it is proportional to the sine of that angle[1]. This general law of emission of heat which different observations had already indicated, is a necessary consequence of the principle of the equilibrium of temperature and of the laws of propagation of heat in solid bodies.

Such are the chief problems which have been discussed in this work; they are all directed to one object only, that is to establish clearly the mathematical principles of the theory of heat, and to keep up in this way with the progress of the useful arts, and of the study of nature.

17. From what precedes it is evident that a very extensive class of phenomena exists, not produced by mechanical forces, but resulting simply from the presence and accumulation of heat. This part of natural philosophy cannot be connected with dynamical theories, it has principles peculiar to itself, and is founded on a method similar to that of other exact sciences. The solar heat, for example, which penetrates the interior of the globe, distributes itself therein according to a regular law which does not depend on the laws of motion, and cannot be determined by the principles of mechanics. The dilatations which the repulsive force of heat produces, observation of which serves to measure temperatures, are in truth dynamical effects; but it is not these dilatations which we calculate, when we investigate the laws of the propagation of heat.

18. There are other more complex natural effects, which depend at the same time on the influence of heat, and of attractive forces: thus, the variations of temperatures which the movements of the sun occasion in the atmosphere and in the ocean, change continually the density of the different parts of the air and the waters. The effect of the forces which these masses obey is modified at every instant by a new distribution of heat, and it cannot be doubted that this cause produces the regular winds, and the chief currents of the sea; the solar and lunar attractions occasioning in the atmosphere effects but slightly sensible, and not general displacements. It was therefore necessary, in order to

[1] *Mém. Acad. d. Sc. Tome V.* Paris, 1826, pp. 179—213. [A. F.]

submit these grand phenomena to calculation, to discover the mathematical laws of the propagation of heat in the interior of masses.

19. It will be perceived, on reading this work, that heat attains in bodies a regular disposition independent of the original distribution, which may be regarded as arbitrary.

In whatever manner the heat was at first distributed, the system of temperatures altering more and more, tends to coincide sensibly with a definite state which depends only on the form of the solid. In the ultimate state the temperatures of all the points are lowered in the same time, but preserve amongst each other the same ratios: in order to express this property the analytical formulæ contain terms composed of exponentials and of quantities analogous to trigonometric functions.

Several problems of mechanics present analogous results, such as the isochronism of oscillations, the multiple resonance of sonorous bodies. Common experiments had made these results remarked, and analysis afterwards demonstrated their true cause. As to those results which depend on changes of temperature, they could not have been recognised except by very exact experiments; but mathematical analysis has outrun observation, it has supplemented our senses, and has made us in a manner witnesses of regular and harmonic vibrations in the interior of bodies.

20. These considerations present a singular example of the relations which exist between the abstract science of numbers and natural causes.

When a metal bar is exposed at one end to the constant action of a source of heat, and every point of it has attained its highest temperature, the system of fixed temperatures corresponds exactly to a table of logarithms; the numbers are the elevations of thermometers placed at the different points, and the logarithms are the distances of these points from the source. In general heat distributes itself in the interior of solids according to a simple law expressed by a partial differential equation common to physical problems of different order. The irradiation of heat has an evident relation to the tables of sines, for the rays which depart from the same point of a heated surface, differ very much from each other, and their intensity is rigorously proportional to the sine of the angle which the direction of each ray makes with the element of surface.

If we could observe the changes of temperature for every instant at every point of a solid homogeneous mass, we should discover in these series of observations the properties of recurring series, as of sines and logarithms; they would be noticed for example in the diurnal or annual variations of temperature of different points of the earth near its surface.

We should recognise again the same results and all the chief elements of general analysis in the vibrations of elastic media, in the properties of lines or of curved surfaces, in the movements of the stars, and those of light or of fluids. Thus the functions obtained by successive differentiations, which are employed in the development of infinite series and in the solution of numerical equations, correspond also to physical properties. The first of these functions, or the fluxion properly so called, expresses in geometry the inclination of the tangent of a curved line, and in dynamics the velocity of a moving body when the motion varies; in the theory of heat it measures the quantity of heat which flows at each point of a body across a given surface. Mathematical analysis has therefore necessary relations with sensible phenomena; its object is not created by human intelligence; it is a pre-existent element of the universal order, and is not in any way contingent or fortuitous; it is imprinted throughout all nature.

21. Observations more exact and more varied will presently ascertain whether the effects of heat are modified by causes which have not yet been perceived, and the theory will acquire fresh perfection by the continued comparison of its results with the results of experiment; it will explain some important phenomena which we have not yet been able to submit to calculation; it will shew how to determine all the thermometric effects of the solar rays, the fixed or variable temperature which would be observed at different distances from the equator, whether in the interior of the earth or beyond the limits of the atmosphere, whether in the ocean or in different regions of the air. From it will be derived the mathematical knowledge of the great movements which result from the influence of heat combined with that of gravity. The

same principles will serve to measure the conducibilities, proper or relative, of different bodies, and their specific capacities, to distinguish all the causes which modify the emission of heat at the surface of solids, and to perfect thermometric instruments.

The theory of heat will always attract the attention of mathematicians, by the rigorous exactness of its elements and the analytical difficulties peculiar to it, and above all by the extent and usefulness of its applications; for all its consequences concern at the same time general physics, the operations of the arts, domestic uses and civil economy.

SECTION II.

Preliminary definitions and general notions.

22. Of the nature of heat uncertain hypotheses only could be formed, but the knowledge of the mathematical laws to which its effects are subject is independent of all hypothesis; it requires only an attentive examination of the chief facts which common observations have indicated, and which have been confirmed by exact experiments.

It is necessary then to set forth, in the first place, the general results of observation, to give exact definitions of all the elements of the analysis, and to establish the principles upon which this analysis ought to be founded.

The action of heat tends to expand all bodies, solid, liquid or gaseous; this is the property which gives evidence of its presence. Solids and liquids increase in volume, if the quantity of heat which they contain increases; they contract if it diminishes.

When all the parts of a solid homogeneous body, for example those of a mass of metal, are equally heated, and preserve without any change the same quantity of heat, they have also and retain the same density. This state is expressed by saying that throughout the whole extent of the mass the molecules have a common and permanent temperature.

23. The thermometer is a body whose smallest changes of volume can be appreciated; it serves to measure temperatures by

the dilatation of a fluid or of air. We assume the construction, use and properties of this instrument to be accurately known. The temperature of a body equally heated in every part, and which keeps its heat, is that which the thermometer indicates when it is and remains in *perfect contact* with the body in question.

Perfect contact is when the thermometer is completely immersed in a fluid mass, and, in general, when there is no point of the external surface of the instrument which is not touched by one of the points of the solid or liquid mass whose temperature is to be measured. In experiments it is not always necessary that this condition should be rigorously observed; but it ought to be assumed in order to make the definition exact.

24. Two fixed temperatures are determined on, namely: the temperature of melting ice which is denoted by 0, and the temperature of boiling water which we will denote by 1: the water is supposed to be boiling under an atmospheric pressure represented by a certain height of the barometer (76 centimetres), the mercury of the barometer being at the temperature 0.

25. Different quantities of heat are measured by determining how many times they contain a fixed quantity which is taken as the unit. Suppose a mass of ice having a definite weight (a kilogramme) to be at temperature 0, and to be converted into water at the same temperature 0 by the addition of a certain quantity of heat: the quantity of heat thus added is taken as the unit of measure. Hence the quantity of heat expressed by a number C contains C times the quantity required to ~~dissolve~~ a kilogramme of ice at the temperature zero into a mass of water at the same zero temperature.

26. To raise a metallic mass having a certain weight, a kilogramme of iron for example, from the temperature 0 to the temperature 1, a new quantity of heat must be added to that which is already contained in the mass. The number C which denotes this additional quantity of heat, is the specific capacity of iron for heat; the number C has very different values for different substances.

27. If a body of definite nature and weight (a kilogramme of mercury) occupies a volume V at temperature 0, it will occupy a greater volume $V+\Delta$, when it has acquired the temperature 1, that is to say, when the heat which it contained at the temperature 0 has been increased by a new quantity C, equal to the specific capacity of the body for heat. But if, instead of adding this quantity C, a quantity zC is added (z being a number positive or negative) the new volume will be $V+\delta$ instead of $V+\Delta$. Now experiments shew that if z is equal to $\frac{1}{2}$, the increase of volume δ is only half the total increment Δ, and that in general the value of δ is $z\Delta$, when the quantity of heat added is zC.

28. The ratio z of the two quantities zC and C of heat added, which is the same as the ratio of the two increments of volume δ and Δ, is that which is called the *temperature;* hence the quantity which expresses the actual temperature of a body represents the excess of its actual volume over the volume which it would occupy at the temperature of melting ice, unity representing the whole excess of volume which corresponds to the boiling point of water, over the volume which corresponds to the melting point of ice.

29. The increments of volume of bodies are in general proportional to the increments of the quantities of heat which produce the dilatations, but it must be remarked that this proportion is exact only in the case where the bodies in question are subjected to temperatures remote from those which determine their change of state. The application of these results to all liquids must not be relied on; and with respect to water in particular, dilatations do not always follow augmentations of heat.

In general the temperatures are numbers proportional to the quantities of heat added, and in the cases considered by us, these numbers are proportional also to the increments of volume.

30. Suppose that a body bounded by a plane surface having a certain area (a square metre) is maintained in any manner

whatever at constant temperature 1, common to all its points, and that the surface in question is in contact with air maintained at temperature 0: the heat which escapes continuously at the surface and passes into the surrounding medium will be replaced always by the heat which proceeds from the constant cause to whose action the body is exposed; thus, a certain quantity of heat denoted by h will flow through the surface in a definite time (a minute).

This amount h, of a flow continuous and always similar to itself, which takes place at a unit of surface at a fixed temperature, is the measure of the external conducibility of the body, that is to say, of the facility with which its surface transmits heat to the atmospheric air.

The air is supposed to be continually displaced with a given uniform velocity: but if the velocity of the current increased, the quantity of heat communicated to the medium would vary also: the same would happen if the density of the medium were increased.

31. If the excess of the constant temperature of the body over the temperature of surrounding bodies, instead of being equal to 1, as has been supposed, had a less value, the quantity of heat dissipated would be less than h. The result of observation is, as we shall see presently, that this quantity of heat lost may be regarded as sensibly proportional to the excess of the temperature of the body over that of the air and surrounding bodies. Hence the quantity h having been determined by one experiment in which the surface heated is at temperature 1, and the medium at temperature 0; we conclude that hz would be the quantity, if the temperature of the surface were z, all the other circumstances remaining the same. This result must be admitted when z is a small fraction.

32. The value h of the quantity of heat which is dispersed across a heated surface is different for different bodies; and it varies for the same body according to the different states of the surface. The effect of irradiation diminishes as the surface becomes more polished; so that by destroying the polish of the surface the value of h is considerably increased. A heated

metallic body will be more quickly cooled if its external surface is covered with a black coating such as will entirely tarnish its metallic lustre.

33. The rays of heat which escape from the surface of a body pass freely through spaces void of air; they are propagated also in atmospheric air: their directions are not disturbed by agitations in the intervening air: they can be reflected by metal mirrors and collected at their foci. Bodies at a high temperature, when plunged into a liquid, heat directly only those parts of the mass with which their surface is in contact. The molecules whose distance from this surface is not extremely small, receive no direct heat; it is not the same with aëriform fluids; in these the rays of heat are borne with extreme rapidity to considerable distances, whether it be that part of these rays traverses freely the layers of air, or whether these layers transmit the rays suddenly without altering their direction.

34. When the heated body is placed in air which is maintained at a sensibly constant temperature, the heat communicated to the air makes the layer of the fluid nearest to the surface of the body lighter; this layer rises more quickly the more intensely it is heated, and is replaced by another mass of cool air. A current is thus established in the air whose direction is vertical, and whose velocity is greater as the temperature of the body is higher. For this reason if the body cooled itself gradually the velocity of the current would diminish with the temperature, and the law of cooling would not be exactly the same as if the body were exposed to a current of air at a constant velocity.

35. When bodies are sufficiently heated to diffuse a vivid light, part of their radiant heat mixed with that light can traverse transparent solids or liquids, and is subject to the force which produces refraction. The quantity of heat which possesses this faculty becomes less as the bodies are less inflamed; it is, we may say, insensible for very opaque bodies however highly they may be heated. A thin transparent plate intercepts almost all the direct heat which proceeds from an ardent mass of metal; but it becomes heated in proportion as the intercepted rays are accumulated in

it; whence, if it is formed of ice, it becomes liquid; but if this plate of ice is exposed to the rays of a torch it allows a sensible amount of heat to pass through with the light.

36. We have taken as the measure of the external conducibility of a solid body a coefficient h, which denotes the quantity of heat which would pass, in a definite time (a minute), from the surface of this body, into atmospheric air, supposing that the surface had a definite extent (a square metre), that the constant temperature of the body was 1, and that of the air 0, and that the heated surface was exposed to a current of air of a given invariable velocity. This value of h is determined by observation. The quantity of heat expressed by the coefficient is composed of two distinct parts which cannot be measured except by very exact experiments. One is the heat communicated by way of contact to the surrounding air: the other, much less than the first, is the radiant heat emitted. We must assume, in our first investigations, that the quantity of heat lost does not change when the temperatures of the body and of the medium are augmented by the same sufficiently small quantity.

37. Solid substances differ again, as we have already remarked, by their property of being more or less permeable to heat; this quality is their conducibility proper: we shall give its definition and exact measure, after having treated of the uniform and linear propagation of heat. Liquid substances possess also the property of transmitting heat from molecule to molecule, and the numerical value of their conducibility varies according to the nature of the substances: but this effect is observed with difficulty in liquids, since their molecules change places on change of temperature. The propagation of heat in them depends chiefly on this continual displacement, in all cases where the lower parts of the mass are most exposed to the action of the source of heat. If, on the contrary, the source of heat be applied to that part of the mass which is highest, as was the case in several of our experiments, the transfer of heat, which is very slow, does not produce any displacement, at least when the increase of temperature does not diminish the volume, as is indeed noticed in singular cases bordering on changes of state.

38. To this explanation of the chief results of observation, a general remark must be added on equilibrium of temperatures; which consists in this, that different bodies placed in the same region, all of whose parts are and remain equally heated, acquire also a common and permanent temperature.

Suppose that all the parts of a mass M have a common and constant temperature a, which is maintained by any cause whatever: if a smaller body m be placed in perfect contact with the mass M, it will assume the common temperature a.

In reality this result would not strictly occur except after an infinite time: but the exact meaning of the proposition is that if the body m had the temperature a before being placed in contact, it would keep it without any change. The same would be the case with a multitude of other bodies n, p, q, r each of which was placed separately in perfect contact with the mass M: all would acquire the constant temperature a. Thus a thermometer if successively applied to the different bodies m, n, p, q, r would indicate the same temperature.

39. The effect in question is independent of contact, and would still occur, if every part of the body m were enclosed in the solid M, as in an enclosure, without touching any of its parts. For example, if the solid were a spherical envelope of a certain thickness, maintained by some external cause at a temperature a, and containing a space entirely deprived of air, and if the body m could be placed in any part whatever of this spherical space, without touching any point of the internal surface of the enclosure, it would acquire the common temperature a, or rather, it would preserve it if it had it already. The result would be the same for all the other bodies n, p, q, r, whether they were placed separately or all together in the same enclosure, and whatever also their substance and form might be.

40. Of all modes of presenting to ourselves the action of heat, that which seems simplest and most conformable to observation, consists in comparing this action to that of light. Molecules separated from one another reciprocally communicate, across empty space, their rays of heat, just as shining bodies transmit their light.

If within an enclosure closed in all directions, and maintained by some external cause at a fixed temperature a, we suppose different bodies to be placed without touching any part of the boundary, different effects will be observed according as the bodies, introduced into this space free from air, are more or less heated. If, in the first instance, we insert only one of these bodies, at the same temperature as the enclosure, it will send from all points of its surface as much heat as it receives from the solid which surrounds it, and is maintained in its original state by this exchange of equal quantities.

If we insert a second body whose temperature b is less than a, it will at first receive from the surfaces which surround it on all sides without touching it, a quantity of heat greater than that which it gives out: it will be heated more and more and will absorb through its surface more heat than in the first instance.

The initial temperature b continually rising, will approach without ceasing the fixed temperature a, so that after a certain time the difference will be almost insensible. The effect would be opposite if we placed within the same enclosure a third body whose temperature was greater than a.

41. All bodies have the property of emitting heat through their surface; the hotter they are the more they emit; the intensity of the emitted rays changes very considerably with the state of the surface.

42. Every surface which receives rays of heat from surrounding bodies reflects part and admits the rest: the heat which is not reflected, but introduced through the surface, accumulates within the solid; and so long as it exceeds the quantity dissipated by irradiation, the temperature rises.

43. The rays which tend to go out of heated bodies are arrested at the surface by a force which reflects part of them into the interior of the mass. The cause which hinders the incident rays from traversing the surface, and which divides these rays into two parts, of which one is reflected and the other admitted, acts in the same manner on the rays which are directed from the interior of the body towards external space.

If by modifying the state of the surface we increase the force by which it reflects the incident rays, we increase at the same time the power which it has of reflecting towards the interior of the body rays which are tending to go out. The incident rays introduced into the mass, and the rays emitted through the surface, are equally diminished in quantity.

44. If within the enclosure above mentioned a number of bodies were placed at the same time, separate from each other and unequally heated, they would receive and transmit rays of heat so that at each exchange their temperatures would continually vary, and would all tend to become equal to the fixed temperature of the enclosure.

This effect is precisely the same as that which occurs when heat is propagated within solid bodies; for the molecules which compose these bodies are separated by spaces void of air, and have the property of receiving, accumulating and emitting heat. Each of them sends out rays on all sides, and at the same time receives other rays from the molecules which surround it.

45. The heat given out by a point situated in the interior of a solid mass can pass directly to an extremely small distance only; it is, we may say, intercepted by the nearest particles; these particles only receive the heat directly and act on more distant points. It is different with gaseous fluids; the direct effects of radiation become sensible in them at very considerable distances.

46. Thus the heat which escapes in all directions from a part of the surface of a solid, passes on in air to very distant points; but is emitted only by those molecules of the body which are extremely near the surface. A point of a heated mass situated at a very small distance from the plane superficies which separates the mass from external space, sends to that space an infinity of rays, but they do not all arrive there; they are diminished by all that quantity of heat which is arrested by the intermediate molecules of the solid. The part of the ray actually dispersed into space becomes less according as it traverses a longer path within the mass. Thus the ray which escapes perpendicular to the surface has greater intensity than that which, departing from the same point, follows

an oblique direction, and the most oblique rays are wholly intercepted.

The same consequences apply to all the points which are near enough to the surface to take part in the emission of heat, from which it necessarily follows that the whole quantity of heat which escapes from the surface in the normal direction is very much greater than that whose direction is oblique. We have submitted this question to calculation, and our analysis proves that the intensity of the ray is proportional to the sine of the angle which the ray makes with the element of surface. Experiments had already indicated a similar result.

47. This theorem expresses a general law which has a necessary connection with the equilibrium and mode of action of heat. If the rays which escape from a heated surface had the same intensity in all directions, a thermometer placed at one of the points of a space bounded on all sides by an enclosure maintained at a constant temperature would indicate a temperature incomparably greater than that of the enclosure[1]. Bodies placed within this enclosure would not take a common temperature, as is always noticed; the temperature acquired by them would depend on the place which they occupied, or on their form, or on the forms of neighbouring bodies.

The same results would be observed, or other effects equally opposed to common experience, if between the rays which escape from the same point any other relations were admitted different from those which we have enunciated. We have recognised this law as the only one compatible with the general fact of the equilibrium of radiant heat.

48. If a space free from air is bounded on all sides by a solid enclosure whose parts are maintained at a common and constant temperature a, and if a thermometer, having the actual temperature a, is placed at any point whatever of the space, its temperature will continue without any change. It will receive therefore at each instant from the inner surface of the enclosure as much heat as it gives out to it. This effect of the rays of heat in a given space is, properly speaking, the measure of the temperature: but

[1] See proof by M. Fourier, *Ann. d. Ch. et Ph.* Ser. 2, IV. p. 128. [A. F.]

this consideration presupposes the mathematical theory of radiant heat.

If now between the thermometer and a part of the surface of the enclosure a body *M* be placed whose temperature is *a*, the thermometer will cease to receive rays from one part of the inner surface, but the rays will be replaced by those which it will receive from the interposed body *M*. An easy calculation proves that the compensation is exact, so that the state of the thermometer will be unchanged. It is not the same if the temperature of the body *M* is different from that of the enclosure. When it is greater, the rays which the interposed body *M* sends to the thermometer and which replace the intercepted rays convey more heat than the latter; the temperature of the thermometer must therefore rise.

If, on the contrary, the intervening body has a temperature less than *a*, that of the thermometer must fall; for the rays which this body intercepts are replaced by those which it gives out, that is to say, by rays cooler than those of the enclosure; thus the thermometer does not receive all the heat necessary to maintain its temperature *a*.

40. Up to this point abstraction has been made of the power which all surfaces have of reflecting part of the rays which are sent to them. If this property were disregarded we should have only a very incomplete idea of the equilibrium of radiant heat.

Suppose then that on the inner surface of the enclosure, maintained at a constant temperature, there is a portion which enjoys, in a certain degree, the power in question; each point of the reflecting surface will send into space two kinds of rays; the one go out from the very interior of the substance of which the enclosure is formed, the others are merely reflected by the same surface against which they had been sent. But at the same time that the surface repels on the outside part of the incident rays, it retains in the inside part of its own rays. In this respect an exact compensation is established, that is to say, every one of its own rays which the surface hinders from going out is replaced by a reflected ray of equal intensity.

The same result would happen, if the power of reflecting rays affected in any degree whatever other parts of the enclosure, or the surface of bodies placed within the same space and already at the common temperature.

Thus the reflection of heat does not disturb the equilibrium of temperatures, and does not introduce, whilst that equilibrium exists, any change in the law according to which the intensity of rays which leave the same point decreases proportionally to the sine of the angle of emission.

50. Suppose that in the same enclosure, all of whose parts maintain the temperature a, we place an isolated body M, and a polished metal surface R, which, turning its concavity towards the body, reflects great part of the rays which it received from the body; if we place a thermometer between the body M and the reflecting surface R, at the focus of this mirror, three different effects will be observed according as the temperature of the body M is equal to the common temperature a, or is greater or less.

In the first case, the thermometer preserves the temperature a; it receives 1°, rays of heat from all parts of the enclosure not hidden from it by the body M or by the mirror; 2°, rays given out by the body; 3°, those which the surface R sends out to the focus, whether they come from the mass of the mirror itself, or whether its surface has simply reflected them; and amongst the last we may distinguish between those which have been sent to the mirror by the mass M, and those which it has received from the enclosure. All the rays in question proceed from surfaces which, by hypothesis, have a common temperature a, so that the thermometer is precisely in the same state as if the space bounded by the enclosure contained no other body but itself.

In the second case, the thermometer placed between the heated body M and the mirror, must acquire a temperature greater than a. In reality, it receives the same rays as in the first hypothesis; but with two remarkable differences: one arises from the fact that the rays sent by the body M to the mirror, and reflected upon the thermometer, contain more heat than in the first case. The other difference depends on the fact that the rays sent directly by the body M to the thermometer contain more heat than formerly. Both causes, and chiefly the first, assist in raising the temperature of the thermometer.

In the third case, that is to say, when the temperature of the mass M is less than a, the temperature must assume also a temperature less than a. In fact, it receives again all the varieties of rays which we distinguished in the first case: but there are two kinds of them which contain less heat than in this first hypothesis, that is to say, those which, being sent out by the body M, are reflected by the mirror upon the thermometer, and those which the same body M sends to it directly. Thus the thermometer does not receive all the heat which it requires to preserve its original temperature a. It gives out more heat than it receives. It is inevitable then that its temperature must fall to the point at which the rays which it receives suffice to compensate those which it loses. This last effect is what is called the reflection of cold, and which, properly speaking, consists in the reflection of too feeble heat. The mirror intercepts a certain quantity of heat, and replaces it by a less quantity.

51. If in the enclosure, maintained at a constant temperature a, a body M be placed, whose temperature a' is less than a, the presence of this body will lower the thermometer exposed to its rays, and we may remark that the rays sent to the thermometer from the surface of the body M, are in general of two kinds, namely, those which come from inside the mass M, and those which, coming from different parts of the enclosure, meet the surface M and are reflected upon the thermometer. The latter rays have the common temperature a, but those which belong to the body M contain less heat, and these are the rays which cool the thermometer. If now, by changing the state of the surface of the body M, for example, by destroying the polish, we diminish the power which it has of reflecting the incident rays, the thermometer will fall still lower, and will assume a temperature a'' less than a. In fact all the conditions would be the same as in the preceding case, if it were not that the body M gives out a greater quantity of its own rays and reflects a less quantity of the rays which it receives from the enclosure; that is to say, these last rays, which have the common temperature, are in part replaced by cooler rays. Hence the thermometer no longer receives so much heat as formerly.

If, independently of the change in the surface of the body M, we place a metal mirror adapted to reflect upon the thermometer the rays which have left M, the temperature will assume a value a''' less than a''. The mirror, in fact, intercepts from the thermometer part of the rays of the enclosure which all have the temperature a, and replaces them by three kinds of rays; namely, 1°, those which come from the interior of the mirror itself, and which have the common temperature; 2°, those which the different parts of the enclosure send to the mirror with the same temperature, and which are reflected to the focus; 3°, those which, coming from the interior of the body M, fall upon the mirror, and are reflected upon the thermometer. The last rays have a temperature less than a; hence the thermometer no longer receives so much heat as it received before the mirror was set up.

Lastly, if we proceed to change also the state of the surface of the mirror, and by giving it a more perfect polish, increase its power of reflecting heat, the thermometer will fall still lower. In fact, all the conditions exist which occurred in the preceding case. Only, it happens that the mirror gives out a less quantity of its own rays, and replaces them by those which it reflects. Now, amongst these last rays, all those which proceed from the interior of the mass M are less intense than if they had come from the interior of the metal mirror; hence the thermometer receives still less heat than formerly: it will assume therefore a temperature a'''' less than a'''.

By the same principles all the known facts of the radiation of heat or of cold are easily explained.

52. The effects of heat can by no means be compared with those of an elastic fluid whose molecules are at rest.

It would be useless to attempt to deduce from this hypothesis the laws of propagation which we have explained in this work, and which all experience has confirmed. The free state of heat is the same as that of light; the active state of this element is then entirely different from that of gaseous substances. Heat acts in the same manner in a vacuum, in elastic fluids, and in liquid or solid masses, it is propagated only by way of radiation, but its sensible effects differ according to the nature of bodies.

53. Heat is the origin of all elasticity; it is the repulsive force which preserves the form of solid masses, and the volume of

liquids. In solid masses, neighbouring molecules would yield to their mutual attraction, if its effect were not destroyed by the heat which separates them.

This elastic force is greater according as the temperature is higher; which is the reason why bodies dilate or contract when their temperature is raised or lowered.

54. The equilibrium which exists, in the interior of a solid mass, between the repulsive force of heat and the molecular attraction, is stable; that is to say, it re-establishes itself when disturbed by an accidental cause. If the molecules are arranged at distances proper for equilibrium, and if an external force begins to increase this distance without any change of temperature, the effect of attraction begins by surpassing that of heat, and brings back the molecules to their original position, after a multitude of oscillations which become less and less sensible.

A similar effect is exerted in the opposite sense when a mechanical cause diminishes the primitive distance of the molecules; such is the origin of the vibrations of sonorous or flexible bodies, and of all the effects of their elasticity.

55. In the liquid or gaseous state of matter, the external pressure is additional or supplementary to the molecular attraction, and, acting on the surface, does not oppose change of form, but only change of the volume occupied. Analytical investigation will best shew how the repulsive force of heat, opposed to the attraction of the molecules or to the external pressure, assists in the composition of bodies, solid or liquid, formed of one or more elements, and determines the elastic properties of gaseous fluids; but these researches do not belong to the object before us, and appear in dynamic theories.

56. It cannot be doubted that the mode of action of heat always consists, like that of light, in the reciprocal communication of rays, and this explanation is at the present time adopted by the majority of physicists; but it is not necessary to consider the phenomena under this aspect in order to establish the theory of heat. In the course of this work it will be seen how the laws of equilibrium and propagation of radiant heat, in solid or liquid masses,

can be rigorously demonstrated, independently of any physical explanation, as the necessary consequences of common observations.

SECTION III.

Principle of the communication of heat.

57. We now proceed to examine what experiments teach us concerning the communication of heat.

If two equal molecules are formed of the same substance and have the same temperature, each of them receives from the other as much heat as it gives up to it; their mutual action may then be regarded as null, since the result of this action can bring about no change in the state of the molecules. If, on the contrary, the first is hotter than the second, it sends to it more heat than it receives from it; the result of the mutual action is the difference of these two quantities of heat. In all cases we make abstraction of the two equal quantities of heat which any two material points reciprocally give up; we conceive that the point most heated acts only on the other, and that, in virtue of this action, the first loses a certain quantity of heat which is acquired by the second. Thus the action of two molecules, or the quantity of heat which the hottest communicates to the other, is the difference of the two quantities which they give up to each other.

58. Suppose that we place in air a solid homogeneous body, whose different points have unequal actual temperatures; each of the molecules of which the body is composed will begin to receive heat from those which are at extremely small distances, or will communicate it to them. This action exerted during the same instant between all points of the mass, will produce an infinitesimal resultant change in all the temperatures: the solid will experience at each instant similar effects, so that the variations of temperature will become more and more sensible.

Consider only the system of two molecules, m and n, equal and extremely near, and let us ascertain what quantity of heat the first can receive from the second during one instant: we may then apply the same reasoning to all the other points which are near enough to the point m, to act directly on it during the first instant.

The quantity of heat communicated by the point n to the point m depends on the duration of the instant, on the very small distance between these points, on the actual temperature of each point, and on the nature of the solid substance; that is to say, if one of these elements happened to vary, all the other remaining the same, the quantity of heat transmitted would vary also. Now experiments have disclosed, in this respect, a general result: it consists in this, that all the other circumstances being the same, the quantity of heat which one of the molecules receives from the other is proportional to the difference of temperature of the two molecules. Thus the quantity would be double, triple, quadruple, if everything else remaining the same, the difference of the temperature of the point n from that of the point m became double, triple, or quadruple. To account for this result, we must consider that the action of n on m is always just as much greater as there is a greater difference between the temperatures of the two points: it is null, if the temperatures are equal, but if the molecule n contains more heat than the equal molecule m, that is to say, if the temperature of m being v, that of n is $v + \Delta$, a portion of the exceeding heat will pass from n to m. Now, if the excess of heat were double, or, which is the same thing, if the temperature of n were $v + 2\Delta$, the exceeding heat would be composed of two equal parts corresponding to the two halves of the whole difference of temperature 2Δ; each of these parts would have its proper effect as if it alone existed: thus the quantity of heat communicated by n to m would be twice as great as when the difference of temperature is only Δ. This simultaneous action of the different parts of the exceeding heat is that which constitutes the principle of the communication of heat. It follows from it that the sum of the partial actions, or the total quantity of heat which m receives from n is proportional to the difference of the two temperatures.

59. Denoting by v and v' the temperatures of two equal molecules m and n, by p, their extremely small distance, and by dt, the infinitely small duration of the instant, the quantity of heat which m receives from n during this instant will be expressed by $(v' - v)\, \phi(p) \,.\, dt$. We denote by $\phi(p)$ a certain function of the

distance p which, in solid bodies and in liquids, becomes nothing when p has a sensible magnitude. The function is the same for every point of the same given substance; it varies with the nature of the substance.

60. The quantity of heat which bodies lose through their surface is subject to the same principle. If we denote by σ the area, finite or infinitely small, of the surface, all of whose points have the temperature v, and if a represents the temperature of the atmospheric air, the coefficient h being the measure of the external conducibility, we shall have $\sigma h (v - a)\, dt$ as the expression for the quantity of heat which this surface σ transmits to the air during the instant dt.

When the two molecules, one of which transmits to the other a certain quantity of heat, belong to the same solid, the exact expression for the heat communicated is that which we have given in the preceding article; and since the molecules are extremely near, the difference of the temperatures is extremely small. It is not the same when heat passes from a solid body into a gaseous medium. But the experiments teach us that if the difference is a quantity sufficiently small, the heat transmitted is sensibly proportional to that difference, and that the number h may, in these first researches[1], be considered as having a constant value, proper to each state of the surface, but independent of the temperature.

61. These propositions relative to the quantity of heat communicated have been derived from different observations. We see first, as an evident consequence of the expressions in question, that if we increased by a common quantity all the initial temperatures of the solid mass, and that of the medium in which it is placed, the successive changes of temperature would be exactly the same as if this increase had not been made. Now this result is sensibly in accordance with experiment; it has been admitted by the physicists who first have observed the effects of heat.

[1] More exact laws of cooling investigated experimentally by Dulong and Petit will be found in the *Journal de l'Ecole Polytechnique*, Tome XI. pp. 234—294, Paris, 1820, or in *Jamin, Cours de Physique, Leçon* 47. [A. F.]

62. If the medium is maintained at a constant temperature, and if the heated body which is placed in that medium has dimensions sufficiently small for the temperature, whilst falling more and more, to remain sensibly the same at all points of the body, it follows from the same propositions, that a quantity of heat will escape at each instant through the surface of the body proportional to the excess of its actual temperature over that of the medium. Whence it is easy to conclude, as will be seen in the course of this work, that the line whose abscissæ represent the times elapsed, and whose ordinates represent the temperatures corresponding to those times, is a logarithmic curve: now, observations also furnish the same result, when the excess of the temperature of the solid over that of the medium is a sufficiently small quantity.

63. Suppose the medium to be maintained at the constant temperature 0, and that the initial temperatures of different points a, b, c, d &c. of the same mass are $\alpha, \beta, \gamma, \delta$ &c., that at the end of the first instant they have become $\alpha', \beta', \gamma', \delta'$ &c., that at the end of the second instant they have become $\alpha'', \beta'', \gamma'', \delta''$ &c., and so on. We may easily conclude from the propositions enunciated, that if the initial temperatures of the same points had been $g\alpha, g\beta, g\gamma, g\delta$ &c. (g being any number whatever), they would have become, at the end of the first instant, by virtue of the action of the different points, $g\alpha', g\beta', g\gamma', g\delta'$ &c., and at the end of the second instant, $g\alpha'', g\beta'', g\gamma'', g\delta''$ &c., and so on. For instance, let us compare the case when the initial temperatures of the points, a, b, c, d &c. were $\alpha, \beta, \gamma, \delta$ &c. with that in which they are $2\alpha, 2\beta, 2\gamma, 2\delta$ &c., the medium preserving in both cases the temperature 0. In the second hypothesis, the difference of the temperatures of any two points whatever is double what it was in the first, and the excess of the temperature of each point, over that of each molecule of the medium, is also double; consequently the quantity of heat which any molecule whatever sends to any other, or that which it receives, is, in the second hypothesis, double of that which it was in the first. The change of temperature which each point suffers being proportional to the quantity of heat acquired, it follows that, in the second case, this change is double what it was in the first case. Now we have

supposed that the initial temperature of the first point, which was a, became a' at the end of the first instant; hence if this initial temperature had been $2a$, and if all the other temperatures had been doubled, it would have become $2a'$. The same would be the case with all the other molecules b, c, d, and a similar result would be derived, if the ratio instead of being 2, were any number whatever g. It follows then, from the principle of the communication of heat, that if we increase or diminish in any given ratio all the initial temperatures, we increase or diminish in the same ratio all the successive temperatures.

This, like the two preceding results, is confirmed by observation. It could not have existed if the quantity of heat which passes from one molecule to another had not been, actually, proportional to the difference of the temperatures.

64. Observations have been made with accurate instruments, on the permanent temperatures at different points of a bar or of a metallic ring, and on the propagation of heat in the same bodies and in several other solids of the form of spheres or cubes. The results of these experiments agree with those which are derived from the preceding propositions. They would be entirely different if the quantity of heat transmitted from one solid molecule to another, or to a molecule of air, were not proportional to the excess of temperature. It is necessary first to know all the rigorous consequences of this proposition; by it we determine the chief part of the quantities which are the object of the problem. By comparing then the calculated values with those given by numerous and very exact experiments, we can easily measure the variations of the coefficients, and perfect our first researches.

SECTION IV.

On the uniform and linear movement of heat.

65. We shall consider, in the first place, the uniform movement of heat in the simplest case, which is that of an infinite solid enclosed between two parallel planes.

We suppose a solid body formed of some homogeneous substance to be enclosed between two parallel and infinite planes;

the lower plane A is maintained, by any cause whatever, at a constant temperature a; we may imagine for example that the mass is prolonged, and that the plane A is a section common to the solid and to the enclosed mass, and is heated at all its points by a constant source of heat; the upper plane B is also maintained by a similar cause at a fixed temperature b, whose value is less than that of a; the problem is to determine what would be the result of this hypothesis if it were continued for an infinite time.

If we suppose the initial temperature of all parts of this body to be b, it is evident that the heat which leaves the source A will be propagated farther and farther and will raise the temperature of the molecules included between the two planes: but the temperature of the upper plane being unable, according to hypothesis to rise above b, the heat will be dispersed within the cooler mass, contact with which keeps the plane B at the constant temperature b. The system of temperatures will tend more and more to a final state, which it will never attain, but which would have the property, as we shall proceed to shew, of existing and keeping itself up without any change if it were once formed.

In the final and fixed state, which we are considering, the permanent temperature of a point of the solid is evidently the same at all points of the same section parallel to the base; and we shall prove that this fixed temperature, common to all the points of an intermediate section, decreases in arithmetic progression from the base to the upper plane, that is to say, if we represent the constant temperatures a and b by the ordinates $A\alpha$ and $B\beta$

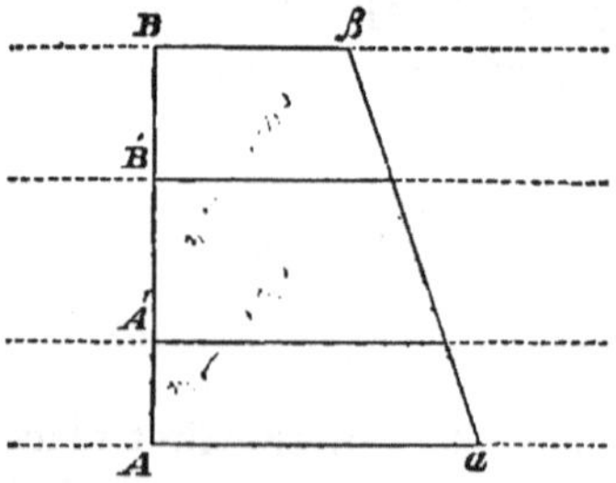

Fig. 1.

(see Fig. 1), raised perpendicularly to the distance AB between the two planes, the fixed temperatures of the intermediate layers will be represented by the ordinates of the straight line $\alpha\beta$ which joins the extremities α and β; thus, denoting by z the height of an intermediate section or its perpendicular distance from the plane A, by e the whole height or distance AB, and by v the temperature of the section whose height is z, we must have the equation $v = a + \frac{b-a}{e} z$.

In fact, if the temperatures were at first established in accordance with this law, and if the extreme surfaces A and B were always kept at the temperatures a and b, no change would happen in the state of the solid. To convince ourselves of this, it will be sufficient to compare the quantity of heat which would traverse an intermediate section A' with that which, during the same time, would traverse another section B'.

Bearing in mind that the final state of the solid is formed and continues, we see that the part of the mass which is below the plane A' must communicate heat to the part which is above that plane, since this second part is cooler than the first.

Imagine two points of the solid, m and m', very near to each other, and placed in any manner whatever, the one m below the plane A', and the other m' above this plane, to be exerting their action during an infinitely small instant: m the hottest point will communicate to m' a certain quantity of heat which will cross the plane A'. Let x, y, z be the rectangular coordinates of the point m, and x', y', z' the coordinates of the point m': consider also two other points n and n' very near to each other, and situated with respect to the plane B', in the same manner in which m and m' are placed with respect to the plane A': that is to say, denoting by ζ the perpendicular distance of the two sections A' and B', the coordinates of the point n will be $x, y, z+\zeta$ and those of the point n', x', y', $z'+\zeta$; the two distances mm' and nn' will be equal: further, the difference of the temperature v of the point m above the temperature v' of the point m' will be the same as the difference of temperature of the two points n and n'. In fact the former difference will be determined by substituting first z and then z' in the general equation

$$v = a + \frac{b-a}{e} z,$$

and subtracting the second equation from the first, whence the result $v - v' = \frac{b-a}{e}(z - z')$. We shall then find, by the substitution of $z + \zeta$ and $z' + \zeta$, that the excess of temperature of the point n over that of the point n' is also expressed by

$$\frac{b-a}{e}(z - z').$$

It follows from this that the quantity of heat sent by the point m to the point m' will be the same as the quantity of heat sent by the point n to the point n', for all the elements which concur in determining this quantity of transmitted heat are the same.

It is manifest that we can apply the same reasoning to every system of two molecules which communicate heat to each other across the section A' or the section B'; whence, if we could sum up the whole quantity of heat which flows, during the same instant, across the section A' or the section B', we should find this quantity to be the same for both sections.

From this it follows that the part of the solid included between A' and B' receives always as much heat as it loses, and since this result is applicable to any portion whatever of the mass included between two parallel sections, it is evident that no part of the solid can acquire a temperature higher than that which it has at present. Thus, it has been rigorously demonstrated that the state of the prism will continue to exist just as it was at first.

Hence, the permanent temperatures of different sections of a solid enclosed between two parallel infinite planes, are represented by the ordinates of a straight line $\alpha\beta$, and satisfy the linear equation $v = a + \frac{b-a}{e} z$.

66. By what precedes we see distinctly what constitutes the propagation of heat in a solid enclosed between two parallel and infinite planes, each of which is maintained at a constant temperature. Heat penetrates the mass gradually across the lower plane: the temperatures of the intermediate sections are raised, but can never exceed nor even quite attain a certain limit which they approach nearer and nearer: this limit or final temperature is different for different intermediate layers, and

decreases in arithmetic progression from the fixed temperature of the lower plane to the fixed temperature of the upper plane.

The final temperatures are those which would have to be given to the solid in order that its state might be permanent; the variable state which precedes it may also be submitted to analysis, as we shall see presently: but we are now considering only the system of final and permanent temperatures. In the last state, during each division of time, across a section parallel to the base, or a definite portion of that section, a certain quantity of heat flows, which is constant if the divisions of time are equal. This uniform flow is the same for all the intermediate sections; it is equal to that which proceeds from the source, and to that which is lost during the same time, at the upper surface of the solid, by virtue of the cause which keeps the temperature constant.

67. The problem now is to measure that quantity of heat which is propagated uniformly within the solid, during a given time, across a definite part of a section parallel to the base: it depends, as we shall see, on the two extreme temperatures a and b, and on the distance e between the two sides of the solid; it would vary if any one of these elements began to change, the other remaining the same. Suppose a second solid to be formed of the same substance as the first, and enclosed between two

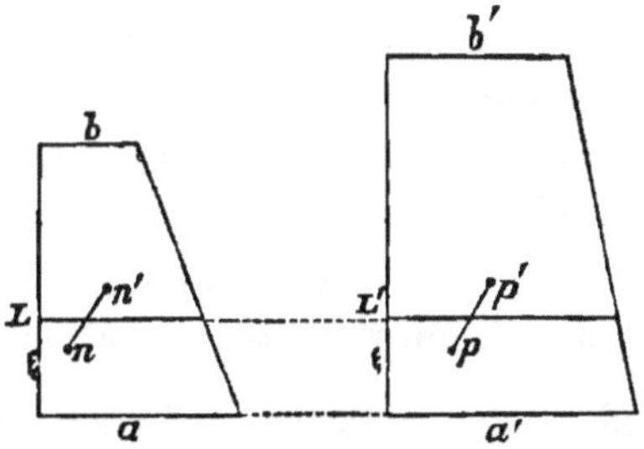

Fig. 2.

infinite parallel planes, whose perpendicular distance is e' (see fig. 2): the lower side is maintained at a fixed temperature a', and the upper side at the fixed temperature b'; both solids are considered to be in that final and permanent state which has the property of maintaining itself as soon as it has been formed.

Thus the law of the temperatures is expressed for the first body by the equation $v=a+\frac{b-a}{e}z$, and for the second, by the equation $u=a'+\frac{b'-a'}{e'}z$, v in the first solid, and u in the second, being the temperature of the section whose height is z.

This arranged, we will compare the quantity of heat which, during the unit of time traverses a unit of area taken on an intermediate section L of the first solid, with that which during the same time traverses an equal area taken on the section L' of the second, ϵ being the height common to the two sections, that is to say, the distance of each of them from their own base. We shall consider two very near points n and n' in the first body, one of which n is below the plane L and the other n' above this plane: x, y, z are the co-ordinates of n: and x', y', z' the co-ordinates of n', ϵ being less than z', and greater than z.

We shall consider also in the second solid the instantaneous action of two points p and p', which are situated, with respect to the section L', in the same manner as the points n and n' with respect to the section L of the first solid. Thus the same co-ordinates x, y, z, and x', y', z' referred to three rectangular axes in the second body, will fix also the position of the points p and p'.

Now, the distance from the point n to the point n' is equal to the distance from the point p to the point p', and since the two bodies are formed of the same substance, we conclude, according to the principle of the communication of heat, that the action of n on n', or the quantity of heat given by n to n', and the action of p on p', are to each other in the same ratio as the differences of the temperature $v-v'$ and $u-u'$.

Substituting v and then v' in the equation which belongs to the first solid, and subtracting, we find $v-v'=\frac{b-a}{e}(z-z')$; we have also by means of the second equation $u-u'=\frac{b'-a'}{e'}(z-z')$, whence the ratio of the two actions in question is that of $\frac{a-b}{e}$ to $\frac{a'-b'}{e'}$.

We may now imagine many other systems of two molecules, the first of which sends to the second across the plane L, a certain quantity of heat, and each of these systems, chosen in the first solid, may be compared with a homologous system situated in the second, and whose action is exerted across the section L'; we can then apply again the previous reasoning to prove that the ratio of the two actions is always that of $\frac{a-b}{e}$ to $\frac{a'-b'}{e'}$.

Now, the whole quantity of heat which, during one instant, crosses the section L, results from the simultaneous action of a multitude of systems each of which is formed of two points; hence this quantity of heat and that which, in the second solid, crosses during the same instant the section L', are also to each other in the ratio of $\frac{a-b}{e}$ to $\frac{a'-b'}{e'}$.

It is easy then to compare with each other the intensities of the constant flows of heat which are propagated uniformly in the two solids, that is to say, the quantities of heat which, during unit of time, cross unit of surface of each of these bodies. The ratio of these intensities is that of the two quotients $\frac{a-b}{e}$ and $\frac{a'-b'}{e'}$. If the two quotients are equal, the flows are the same, whatever in other respects the values a, b, e, a', b', e', may be; in general, denoting the first flow by F and the second by F', we shall have $\frac{F}{F'} = \frac{a-b}{e} \div \frac{a'-b'}{e'}$.

68. Suppose that in the second solid, the permanent temperature a' of the lower plane is that of boiling water, 1; that the temperature e' of the upper plane is that of melting ice, 0; that the distance e' of the two planes is the unit of measure (a metre); let us denote by K the constant flow of heat which, during unit of time (a minute) would cross unit of surface in this last solid, if it were formed of a given substance; K expressing a certain number of units of heat, that is to say a certain number of times the heat necessary to convert a kilogramme of ice into water: we shall have, in general, to determine the

constant flow F, in a solid formed of the same substance, the equation $\frac{F}{K}=\frac{a-b}{e}$ or $F=K\frac{a-b}{e}$.

The value of F denotes the quantity of heat which, during the unit of time, passes across a unit of area of the surface taken on a section parallel to the base.

Thus the thermometric state of a solid enclosed between two parallel infinite plane sides whose perpendicular distance is e, and which are maintained at fixed temperatures a and b, is represented by the two equations:

$$v=a+\frac{b-a}{e}z, \text{ and } F=K\frac{a-b}{e} \text{ or } F=-K\frac{dv}{dz}.$$

The first of these equations expresses the law according to which the temperatures decrease from the lower side to the opposite side, the second indicates the quantity of heat which, during a given time, crosses a definite part of a section parallel to the base.

69. We have taken this coefficient K, which enters into the second equation, to be the measure of the specific conducibility of each substance; this number has very different values for different bodies.

It represents, in general, the quantity of heat which, in a homogeneous solid formed of a given substance and enclosed between two infinite parallel planes, flows, during one minute, across a surface of one square metre taken on a section parallel to the extreme planes, supposing that these two planes are maintained, one at the temperature of boiling water, the other at the temperature of melting ice, and that all the intermediate planes have acquired and retain a permanent temperature.

We might employ another definition of conducibility, since we could estimate the capacity for heat by referring it to unit of volume, instead of referring it to unit of mass. All these definitions are equally good provided they are clear and precise.

We shall shew presently how to determine by observation the value K of the conducibility or conductibility in different substances.

70. In order to establish the equations which we have cited in Article 68, it would not be necessary to suppose the points which exert their action across the planes to be at extremely small distances.

The results would still be the same if the distances of these points had any magnitude whatever; they would therefore apply also to the case where the direct action of heat extended within the interior of the mass to very considerable distances, all the circumstances which constitute the hypothesis remaining in other respects the same.

We need only suppose that the cause which maintains the temperatures at the surface of the solid, affects not only that part of the mass which is extremely near to the surface, but that its action extends to a finite depth. The equation $v = a - \frac{a-b}{e} z$ will still represent in this case the permanent temperatures of the solid. The true sense of this proposition is that, if we give to all points of the mass the temperatures expressed by the equation, and if besides any cause whatever, acting on the two extreme laminæ, retained always every one of their molecules at the temperature which the same equation assigns to them, the interior points of the solid would preserve without any change their initial state.

If we supposed that the action of a point of the mass could extend to a finite distance ϵ, it would be necessary that the thickness of the extreme laminæ, whose state is maintained by the external cause, should be at least equal to ϵ. But the quantity ϵ having in fact, in the natural state of solids, only an inappreciable value, we may make abstraction of this thickness; and it is sufficient for the external cause to act on each of the two layers, extremely thin, which bound the solid. This is always what must be understood by the expression, *to maintain the temperature of the surface constant.*

71. We proceed further to examine the case in which the same solid would be exposed, at one of its faces, to atmospheric air maintained at a constant temperature.

Suppose then that the lower plane preserves the fixed temperature a, by virtue of any external cause whatever, and that the upper plane, instead of being maintained as formerly at a less temperature b, is exposed to atmospheric air maintained at that temperature b, the perpendicular distance of the two planes being denoted always by e: the problem is to determine the final temperatures.

Assuming that in the initial state of the solid, the common temperature of its molecules is b or less than b, we can readily imagine that the heat which proceeds incessantly from the source A penetrates the mass, and raises more and more the temperatures of the intermediate sections; the upper surface is gradually heated, and permits part of the heat which has penetrated the solid to escape into the air. The system of temperatures continually approaches a final state which would exist of itself if it were once formed; in this final state, which is that which we are considering, the temperature of the plane B has a fixed but unknown value, which we will denote by β, and since the lower plane A preserves also a permanent temperature a, the system of temperatures is represented by the general equation $v = a + \frac{\beta - a}{e} z$, v denoting always the fixed temperature of the section whose height is z. The quantity of heat which flows during unit of time across a unit of surface taken on any section whatever is $k \frac{a - \beta}{e}$, k denoting the interior conducibility.

We must now consider that the upper surface B, whose temperature is β, permits the escape into the air of a certain quantity of heat which must be exactly equal to that which crosses any section whatever L of the solid. If it were not so, the part of the mass included between this section L and the plane B would not receive a quantity of heat equal to that which it loses; hence it would not maintain its state, which is contrary to hypothesis; the constant flow at the surface is therefore equal to that which traverses the solid: now, the quantity of heat which escapes, during unit of time, from unit of surface taken on the plane B, is expressed by $h(\beta - b)$, b being the fixed temperature of the air, and h the measure of the conducibility of the surface B; we must therefore have the equation $k \frac{a - \beta}{e} = h(\beta - b)$, which will determine the value of β.

From this may be derived $a-\beta=\frac{he(a-b)}{he+k}$, an equation whose second member is known; for the temperatures a and b are given, as are also the quantities h, k, e.

Introducing this value of $a-\beta$ into the general equation $v=a+\frac{\beta-a}{e}z$, we shall have, to express the temperatures of any section of the solid, the equation $a-v=\frac{hz(a-b)}{he+k}$, in which known quantities only enter with the corresponding variables v and z.

72. So far we have determined the final and permanent state of the temperatures in a solid enclosed between two infinite and parallel plane surfaces, maintained at unequal temperatures. This first case is, properly speaking, the case of the linear and uniform propagation of heat, for there is no transfer of heat in the plane parallel to the sides of the solid; that which traverses the solid flows uniformly, since the value of the flow is the same for all instants and for all sections.

We will now restate the three chief propositions which result from the examination of this problem; they are susceptible of a great number of applications, and form the first elements of our theory.

1st. If at the two extremities of the thickness e of the solid we erect perpendiculars to represent the temperatures a and b of the two sides, and if we draw the straight line which joins the extremities of these two first ordinates, all the intermediate temperatures will be proportional to the ordinates of this straight line; they are expressed by the general equation $a-v=\frac{a-b}{e}z$, v denoting the temperature of the section whose height is z.

2nd. The quantity of heat which flows uniformly, during unit of time, across unit of surface taken on any section whatever parallel to the sides, all other things being equal, is directly proportional to the difference $a-b$ of the extreme temperatures, and inversely proportional to the distance e which separates these sides. The quantity of heat is expressed by $K\frac{a-b}{e}$, or

$-K\frac{dv}{dz}$, if we derive from the general equation the value of $\frac{dv}{dz}$ which is constant; this uniform flow may always be represented, for a given substance and in the solid under examination, by the tangent of the angle included between the perpendicular e and the straight line whose ordinates represent the temperatures.

3rd. One of the extreme surfaces of the solid being submitted always to the temperature a, if the other plane is exposed to air maintained at a fixed temperature b; the plane in contact with the air acquires, as in the preceding case, a fixed temperature β, greater than b, and it permits a quantity of heat to escape into the air across unit of surface, during unit of time, which is expressed by $h(\beta - b)$, h denoting the external conducibility of the plane.

The same flow of heat $h(\beta - b)$ is equal to that which traverses the prism and whose value is $K(a - \beta)$; we have therefore the equation $h(\beta - b) = K\frac{a - \beta}{e}$, which gives the value of β.

SECTION V.

Law of the permanent temperatures in a prism of small thickness.

73. We shall easily apply the principles which have just been explained to the following problem, very simple in itself, but one whose solution it is important to base on exact theory.

A metal bar, whose form is that of a rectangular parallelopiped infinite in length, is exposed to the action of a source of heat which produces a constant temperature at all points of its extremity A. It is required to determine the fixed temperatures at the different sections of the bar.

The section perpendicular to the axis is supposed to be a square whose side $2l$ is so small that we may without sensible error consider the temperatures to be equal at different points of the same section. The air in which the bar is placed is maintained at a constant temperature 0, and carried away by a current with uniform velocity.

Within the interior of the solid, heat will pass successively all the parts situate to the right of the source, and not exposed directly to its action; they will be heated more and more, but the temperature of each point will not increase beyond a certain limit. This maximum temperature is not the same for every section; it in general decreases as the distance of the section from the origin increases: we shall denote by v the fixed temperature of a section perpendicular to the axis, and situate at a distance x from the origin A.

Before every point of the solid has attained its highest degree of heat, the system of temperatures varies continually, and approaches more and more to a fixed state, which is that which we consider. This final state is kept up of itself when it has once been formed. In order that the system of temperatures may be permanent, it is necessary that the quantity of heat which, during unit of time, crosses a section made at a distance x from the origin, should balance exactly all the heat which, during the same time, escapes through that part of the external surface of the prism which is situated to the right of the same section. The lamina whose thickness is dx, and whose external surface is $8ldx$, allows the escape into the air, during unit of time, of a quantity of heat expressed by $8hlv \,.\, dx$, h being the measure of the external conducibility of the prism. Hence taking the integral $\int 8hlv \,.\, dx$ from $x=0$ to $x=\infty$, we shall find the quantity of heat which escapes from the whole surface of the bar during unit of time; and if we take the same integral from $x=0$ to $x=x$, we shall have the quantity of heat lost through the part of the surface included between the source of heat and the section made at the distance x. Denoting the first integral by C, whose value is constant, and the variable value of the second by $\int 8hlv \,.\, dx$; the difference $C - \int 8hlv \,.\, dx$ will express the whole quantity of heat which escapes into the air across the part of the surface situate to the right of the section. On the other hand, the lamina of the solid, enclosed between two sections infinitely near at distances x and $x+dx$, must resemble an infinite solid, bounded by two parallel planes, subject to fixed temperatures v and $v+dv$, since, by hypothesis, the temperature does not vary throughout the whole extent of the same section. The thickness of the solid is dx, and the area of the section is $4l^2$: hence the quantity of heat which flows uniformly, during unit of time, across a section of this solid, is, according to the preceding principles, $-4l^2k\frac{dv}{dx}$, k being the specific internal conducibility: we must therefore have the equation

$$-4l^2k\frac{dv}{dx} = C - \int 8hlv\,.\,dx,$$

whence

$$kl\frac{d^2v}{dx^2} = 2hv.$$

74. We should obtain the same result by considering the equilibrium of heat in a single lamina infinitely thin, enclosed between two sections at distances x and $x + dx$. In fact, the quantity of heat which, during unit of time, crosses the first section situate at distance x, is $-4l^2k\frac{dv}{dx}$. To find that which flows during the same time across the successive section situate at distance $x + dx$, we must in the preceding expression change x into $x + dx$, which gives $-4l^2k\,.\left[\frac{dv}{dx} + d\left(\frac{dv}{dx}\right)\right]$. If we subtract the second expression from the first we shall find how much heat is acquired by the lamina bounded by these two sections during unit of time; and since the state of the lamina is permanent, it follows that all the heat acquired is dispersed into the air across the external surface $8ldx$ of the same lamina: now the last quantity of heat is $8hlvdx$: we shall obtain therefore the same equation

$$8hlvdx = 4l^2kd\left(\frac{dv}{dx}\right), \text{ whence } \frac{d^2v}{dx^2} = \frac{2h}{kl}v.$$

75. In whatever manner this equation is formed, it is necessary to remark that the quantity of heat which passes into the lamina whose thickness is dx, has a finite value, and that its exact expression is $-4l^2k\frac{dv}{dx}$. The lamina being enclosed between two surfaces the first of which has a temperature v,

and the second a lower temperature v', we see that the quantity of heat which it receives through the first surface depends on the difference $v-v'$, and is proportional to it: but this remark is not sufficient to complete the calculation. The quantity in question is not a differential: it has a finite value, since it is equivalent to all the heat which escapes through that part of the external surface of the prism which is situate to the right of the section. To form an exact idea of it, we must compare the lamina whose thickness is dx, with a solid terminated by two parallel planes whose distance is e, and which are maintained at unequal temperatures a and b. The quantity of heat which passes into such a prism across the hottest surface, is in fact proportional to the difference $a-b$ of the extreme temperatures, but it does not depend only on this difference: all other things being equal, it is less when the prism is thicker, and in general it is proportional to $\frac{a-b}{e}$. This is why the quantity of heat which passes through the first surface into the lamina, whose thickness is dx, is proportional to $\frac{v-v'}{dx}$.

We lay stress on this remark because the neglect of it has been the first obstacle to the establishment of the theory. If we did not make a complete analysis of the elements of the problem, we should obtain an equation not homogeneous, and, *a fortiori*, we should not be able to form the equations which express the movement of heat in more complex cases.

It was necessary also to introduce into the calculation the dimensions of the prism, in order that we might not regard, as general, consequences which observation had furnished in a particular case. Thus, it was discovered by experiment that a bar of iron, heated at one extremity, could not acquire, at a distance of six feet from the source, a temperature of one degree (octogesimal[1]); for to produce this effect, it would be necessary for the heat of the source to surpass considerably the point of fusion of iron; but this result depends on the thickness of the prism employed. If it had been greater, the heat would have been propagated to a greater distance, that is to say, the point of the bar which acquires a fixed temperature of one degree is

[1] Reaumur's Scale of Temperature. [A. F.]

much more remote from the source when the bar is thicker, all other conditions remaining the same. We can always raise by one degree the temperature of one end of a bar of iron, by heating the solid at the other end; we need only give the radius of the base a sufficient length: which is, we may say, evident, and of which besides a proof will be found in the solution of the problem (Art. 78).

76. The integral of the preceding equation is

$$v = Ae^{-x\sqrt{\frac{2h}{kl}}} + Be^{+x\sqrt{\frac{2h}{kl}}},$$

A and B being two arbitrary constants; now, if we suppose the distance x infinite, the value of the temperature v must be infinitely small; hence the term $Be^{+x\sqrt{\frac{2h}{kl}}}$ does not exist in the integral: thus the equation $v = Ae^{-x\sqrt{\frac{2h}{kl}}}$ represents the permanent state of the solid; the temperature at the origin is denoted by the constant A, since that is the value of v when x is zero.

This law according to which the temperatures decrease is the same as that given by experiment; several physicists have observed the fixed temperatures at different points of a metal bar exposed at its extremity to the constant action of a source of heat, and they have ascertained that the distances from the origin represent logarithms, and the temperatures the corresponding numbers.

77. The numerical value of the constant quotient of two consecutive temperatures being determined by observation, we easily deduce the value of the ratio $\frac{h}{k}$; for, denoting by v_1, v_2 the temperatures corresponding to the distances x_1, x_2, we have

$$\frac{v_1}{v_2} = e^{-(x_1 - x_2)\sqrt{\frac{2h}{kl}}}, \text{ whence } \sqrt{\frac{2h}{k}} = \frac{\log v_1 - \log v_2}{x_2 - x_1}\sqrt{l}.$$

As for the separate values of h and k, they cannot be determined by experiments of this kind: we must observe also the varying motion of heat.

78. Suppose two bars of the same material and different dimensions to be submitted at their extremities to the same tem-

perature A; let l_1 be the side of a section in the first bar, and l_2 in the second, we shall have, to express the temperatures of these two solids, the equations

$$v_1 = Ae^{-x_1\sqrt{\frac{2h}{kl_1}}} \text{ and } v_2 = Ae^{-x_2\sqrt{\frac{2h}{kl_2}}},$$

v_1, in the first solid, denoting the temperature of a section made at distance x_1, and v_2, in the second solid, the temperature of a section made at distance x_2.

When these two bars have arrived at a fixed state, the temperature of a section of the first, at a certain distance from the source, will not be equal to the temperature of a section of the second at the same distance from the focus; in order that the fixed temperatures may be equal, the distances must be different. If we wish to compare with each other the distances x_1 and x_2 from the origin up to the points which in the two bars attain the same temperature, we must equate the second members of these equations, and from them we conclude that $\frac{x_1^2}{x_2^2} = \frac{l_1}{l_2}$. Thus the distances in question are to each other as the square roots of the thicknesses.

79. If two metal bars of equal dimensions, but formed of different substances, are covered with the same coating, which gives them the same external conducibility[1], and if they are submitted at their extremities to the same temperature, heat will be propagated most easily and to the greatest distance from the origin in that which has the greatest conducibility. To compare with each other the distances x_1 and x_2 from the common origin up to the points which acquire the same fixed temperature, we must, after denoting the respective conducibilities of the two substances by k_1 and k_2, write the equation

$$e^{-x_1\sqrt{\frac{2h}{k_1 l}}} = e^{-x_2\sqrt{\frac{2h}{k_2 l}}}, \text{ whence } \frac{x_1^2}{x_2^2} = \frac{k_1}{k_2}.$$

Thus the ratio of the two conducibilities is that of the squares of the distances from the common origin to the points which attain the same fixed temperature.

[1] Ingenhousz (1789), *Sur les métaux comme conducteurs de la chaleur. Journal de Physique*, XXXIV., 68, 380. Gren's *Journal der Physik*, Bd. I. [A. F.]

80. It is easy to ascertain how much heat flows during unit of time through a section of the bar arrived at its fixed state: this quantity is expressed by $-4kl^2\frac{dv}{dx}$, or $4A\sqrt{2khl^3}.e^{-x\sqrt{\frac{2h}{kl}}}$, and if we take its value at the origin, we shall have $4A\sqrt{2khl^3}$ as the measure of the quantity of heat which passes from the source into the solid during unit of time; thus the expenditure of the source of heat is, all other things being equal, proportional to the square root of the cube of the thickness.

We should obtain the same result on taking the integral $\int 8hlv\,.\,dx$ from x nothing to x infinite.

SECTION VI.

On the heating of closed spaces.

81. We shall again make use of the theorems of Article 72 in the following problem, whose solution offers useful applications; it consists in determining the extent of the heating of closed spaces.

Imagine a closed space, of any form whatever, to be filled with atmospheric air and closed on all sides, and that all parts of the boundary are homogeneous and have a common thickness e, so small that the ratio of the external surface to the internal surface differs little from unity. The space which this boundary terminates is heated by a source whose action is constant; for example, by means of a surface whose area is σ maintained at a constant temperature α.

We consider here only the mean temperature of the air contained in the space, without regard to the unequal distribution of heat in this mass of air; thus we suppose that the existing causes incessantly mingle all the portions of air, and make their temperatures uniform.

We see first that the heat which continually leaves the source spreads itself in the surrounding air and penetrates the mass of which the boundary is formed, is partly dispersed at the surface, and passes into the external air, which we suppose to be maintained at a lower and permanent temperature n. The inner air is heated more and more: the same is the case with the solid boundary: the system of temperatures steadily approaches a final state which is the object of the problem, and has the property of existing by itself and of being kept up unchanged, provided the surface of the source σ be maintained at the temperature α, and the external air at the temperature n.

In the permanent state which we wish to determine the air preserves a fixed temperature m; the temperature of the inner surface s of the solid boundary has also a fixed value a; lastly, the outer surface s, which terminates the enclosure, preserves a fixed temperature b less than a, but greater than n. The quantities σ, α, s, e and n are known, and the quantities m, a and b are unknown.

The degree of heating consists in the excess of the temperature m over n, the temperature of the external air; this excess evidently depends on the area σ of the heating surface and on its temperature α; it depends also on the thickness e of the enclosure, on the area s of the surface which bounds it, on the facility with which heat penetrates the inner surface or that which is opposite to it; finally, on the specific conducibility of the solid mass which forms the enclosure: for if any one of these elements were to be changed, the others remaining the same, the degree of the heating would vary also. The problem is to determine how all these quantities enter into the value of $m - n$.

82. The solid boundary is terminated by two equal surfaces, each of which is maintained at a fixed temperature; every prismatic element of the solid enclosed between two opposite portions of these surfaces, and the normals raised round the contour of the bases, is therefore in the same state as if it belonged to an infinite solid enclosed between two parallel planes, maintained at unequal temperatures. All the prismatic elements which compose the boundary touch along their whole length. The points of the mass which are equidistant from the inner surface have equal temperatures, to whatever prism they belong; consequently there cannot be any transfer of heat in the direction perpendicular to the length of these prisms. The case is, therefore, the same

as that of which we have already treated, and we must apply to it the linear equations which have been stated in former articles.

83. Thus in the permanent state which we are considering, the flow of heat which leaves the surface σ during a unit of time, is equal to that which, during the same time, passes from the surrounding air into the inner surface of the enclosure; it is equal also to that which, in a unit of time, crosses an intermediate section made within the solid enclosure by a surface equal and parallel to those which bound this enclosure; lastly, the same flow is again equal to that which passes from the solid enclosure across its external surface, and is dispersed into the air. If these four quantities of flow of heat were not equal, some variation would necessarily occur in the state of the temperatures, which is contrary to the hypothesis.

The first quantity is expressed by $\sigma(\alpha - m)g$, denoting by g the external conducibility of the surface σ, which belongs to the source of heat.

The second is $s(m-a)h$, the coefficient h being the measure of the external conducibility of the surface s, which is exposed to the action of the source of heat.

The third is $s\dfrac{a-b}{e}K$, the coefficient K being the measure of the conducibility proper to the homogeneous substance which forms the boundary.

The fourth is $s(b-n)H$, denoting by H the external conducibility of the surface s, which the heat quits to be dispersed into the air. The coefficients h and H may have very unequal values on account of the difference of the state of the two surfaces which bound the enclosure; they are supposed to be known, as also the coefficient K: we shall have then, to determine the three unknown quantities m, a and b, the three equations:

$$\sigma(\alpha - m)g = s(m-a)h,$$

$$\sigma(\alpha - m)g = s\frac{a-b}{e}K,$$

$$\sigma(\alpha - m)g = s(b-n)H.$$

84. The value of m is the special object of the problem. It may be found by writing the equations in the form

$$m - a = \frac{\sigma}{s}\frac{g}{h}(\alpha - m),$$

$$a - b = \frac{\sigma}{s}\frac{ge}{K}(\alpha - m),$$

$$b - n = \frac{\sigma}{s}\frac{g}{H}(\alpha - m);$$

adding, we have $m - n = (\alpha - m) P,$

denoting by P the known quantity $\frac{\sigma}{s}\left(\frac{g}{h} + \frac{ge}{K} + \frac{g}{H}\right)$;

whence we conclude

$$m - n = (\alpha - n)\frac{P}{1+P} = \frac{(\alpha - n)\frac{\sigma}{s}\left(\frac{g}{h} + \frac{ge}{K} + \frac{g}{H}\right)}{1 + \frac{\sigma}{s}\left(\frac{g}{h} + \frac{ge}{K} + \frac{g}{H}\right)}.$$

85. The result shews how $m - n$, the extent of the heating, depends on given quantities which constitute the hypothesis. We will indicate the chief results to be derived from it[1].

1st. The extent of the heating $m - n$ is directly proportional to the excess of the temperature of the source over that of the external air.

2nd. The value of $m - n$ does not depend on the form of the enclosure nor on its volume, but only on the ratio $\frac{\sigma}{s}$ of the surface from which the heat proceeds to the surface which receives it, and also on e the thickness of the boundary.

If we double σ the surface of the source of heat, the extent of the heating does not become double, but increases according to a certain law which the equation expresses.

[1] These results were stated by the author in a rather different manner in the extract from his original memoir published in the *Bulletin par la Société Philomatique de Paris*, 1818, pp. 1—11. [A. F.]

3rd. All the specific coefficients which regulate the action of the heat, that is to say, g, K, H and h, compose, with the dimension e, in the value of $m - n$ a single element $\frac{g}{h} + \frac{ge}{K} + \frac{g}{H}$, whose value may be determined by observation.

If we doubled e the thickness of the boundary, we should have the same result as if, in forming it, we employed a substance whose conducibility proper was twice as great. Thus the employment of substances which are bad conductors of heat permits us to make the thickness of the boundary small; the effect which is obtained depends only on the ratio $\frac{e}{K}$.

4th. If the conducibility K is nothing, we find $m - n = \alpha$; that is to say, the inner air assumes the temperature of the source: the same is the case if H is zero, or h zero. These consequences are otherwise evident, since the heat cannot then be dispersed into the external air.

5th. The values of the quantities g, H, h, K and α, which we supposed known, may be measured by direct experiments, as we shall shew in the sequel; but in the actual problem, it will be sufficient to notice the value of $m - n$ which corresponds to given values of σ and of α, and this value may be used to determine the whole coefficient $\frac{g}{h} + \frac{ge}{K} + \frac{g}{H}$, by means of the equation $m - n = (\alpha - n)\frac{\sigma}{s} p \div \left(1 + \frac{\sigma}{s} p\right)$ in which p denotes the coefficient sought. We must substitute in this equation, instead of $\frac{\sigma}{s}$ and $\alpha - n$, the values of those quantities, which we suppose given, and that of $m - n$ which observation will have made known. From it may be derived the value of p, and we may then apply the formula to any number of other cases.

6th. The coefficient H enters into the value of $m - n$ in the same manner as the coefficient h; consequently the state of the surface, or that of the envelope which covers it, produces the same effect, whether it has reference to the inner or outer surface.

We should have considered it useless to take notice of these different consequences, if we were not treating here of entirely new problems, whose results may be of direct use.

86. We know that animated bodies retain a temperature sensibly fixed, which we may regard as independent of the temperature of the medium in which they live. These bodies are, after some fashion, constant sources of heat, just as inflamed substances are in which the combustion has become uniform. We may then, by aid of the preceding remarks, foresee and regulate exactly the rise of temperature in places where a great number of men are collected together. If we there observe the height of the thermometer under given circumstances, we shall determine in advance what that height would be, if the number of men assembled in the same space became very much greater.

In reality, there are several accessory circumstances which modify the results, such as the unequal thickness of the parts of the enclosure, the difference of their aspect, the effects which the outlets produce, the unequal distribution of heat in the air. We cannot therefore rigorously apply the rules given by analysis; nevertheless these rules are valuable in themselves, because they contain the true principles of the matter: they prevent vague reasonings and useless or confused attempts.

87. If the same space were heated by two or more sources of different kinds, or if the first inclosure were itself contained in a second enclosure separated from the first by a mass of air, we might easily determine in like manner the degree of heating and the temperature of the surfaces.

If we suppose that, besides the first source σ, there is a second heated surface π, whose constant temperature is β, and external conducibility j, we shall find, all the other denominations being retained, the following equation:

$$m - n = \frac{\dfrac{(\alpha - n)\,\sigma g + (\beta - n)\,\pi j}{s}\left(\dfrac{e}{K} + \dfrac{1}{H} + \dfrac{1}{h}\right)}{1 + \dfrac{\sigma g + \pi j}{s}\left(\dfrac{e}{K} + \dfrac{1}{H} + \dfrac{1}{h}\right)}.$$

If we suppose only one source σ, and if the first enclosure is itself contained in a second, s', h', K', H', e', representing the elements of the second enclosure which correspond to those of the first which were denoted by s, h, K, H, e; we shall find, p denoting the temperature of the air which surrounds the external surface of the second enclosure, the following equation:

$$m-p=\frac{(a-p)\,P}{1+P}.$$

The quantity P represents

$$\frac{\sigma}{s}\left(\frac{g}{h}+\frac{ge}{K}+\frac{g}{H}\right)+\frac{\sigma}{s'}\left(\frac{g}{h'}+\frac{ge'}{K'}+\frac{g}{H'}\right).$$

We should obtain a similar result if we had three or a greater number of successive enclosures; and from this we conclude that these solid envelopes, separated by air, assist very much in increasing the degree of heating, however small their thickness may be.

88. To make this remark more evident, we will compare the quantity of heat which escapes from the heated surface, with that which the same body would lose, if the surface which envelopes it were separated from it by an interval filled with air.

If the body A be heated by a constant cause, so that its surface preserves a fixed temperature b, the air being maintained at a less temperature a, the quantity of heat which escapes into the air in the unit of time across a unit of surface will be expressed by $h(b-a)$, h being the measure of the external conducibility. Hence in order that the mass may preserve a fixed temperature b, it is necessary that the source, whatever it may be, should furnish a quantity of heat equal to $hS(b-a)$, S denoting the area of the surface of the solid.

Suppose an extremely thin shell to be detached from the body A and separated from the solid by an interval filled with air; and suppose the surface of the same solid A to be still maintained at the temperature b. We see that the air contained between the shell and the body will be heated and will take a temperature a' greater than a. The shell itself will attain a permanent state and will transmit to the external air whose fixed temperature is a all the heat which the body loses. It follows that the quantity of heat escaping from the solid will be $hS(b-a')$, instead of being $hS(b-a)$, for we suppose that the new surface of the solid and the surfaces which bound the shell have likewise the same external conducibility h. It is evident that the expenditure of the source of heat will be less than it was at first. The problem is to determine the exact ratio of these quantities.

89. Let e be the thickness of the shell, m the fixed temperature of its inner surface, n that of its outer surface, and K its internal conducibility. We shall have, as the expression of the quantity of heat which leaves the solid through its surface, $hS(b-a')$.

As that of the quantity which penetrates the inner surface of the shell, $hS(a'-m)$.

As that of the quantity which crosses any section whatever of the same shell, $KS\frac{m-n}{e}$.

Lastly, as the expression of the quantity which passes through the outer surface into the air, $hS(n-a)$.

All these quantities must be equal, we have therefore the following equations:

$$h(n-a)=\frac{K}{e}(m-n),$$

$$h(n-a)=h(a'-m),$$

$$h(n-a)=h(b-a').$$

If moreover we write down the identical equation

$$h(n-a)=h(n-a),$$

and arrange them all under the forms

$$n-a=n-a,$$

$$m-n=\frac{he}{K}(n-a),$$

$$a'-m=n-a,$$

$$b-a'=n-a,$$

we find, on addition,

$$b-a=(n-a)\left(3+\frac{he}{K}\right).$$

The quantity of heat lost by the solid was $hS(b-a)$, when its surface communicated freely with the air, it is now $hS(b-a')$ or $hS(n-a)$, which is equivalent to $hS\dfrac{b-a}{3+\dfrac{he}{K}}$.

The first quantity is greater than the second in the ratio of $3+\dfrac{he}{K}$ to 1.

In order therefore to maintain at temperature b a solid whose surface communicates directly to the air, more than three times as much heat is necessary than would be required to maintain it at temperature b, when its extreme surface is not adherent but separated from the solid by any small interval whatever filled with air.

If we suppose the thickness e to be infinitely small, the ratio of the quantities of heat lost will be 3, which would also be the value if K were infinitely great.

We can easily account for this result, for the heat being unable to escape into the external air, without penetrating several surfaces, the quantity which flows out must diminish as the number of interposed surfaces increases; but we should have been unable to arrive at any exact judgment in this case, if the problem had not been submitted to analysis.

90. We have not considered, in the preceding article, the effect of radiation across the layer of air which separates the two surfaces; nevertheless this circumstance modifies the problem, since there is a portion of heat which passes directly across the intervening air. We shall suppose then, to make the object of the analysis more distinct, that the interval between the surfaces is free from air, and that the heated body is covered by any number whatever of parallel laminæ separated from each other.

If the heat which escapes from the solid through its plane superficies maintained at a temperature b expanded itself freely in vacuo and was received by a parallel surface maintained at a less temperature a, the quantity which would be dispersed in unit of time across unit of surface would be proportional to $(b-a)$, the difference of the two constant temperatures: this quantity would be represented by $H(b-a)$, H being the value of the relative conducibility which is not the same as h.

The source which maintains the solid in its original state must therefore furnish, in every unit of time, a quantity of heat equal to $HS(b-a)$.

We must now determine the new value of this expenditure in the case where the surface of the body is covered by several successive laminæ separated by intervals free from air, supposing always that the solid is subject to the action of any external cause whatever which maintains its surface at the temperature b.

Imagine the whole system of temperatures to have become fixed; let m be the temperature of the under surface of the first lamina which is consequently opposite to that of the solid, let n be the temperature of the upper surface of the same lamina, e its thickness, and K its specific conducibility; denote also by $m_1, n_1, m_2, n_2, m_3, n_3, m_4, n_4$, &c. the temperatures of the under and upper surfaces of the different laminæ, and by K, e, the conducibility and thickness of the same laminæ; lastly, suppose all these surfaces to be in a state similar to the surface of the solid, so that the value of the coefficient H is common to them.

The quantity of heat which penetrates the under surface of a lamina corresponding to any suffix i is $HS(n_{i-1}-m_i)$, that which crosses this lamina is $\frac{KS}{e}(m_i-n_i)$, and the quantity which escapes from its upper surface is $HS(n_i-m_{i+1})$. These three quantities, and all those which refer to the other laminæ are equal; we may therefore form the equation by comparing all these quantities in question with the first of them, which is $HS(b-m_1)$; we shall thus have, denoting the number of laminæ by j:

$$b-m_1=b-m_1,$$

$$m_1-n_1=\frac{He}{K}(b-m_1),$$

$$n_1-m_2=b-m_1,$$

$$m_2-n_2=\frac{He}{K}(b-m_1),$$

$$m_j - n_j = \frac{He}{K}(b - m_1),$$

$$n_j - a = b - m_1.$$

Adding these equations, we find

$$(b - a) = (b - m_1) j \left(1 + \frac{He}{K}\right) + 1.$$

The expenditure of the source of heat necessary to maintain the surface of the body A at the temperature b is $HS(b-a)$, when this surface sends its rays to a fixed surface maintained at the temperature a. The expenditure is $HS(b-m_1)$ when we place between the surface of the body A, and the fixed surface maintained at temperature a, a number j of isolated laminæ; thus the quantity of heat which the source must furnish is very much less in the second hypotheses than in the first, and the ratio of the two quantities is $\dfrac{1}{j\left(1 + \dfrac{He}{K}\right) + 1}$. If we suppose the thickness e of the laminæ to be infinitely small, the ratio is $\frac{1}{j+1}$. The expenditure of the source is then inversely as the number of laminæ which cover the surface of the solid.

91. The examination of these results and of those which we obtained when the intervals between successive enclosures were occupied by atmospheric air explain clearly why the separation of surfaces and the intervention of air assist very much in retaining heat.

Analysis furnishes in addition analogous consequences when we suppose the source to be external, and that the heat which emanates from it crosses successively different diathermanous envelopes and the air which they enclose. This is what has happened when experimenters have exposed to the rays of the sun thermometers covered by several sheets of glass within which different layers of air have been enclosed.

For similar reasons the temperature of the higher regions of the atmosphere is very much less than at the surface of the earth.

In general the theorems concerning the heating of air in closed spaces extend to a great variety of problems. It would be useful to revert to them when we wish to foresee and regulate temperature with precision, as in the case of green-houses, drying-houses, sheep-folds, work-shops, or in many civil establishments, such as hospitals, barracks, places of assembly.

In these different applications we must attend to accessory circumstances which modify the results of analysis, such as the unequal thickness of different parts of the enclosure, the introduction of air, &c.; but these details would draw us away from our chief object, which is the exact demonstration of general principles.

For the rest, we have considered only, in what has just been said, the permanent state of temperature in closed spaces. We can in addition express analytically the variable state which precedes, or that which begins to take place when the source of heat is withdrawn, and we can also ascertain in this way, how the specific properties of the bodies which we employ, or their dimensions affect the progress and duration of the heating; but these researches require a different analysis, the principles of which will be explained in the following chapters.

SECTION VII.

On the uniform movement of heat in three dimensions.

92. Up to this time we have considered the uniform movement of heat in one dimension only, but it is easy to apply the same principles to the case in which heat is propagated uniformly in three directions at right angles.

Suppose the different points of a solid enclosed by six planes at right angles to have unequal actual temperatures represented by the linear equation $v = A + ax + by + cz$, x, y, z, being the rectangular co-ordinates of a molecule whose temperature is v. Suppose further that any external causes whatever acting on the six faces of the prism maintain every one of the molecules situated on the surface, at its actual temperature expressed by the general equation

$$v = A + ax + by + cz \text{ .... } (a),$$

we shall prove that the same causes which, by hypothesis, keep the outer layers of the solid in their initial state, are sufficient to preserve also the actual temperatures of every one of the inner molecules, so that their temperatures do not cease to be represented by the linear equation.

The examination of this question is an element of the general theory, it will serve to determine the laws of the varied movement of heat in the interior of a solid of any form whatever, for every one of the prismatic molecules of which the body is composed is during an infinitely small time in a state similar to that which the linear equation (a) expresses. We may then, by following the ordinary principles of the differential calculus, easily deduce from the notion of uniform movement the general equations of varied movement.

93. In order to prove that when the extreme layers of the solid preserve their temperatures no change can happen in the interior of the mass, it is sufficient to compare with each other the quantities of heat which, during the same instant, cross two parallel planes.

Let b be the perpendicular distance of these two planes which we first suppose parallel to the horizontal plane of x and y. Let m and m' be two infinitely near molecules, one of which is above the first horizontal plane and the other below it: let x, y, z be the co-ordinates of the first molecule, and x', y', z' those of the second. In like manner let M and M' denote two infinitely near molecules, separated by the second horizontal plane and situated, relatively to that plane, in the same manner as m and m' are relatively to the first plane; that is to say, the co-ordinates of M are $x, y, z+b$, and those of M' are $x', y', z'+b$. It is evident that the distance mm' of the two molecules m and m' is equal to the distance MM' of the two molecules M and M'; further, let v be the temperature of m, and v' that of m', also let V and V' be the temperatures of M and M', it is easy to see that the two differences $v-v'$ and $V-V'$ are equal; in fact, substituting first the co-ordinates of m and m' in the general equation

$$v = A + ax + by + cz,$$

we find $$v - v' = a(x-x') + b(y-y') + c(z-z'),$$

and then substituting the co-ordinates of M and M', we find also $V - V' = a(x - x') + b(y - y') + c(z - z')$. Now the quantity of heat which m sends to m' depends on the distance mm', which separates these molecules, and it is proportional to the difference $v - v'$ of their temperatures. This quantity of heat transferred may be represented by

$$q(v - v')\,dt;$$

the value of the coefficient q depends in some manner on the distance mm', and on the nature of the substance of which the solid is formed, dt is the duration of the instant. The quantity of heat transferred from M to M', or the action of M on M' is expressed likewise by $q(V - V')\,dt$, and the coefficient q is the same as in the expression $q(v - v')\,dt$, since the distance MM' is equal to mm' and the two actions are effected in the same solid: furthermore $V - V'$ is equal to $v - v'$, hence the two actions are equal.

If we choose two other points n and n', very near to each other, which transfer heat across the first horizontal plane, we shall find in the same manner that their action is equal to that of two homologous points N and N' which communicate heat across the second horizontal plane. We conclude then that the whole quantity of heat which crosses the first plane is equal to that which crosses the second plane during the same instant. We should derive the same result from the comparison of two planes parallel to the plane of x and z, or from the comparison of two other planes parallel to the plane of y and z. Hence any part whatever of the solid enclosed between six planes at right angles, receives through each of its faces as much heat as it loses through the opposite face; hence no portion of the solid can change temperature.

94. From this we see that, across one of the planes in question, a quantity of heat flows which is the same at all instants, and which is also the same for all other parallel sections.

In order to determine the value of this constant flow we shall compare it with the quantity of heat which flows uniformly in the most simple case, which has been already discussed. The case is that of an infinite solid enclosed between two infinite planes and maintained in a constant state. We have seen that the temperatures of the different points of the mass are in this case represented by the equation $v = A + cz$; we proceed to prove that the uniform flow of heat propagated in the vertical direction in the infinite solid is equal to that which flows in the same direction across the prism enclosed by six planes at right angles. This equality necessarily exists if the coefficient c in the equation $v = A + cz$, belonging to the first solid, is the same as the coefficient c in the more general equation $v = A + ax + by + cz$ which represents the state of the prism. In fact, denoting by H a plane in this prism perpendicular to z, and by m and μ two molecules very near to each other, the first of which m is below the plane H, and the second above this plane, let v be the temperature of m whose co-ordinates are x, y, z, and w the temperature of μ whose co-ordinates are $x + \alpha$, $y + \beta$, $z + \gamma$. Take a third molecule μ' whose co-ordinates are $x - \alpha$, $y - \beta$, $z + \gamma$, and whose temperature may be denoted by w'. We see that μ and μ' are on the same horizontal plane, and that the vertical drawn from the middle point of the line $\mu\mu'$, which joins these two points, passes through the point m, so that the distances $m\mu$ and $m\mu'$ are equal. The action of m on μ, or the quantity of heat which the first of these molecules sends to the other across the plane H, depends on the difference $v - w$ of their temperatures. The action of m on μ' depends in the same manner on the difference $v - w'$ of the temperatures of these molecules, since the distance of m from μ is the same as that of m from μ'. Thus, expressing by $q(v - w)$ the action of m on μ during the unit of time, we shall have $q(v - w')$ to express the action of m on μ', q being a common unknown factor, depending on the distance $m\mu$ and on the nature of the solid. Hence the sum of the two actions exerted during unit of time is $q(v - w + v - w')$.

If instead of x, y, and z, in the general equation

$$v = A + ax + by + cz,$$

we substitute the co-ordinates of m and then those of μ and μ', we shall find

$$v - w = -a\alpha - b\beta - c\gamma,$$

$$v - w' = +a\alpha + b\beta - c\gamma.$$

The sum of the two actions of m on μ and of m on μ' is therefore $-2q c\gamma$.

Suppose then that the plane H belongs to the infinite solid whose temperature equation is $v = A + cz$, and that we denote also by m, μ and μ' those molecules in this solid whose co-ordinates are x, y, z for the first, $x+\alpha$, $y+\beta$, $z+\gamma$ for the second, and $x-\alpha$, $y-\beta$, $z+\gamma$ for the third: we shall have, as in the preceding case, $v - w + v - w' = -2c\gamma$. Thus the sum of the two actions of m on μ and of m on μ', is the same in the infinite solid as in the prism enclosed between the six planes at right angles.

We should obtain a similar result, if we considered the action of another point n below the plane H on two others ν and ν', situated at the same height above the plane. Hence, the sum of all the actions of this kind, which are exerted across the plane H, that is to say the whole quantity of heat which, during unit of time, passes to the upper side of this surface, by virtue of the action of very near molecules which it separates, is always the same in both solids.

95. In the second of these two bodies, that which is bounded by two infinite planes, and whose temperature equation is $v = A + cz$, we know that the quantity of heat which flows during unit of time across unit of area taken on any horizontal section whatever is $-cK$, c being the coefficient of z, and K the specific conducibility; hence, the quantity of heat which, in the prism enclosed between six planes at right angles, crosses during unit of time, unit of area taken on any horizontal section whatever, is also $-cK$, when the linear equation which represents the temperatures of the prism is

$$v = A + ax + by + cz.$$

In the same way it may be proved that the quantity of heat which, during unit of time, flows uniformly across unit of area taken on any section whatever perpendicular to x, is expressed by $-aK$, and that the whole quantity which, during unit of time, crosses unit of area taken on a section perpendicular to y, is expressed by $-bK$.

The theorems which we have demonstrated in this and the two preceding articles, suppose the direct action of heat in the interior of the mass to be limited to an extremely small distance, but they would still be true, if the rays of heat sent out by each molecule could penetrate directly to a quite appreciable distance, but it would be necessary in this case, as we have remarked in Article 70, to suppose that the cause which maintains the temperatures of the faces of the solid affects a part extending within the mass to a finite depth.

SECTION VIII.

Measure of the movement of heat at a given point of a solid mass.

96. It still remains for us to determine one of the principal elements of the theory of heat, which consists in defining and in measuring exactly the quantity of heat which passes through every point of a solid mass across a plane whose direction is given.

If heat is unequally distributed amongst the molecules of the same body, the temperatures at any point will vary every instant. Denoting by t the time which has elapsed, and by v the temperature attained after a time t by an infinitely small molecule whose co-ordinates are x, y, z; the variable state of the solid will be expressed by an equation similar to the following $v = F(x, y, z, t)$. Suppose the function F to be given, and that consequently we can determine at every instant the temperature of any point whatever; imagine that through the point m we draw a horizontal plane parallel to that of x and y, and that on this plane we trace an infinitely small circle ω, whose centre is at m; it is required to determine what is the quantity of heat which during the instant dt will pass across the circle ω from the part of the solid which is below the plane into the part above it.

All points extremely near to the point m and under the plane exert their action during the infinitely small instant dt, on all those which are above the plane and extremely near to the point m, that is to say, each of the points situated on one side of this plane will send heat to each of those which are situated on the other side.

We shall consider as positive an action whose effect is to transport a certain quantity of heat above the plane, and as negative that which causes heat to pass below the plane. The sum of all the partial actions which are exerted across the circle ω, that is to say the sum of all the quantities of heat which, crossing any point whatever of this circle, pass from the part of the solid below the plane to the part above, compose the flow whose expression is to be found.

It is easy to imagine that this flow may not be the same throughout the whole extent of the solid, and that if at another point m' we traced a horizontal circle ω equal to the former, the two quantities of heat which rise above these planes ω and ω' during the same instant might not be equal: these quantities are comparable with each other and their ratios are numbers which may be easily determined.

97. We know already the value of the constant flow for the case of linear and uniform movement; thus in the solid enclosed between two infinite horizontal planes, one of which is maintained at the temperature a and the other at the temperature b, the flow of heat is the same for every part of the mass; we may regard it as taking place in the vertical direction only. The value corresponding to unit of surface and to unit of time is $K\left(\frac{a-b}{e}\right)$, e denoting the perpendicular distance of the two planes, and K the specific conducibility: the temperatures at the different points of the solid are expressed by the equation $v = a - \left(\frac{a-b}{e}\right) z$.

When the problem is that of a solid comprised between six rectangular planes, pairs of which are parallel, and the temperatures at the different points are expressed by the equation

$$v = A + ax + by + cz,$$

the propagation takes place at the same time along the directions of x, of y, of z; the quantity of heat which flows across a definite portion of a plane parallel to that of x and y is the same throughout the whole extent of the prism; its value corresponding to unit of surface, and to unit of time is $-cK$, in the direction of z, it is $-bK$, in the direction of y, and $-aK$ in that of x.

In general the value of the vertical flow in the two cases which we have just cited, depends only on the coefficient of z and on the specific conducibility K; this value is always equal to $-K\frac{dv}{dz}$.

The expression of the quantity of heat which, during the instant dt, flows across a horizontal circle infinitely small, whose area is ω, and passes in this manner from the part of the solid which is below the plane of the circle to the part above, is, for the two cases in question, $-K\frac{dv}{dz}\omega dt$.

98. It is easy now to generalise this result and to recognise that it exists in every case of the varied movement of heat expressed by the equation $v = F(x, y, z, t)$.

Let us in fact denote by x', y', z', the co-ordinates of this point m, and its actual temperature by v'. Let $x' + \xi$, $y' + \eta$, $z' + \zeta$, be the co-ordinates of a point μ infinitely near to the point m, and whose temperature is w; ξ, η, ζ are quantities infinitely small added to the co-ordinates x', y', z'; they determine the position of molecules infinitely near to the point m, with respect to three rectangular axes, whose origin is at m, parallel to the axes of x, y, and z. Differentiating the equation

$$v = f(x, y, z, t)$$

and replacing the differentials by ξ, η, ζ, we shall have, to express the value of w which is equivalent to $v + dv$, the linear equation $w = v' + \frac{dv'}{dx}\xi + \frac{dv'}{dy}\eta + \frac{dv'}{dz}\zeta$; the coefficients $v', \frac{dv'}{dx}, \frac{dv'}{dy}, \frac{dv'}{dz}$, are functions of x, y, z, t, in which the given and constant values x', y', z', which belong to the point m, have been substituted for x, y, z.

Suppose that the same point m belongs also to a solid enclosed between six rectangular planes, and that the actual temperatures of the points of this prism, whose dimensions are finite, are expressed by the linear equation $w = A + a\xi + b\eta + c\zeta$; and that the molecules situated on the faces which bound the solid are maintained by some external cause at the temperature which is assigned to them by the linear equation. ξ, η, ζ are the rectangular co-ordinates of a molecule of the prism, whose temperature is w, referred to three axes whose origin is at m.

This arranged, if we take as the values of the constant coefficients A, a, b, c, which enter into the equation for the prism, the quantities $v', \frac{dv'}{dx}, \frac{dv'}{dy}, \frac{dv'}{dz}$, which belong to the differential equation; the state of the prism expressed by the equation

$$w = v' + \frac{dv'}{dx}\xi + \frac{dv'}{dy}\eta + \frac{dv'}{dz}\zeta$$

will coincide as nearly as possible with the state of the solid; that is to say, all the molecules infinitely near to the point m will have the same temperature, whether we consider them to be in the solid or in the prism. This coincidence of the solid and the prism is quite analogous to that of curved surfaces with the planes which touch them.

It is evident, from this, that the quantity of heat which flows in the solid across the circle ω, during the instant dt, is the same as that which flows in the prism across the same circle; for all the molecules whose actions concur in one effect or the other, have the same temperature in the two solids. Hence, the flow in question, in one solid or the other, is expressed by $-K\frac{dv}{dz}\omega dt$. It would be $-K\frac{dv}{dy}\omega dt$, if the circle ω, whose centre is m, were perpendicular to the axis of y, and $-K\frac{dv}{dx}\omega dt$, if this circle were perpendicular to the axis of x.

The value of the flow which we have just determined varies in the solid from one point to another, and it varies also with the time. We might imagine it to have, at all the points of a unit of surface, the same value as at the point m, and to preserve this value during unit of time; the flow would then be expressed by $-K\frac{dv}{dz}$, it would be $-K\frac{dv}{dy}$ in the direction of y, and $-K\frac{dv}{dx}$ in that of x. We shall ordinarily employ in calculation this value of the flow thus referred to unit of time and to unit of surface.

99. This theorem serves in general to measure the velocity with which heat tends to traverse a given point of a plane situated in any manner whatever in the interior of a solid whose temperatures vary with the time. Through the given point m, a perpendicular must be raised upon the plane, and at every point of this perpendicular ordinates must be drawn to represent the actual temperatures at its different points. A plane curve will thus be formed whose axis of abscissæ is the perpendicular.

The fluxion of the ordinate of this curve, answering to the point *m*, taken with the opposite sign, expresses the velocity with which heat is transferred across the plane. This fluxion of the ordinate is known to be the tangent of the angle formed by the element of the curve with a parallel to the abscissæ.

The result which we have just explained is that of which the most frequent applications have been made in the theory of heat. We cannot discuss the different problems without forming a very exact idea of the value of the flow at every point of a body whose temperatures are variable. It is necessary to insist on this fundamental notion; an example which we are about to refer to will indicate more clearly the use which has been made of it in analysis.

100. Suppose the different points of a cubic mass, an edge of which has the length π, to have unequal actual temperatures represented by the equation $v = \cos x \cos y \cos z$. The co-ordinates x, y, z are measured on three rectangular axes, whose origin is at the centre of the cube, perpendicular to the faces. The points of the external surface of the solid are at the actual temperature 0, and it is supposed also that external causes maintain at all these points the actual temperature 0. On this hypothesis the body will be cooled more and more, the temperatures of all the points situated in the interior of the mass will vary, and, after an infinite time, they will all attain the temperature 0 of the surface. Now, we shall prove in the sequel, that the variable state of this solid is expressed by the equation

$$v = e^{-gt} \cos x \cos y \cos z,$$

the coefficient g is equal to $\frac{3K}{C.D}$, K is the specific conducibility of the substance of which the solid is formed, D is the density and C the specific heat; t is the time elapsed.

We here suppose that the truth of this equation is admitted, and we proceed to examine the use which may be made of it to find the quantity of heat which crosses a given plane parallel to one of the three planes at the right angles.

If, through the point m, whose co-ordinates are x, y, z, we draw a plane perpendicular to z, we shall find, after the mode of the preceding article, that the value of the flow, at this point and across the plane, is $-K\frac{dv}{dz}$, or $Ke^{-gt}\cos x . \cos y . \sin z$. The quantity of heat which, during the instant dt, crosses an infinitely small rectangle, situated on this plane, and whose sides are dx and dy, is

$$Ke^{-gt}\cos x \cos y \sin z \, dx \, dy \, dt.$$

Thus the whole heat which, during the instant dt, crosses the entire area of the same plane, is

$$Ke^{-gt}\sin z . dt \iint \cos x \cos y \, dx \, dy\,;$$

the double integral being taken from $x = -\frac{1}{2}\pi$ up to $x = \frac{1}{2}\pi$, and from $y = -\frac{1}{2}\pi$ up to $y = \frac{1}{2}\pi$. We find then for the expression of this total heat,

$$4Ke^{-gt}\sin z . dt.$$

If then we take the integral with respect to t, from $t = 0$ to $t = t$, we shall find the quantity of heat which has crossed the same plane since the cooling began up to the actual moment. This integral is $\frac{4K}{g}\sin z\,(1 - e^{-gt})$, its value at the surface is

$$\frac{4K}{g}(1 - e^{-gt}),$$

so that after an infinite time the quantity of heat lost through one of the faces is $\frac{4K}{g}$. The same reasoning being applicable to each of the six faces, we conclude that the solid has lost by its complete cooling a total quantity of heat equal to $\frac{24K}{g}$ or $8CD$, since g is equivalent to $\frac{3K}{CD}$. The total heat which is dissipated during the cooling must indeed be independent of the special conducibility K, which can only influence more or less the velocity of cooling.

100. A. We may determine in another manner the quantity of heat which the solid loses during a given time, and this will serve in some degree to verify the preceding calculation. In fact, the mass of the rectangular molecule whose dimensions are dx, dy, dz, is $D\,dx\,dy\,dz$, consequently the quantity of heat which must be given to it to bring it from the temperature 0 to that of boiling water is $CD\,dx\,dy\,dz$, and if it were required to raise this molecule to the temperature v, the expenditure of heat would be $v\,CD\,dx\,dy\,dz$.

It follows from this, that in order to find the quantity by which the heat of the solid, after time t, exceeds that which it contained at the temperature 0, we must take the multiple integral $\iiint v\,CD\,dx\,dy\,dz$, between the limits $x=-\frac{1}{2}\pi$, $x=\frac{1}{2}\pi$, $y=-\frac{1}{2}\pi$, $y=\frac{1}{2}\pi$, $z=-\frac{1}{2}\pi$, $z=\frac{1}{2}\pi$.

We thus find, on substituting for v its value, that is to say

$$e^{-gt}\cos x\cos y\cos z,$$

that the excess of actual heat over that which belongs to the temperature 0 is $8CD\,(1-e^{-gt})$; or, after an infinite time, $8CD$, as we found before.

We have described, in this introduction, all the elements which it is necessary to know in order to solve different problems relating to the movement of heat in solid bodies, and we have given some applications of these principles, in order to shew the mode of employing them in analysis; the most important use which we have been able to make of them, is to deduce from them the general equations of the propagation of heat, which is the subject of the next chapter.

Note on Art. 76. The researches of J. D. Forbes on the temperatures of a long iron bar heated at one end shew conclusively that the conducting power K is not constant, but diminishes as the temperature increases.—*Transactions of the Royal Society of Edinburgh*, Vol. XXIII. pp. 133—146 and Vol. XXIV. pp. 73—110.

Note on Art. 98. General expressions for the flow of heat within a mass in which the conductibility varies with the direction of the flow are investigated by Lamé in his *Théorie Analytique de la Chaleur*, pp. 1—8. [A. F.]

CHAPTER II.

EQUATIONS OF THE MOVEMENT OF HEAT.

SECTION I.

Equation of the varied movement of heat in a ring.

101. WE might form the general equations which represent the movement of heat in solid bodies of any form whatever, and apply them to particular cases. But this method would often involve very complicated calculations which may easily be avoided. There are several problems which it is preferable to treat in a special manner by expressing the conditions which are appropriate to them; we proceed to adopt this course and examine separately the problems which have been enunciated in the first section of the introduction; we will limit ourselves at first to forming the differential equations, and shall give the integrals of them in the following chapters.

102. We have already considered the uniform movement of heat in a prismatic bar of small thickness whose extremity is immersed in a constant source of heat. This first case offered no difficulties, since there was no reference except to the permanent state of the temperatures, and the equation which expresses them is easily integrated. The following problem requires a more profound investigation; its object is to determine the variable state of a solid ring whose different points have received initial temperatures entirely arbitrary.

The solid ring or armlet is generated by the revolution of a rectangular section about an axis perpendicular to the plane of the ring (see figure 3), l is the perimeter of the section whose area is S, the coefficient h measures the external conducibility, K the internal conducibility, C the specific capacity for heat, D the density. The line $oxx'x''$ represents the mean circumference of the armlet, or that line which passes through the centres of figure of all the sections; the distance of a section from the origin o is measured by the arc whose length is x; R is the radius of the mean circumference.

Fig. 3.

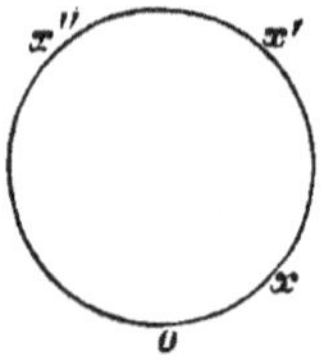

It is supposed that on account of the small dimensions and of the form of the section, we may consider the temperature at the different points of the same section to be equal.

103. Imagine that initial arbitrary temperatures have been given to the different sections of the armlet, and that the solid is then exposed to air maintained at the temperature 0, and displaced with a constant velocity; the system of temperatures will continually vary, heat will be propagated within the ring, and dispersed at the surface: it is required to determine what will be the state of the solid at any given instant.

Let v be the temperature which the section situated at distance x will have acquired after a lapse of time t; v is a certain function of x and t, into which all the initial temperatures also must enter: this is the function which is to be discovered.

104. We will consider the movement of heat in an infinitely small slice, enclosed between a section made at distance x and another section made at distance $x+dx$. The state of this slice for the duration of one instant is that of an infinite solid terminated by two parallel planes maintained at unequal temperatures; thus the quantity of heat which flows during this instant dt across the first section, and passes in this way from the part of the solid which precedes the slice into the slice itself, is measured according to the principles established in the introduction, by the product of four factors, that is to say, the conducibility K, the area of the section S, the ratio $-\frac{dv}{dx}$, and the duration of the instant; its expression is $-KS\frac{dv}{dx}dt$. To determine the quantity of heat

which escapes from the same slice across the second section, and passes into the contiguous part of the solid, it is only necessary to change x into $x+dx$ in the preceding expression, or, which is the same thing, to add to this expression its differential taken with respect to x; thus the slice receives through one of its faces a quantity of heat equal to $-KS\frac{dv}{dx}dt$, and loses through the opposite face a quantity of heat expressed by

$$-KS\frac{dv}{dx}dt - KS\frac{d^2v}{dx^2}dx\,dt.$$

It acquires therefore by reason of its position a quantity of heat equal to the difference of the two preceding quantities, that is

$$KS\frac{d^2v}{dx^2}dx\,dt.$$

On the other hand, the same slice, whose external surface is ldx and whose temperature differs infinitely little from v, allows a quantity of heat equivalent to $hlv\,dx\,dt$ to escape into the air during the instant dt; it follows from this that this infinitely small part of the solid retains in reality a quantity of heat represented by $KS\frac{d^2v}{dx^2}dx\,dt - hlv\,dx\,dt$ which makes its temperature vary. The amount of this change must be examined.

105. The coefficient C expresses how much heat is required to raise unit of weight of the substance in question from temperature 0 up to temperature 1; consequently, multiplying the volume Sdx of the infinitely small slice by the density D, to obtain its weight, and by C the specific capacity for heat, we shall have $CDSdx$ as the quantity of heat which would raise the volume of the slice from temperature 0 up to temperature 1. Hence the increase of temperature which results from the addition of a quantity of heat equal to $KS\frac{d^2v}{dx^2}dx\,dt - hlv\,dx\,dt$ will be found by dividing the last quantity by $CDS\,dx$. Denoting therefore, according to custom, the increase of temperature which takes place during the instant dt by $\frac{dv}{dt}dt$, we shall have the equation

$$\frac{dv}{dt}=\frac{K}{CD}\frac{d^2v}{dx^2}-\frac{hl}{CDS}v \ldots\ldots\ldots (b)$$

We shall explain in the sequel the use which may be made of this equation to determine the complete solution, and what the difficulty of the problem consists in; we limit ourselves here to a remark concerning the permanent state of the armlet.

106. Suppose that, the plane of the ring being horizontal, sources of heat, each of which exerts a constant action, are placed below different points m, n, p, q etc.; heat will be propagated in the solid, and that which is dissipated through the surface being incessantly replaced by that which emanates from the sources, the temperature of every section of the solid will approach more and more to a stationary value which varies from one section to another. In order to express by means of equation (b) the law of the latter temperatures, which would exist of themselves if they were once established, we must suppose that the quantity v does not vary with respect to t; which annuls the term $\frac{dv}{dt}$. We thus have the equation

$$\frac{d^2v}{dx^2}=\frac{hl}{KS}v, \text{ whence } v=Me^{-x\sqrt{\frac{hl}{KS}}}+Ne^{+x\sqrt{\frac{hl}{KS}}},$$

M and N being two constants[1].

[1] This equation is the same as the equation for the steady temperature of a finite bar heated at one end (Art. 76), except that l here denotes the perimeter of a section whose area is S. In the case of the finite bar we can determine two relations between the constants M and N: for, if V be the temperature at the source, where $x=0$, $V=M+N$; and if at the end of the bar remote from the source, where $x=L$ suppose, we make a section at a distance dx from that end, the flow through this section is, in unit of time, $-KS\frac{dv}{dx}$, and this is equal to the waste of heat through the periphery and free end of the slice, $hv\,(ldx+S)$ namely; hence ultimately, dx vanishing,

$$hv+K\frac{dv}{dx}=0, \text{ when } x=L,$$

that is

$$Me^{-L\sqrt{\frac{hl}{KS}}}+Ne^{+L\sqrt{\frac{hl}{KS}}}=\sqrt{\frac{Kl}{hS}}\left(Me^{-L\sqrt{\frac{hl}{KS}}}-Ne^{+L\sqrt{\frac{hl}{KS}}}\right).$$

Cf. Verdet, *Conférences de Physique*, p. 37. [A. F.]

107. Suppose a portion of the circumference of the ring, situated between two successive sources of heat, to be divided into equal parts, and denote by v_1, v_2, v_3, v_4, &c., the temperatures at the points of division whose distances from the origin are x_1, x_2, x_3, x_4, &c.; the relation between v and x will be given by the preceding equation, after that the two constants have been determined by means of the two values of v corresponding to the sources of heat. Denoting by α the quantity $e^{-\sqrt{\frac{hl}{KS}}}$, and by λ the distance $x_2 - x_1$ of two consecutive points of division, we shall have the equations:

$$v_1 = M\alpha^{x_1} + N\alpha^{-x_1},$$

$$v_2 = M\alpha^{\lambda} . \alpha^{x_1} + N\alpha^{-\lambda}\alpha^{-x_1},$$

$$v_3 = M\alpha^{2\lambda}\alpha^{x_1} + N\alpha^{-2\lambda}\alpha^{-x_1},$$

whence we derive the following relation $\frac{v_1 + v_3}{v_2} = \alpha^{\lambda} + \alpha^{-\lambda}$.

We should find a similar result for the three points whose temperatures are v_2, v_3, v_4, and in general for any three consecutive points. It follows from this that if we observed the temperatures v_1, v_2, v_3, v_4, v_5 &c. of several successive points, all situated between the same two sources m and n and separated by a constant interval λ, we should perceive that any three consecutive temperatures are always such that the sum of the two extremes divided by the mean gives a constant quotient $\alpha^{\lambda} + \alpha^{-\lambda}$.

108. If, in the space included between the next two sources of heat n and p, the temperatures of other different points separated by the same interval λ were observed, it would still be found that for any three consecutive points, the sum of the two extreme temperatures, divided by the mean, gives the same quotient $\alpha^{\lambda} + \alpha^{-\lambda}$. The value of this quotient depends neither on the position nor on the intensity of the sources of heat.

109. Let q be this constant value, we have the equation

$$v_3 = qv_2 - v;$$

we see by this that when the circumference is divided into equal parts, the temperatures at the points of division, included between two consecutive sources of heat, are represented by the terms of a recurring series whose scale of relation is composed of two terms q and -1.

Experiments have fully confirmed this result. We have exposed a metallic ring to the permanent and simultaneous action of different sources of heat, and we have observed the stationary temperatures of several points separated by constant intervals; we always found that the temperatures of any three consecutive points, not separated by a source of heat, were connected by the relation in question. Even if the sources of heat be multiplied, and in whatever manner they be disposed, no change can be effected in the numerical value of the quotient $\frac{v_1 + v_3}{v_2}$; it depends only on the dimensions or on the nature of the ring, and not on the manner in which that solid is heated.

110. When we have found, by observation, the value of the constant quotient q or $\frac{v_1 + v_3}{v_2}$, the value of α^λ may be derived from it by means of the equation $\alpha^\lambda + \alpha^{-\lambda} = q$. One of the roots is α^λ, and other root is $\alpha^{-\lambda}$. This quantity being determined, we may derive from it the value of the ratio $\frac{h}{K}$, which is $\frac{S}{l}(\log \alpha)^2$. Denoting α^λ by ω, we shall have $\omega^2 - q\omega + 1 = 0$. Thus the ratio of the two conducibilities is found by multiplying $\frac{S}{l}$ by the square of the hyperbolic logarithm of one of the roots of the equation $\omega^2 - q\omega + 1 = 0$, and dividing the product by λ^2.

SECTION II.

Equation of the varied movement of heat in a solid sphere.

111. A solid homogeneous mass, of the form of a sphere, having been immersed for an infinite time in a medium maintained at a permanent temperature 1, is then exposed to air which is kept at temperature 0, and displaced with constant velocity: it is required to determine the successive states of the body during the whole time of the cooling.

Denote by x the distance of any point whatever from the centre of the sphere, and by v the temperature of the same point, after a time t has elapsed; and suppose, to make the problem more general, that the initial temperature, common to all points situated at the distance x from the centre, is different for different values of x; which is what would have been the case if the immersion had not lasted for an infinite time.

Points of the solid, equally distant from the centre, will not cease to have a common temperature; v is thus a function of x and t. When we suppose $t=0$, it is essential that the value of this function should agree with the initial state which is given, and which is entirely arbitrary.

112. We shall consider the instantaneous movement of heat in an infinitely thin shell, bounded by two spherical surfaces whose radii are x and $x+dx$: the quantity of heat which, during an infinitely small instant dt, crosses the lesser surface whose radius is x, and so passes from that part of the solid which is nearest to the centre into the spherical shell, is equal to the product of four factors which are the conducibility K, the duration dt, the extent $4\pi x^2$ of surface, and the ratio $\frac{dv}{dx}$, taken with the negative sign; it is expressed by $-4K\pi x^2 \frac{dv}{dx} dt$.

To determine the quantity of heat which flows during the same instant through the second surface of the same shell, and passes from this shell into the part of the solid which envelops it, x must be changed into $x+dx$, in the preceding expression: that is to say, to the term $-4K\pi x^2 \frac{dv}{dx} dt$ must be added the differential of this term taken with respect to x. We thus find

$$-4K\pi x^2 \frac{dv}{dx} dt - 4K\pi d\left(x^2 \frac{dv}{dx}\right) . dt$$

as the expression of the quantity of heat which leaves the spherical shell across its second surface; and if we subtract this quantity from that which enters through the first surface, we shall have $4K\pi d\left(x^2 \frac{dv}{dx}\right) dt$. This difference is evidently the quantity of heat which accumulates in the intervening shell, and whose effect is to vary its temperature.

113. The coefficient C denotes the quantity of heat which is necessary to raise, from temperature 0 to temperature 1, a definite unit of weight; D is the weight of unit of volume, $4\pi x^2 dx$ is the volume of the intervening layer, differing from it only by a quantity which may be omitted: hence $4\pi CDx^2dx$ is the quantity of heat necessary to raise the intervening shell from temperature 0 to temperature 1. Hence it is requisite to divide the quantity of heat which accumulates in this shell by $4\pi CDx^2dx$, and we shall then find the increase of its temperature v during the time dt. We thus obtain the equation

$$dv = \frac{K}{CD}\,dt \,.\, \frac{d\left(x^2 \dfrac{dv}{dx}\right)}{x^2dx},$$

or

$$\frac{dv}{dt} = \frac{K}{CD} \cdot \left(\frac{d^2v}{dx^2} + \frac{2}{x}\frac{dv}{dx}\right) \text{ .... } (c).$$

114. The preceding equation represents the law of the movement of heat in the interior of the solid, but the temperatures of points in the surface are subject also to a special condition which must be expressed. This condition relative to the state of the surface may vary according to the nature of the problems discussed: we may suppose for example, that, after having heated the sphere, and raised all its molecules to the temperature of boiling water, the cooling is effected by giving to all points in the surface the temperature 0, and by retaining them at this temperature by any external cause whatever. In this case we may imagine the sphere, whose variable state it is desired to determine, to be covered by a very thin envelope on which the cooling agency exerts its action. It may be supposed, 1°, that this infinitely thin envelope adheres to the solid, that it is of the same substance as the solid and that it forms a part of it, like the other portions of the mass; 2°, that all the molecules of the envelope are subjected to temperature 0 by a cause always in action which prevents the temperature from ever being above or below zero. To express this condition theoretically, the function v, which contains x and t,

must be made to become nul, when we give to x its complete value X equal to the radius of the sphere, whatever else the value of t may be. We should then have, on this hypothesis, if we denote by $\phi(x, t)$ the function of x and t, which expresses the value of v, the two equations

$$\frac{dv}{dt} = \frac{K}{CD}\left(\frac{d^2v}{dx^2} + \frac{2}{x}\frac{dv}{dx}\right), \qquad \text{and} \quad \phi(X, t) = 0.$$

Further, it is necessary that the initial state should be represented by the same function $\phi(x, t)$: we shall therefore have as a second condition $\phi(x, 0) = 1$. Thus the variable state of a solid sphere on the hypothesis which we have first described will be represented by a function v, which must satisfy the three preceding equations. The first is general, and belongs at every instant to all points of the mass; the second affects only the molecules at the surface, and the third belongs only to the initial state.

115. If the solid is being cooled in air, the second equation is different; it must then be imagined that the very thin envelope is maintained by some external cause, in a state such as to produce the escape from the sphere, at every instant, of a quantity of heat equal to that which the presence of the medium can carry away from it.

Now the quantity of heat which, during an infinitely small instant dt, flows within the interior of the solid across the spherical surface situate at distance x, is equal to $-4K\pi x^2 \frac{dv}{dx} dt$; and this general expression is applicable to all values of x. Thus, by supposing $x = X$ we shall ascertain the quantity of heat which in the variable state of the sphere would pass across the very thin envelope which bounds it; on the other hand, the external surface of the solid having a variable temperature, which we shall denote by V, would permit the escape into the air of a quantity of heat proportional to that temperature, and to the extent of the surface, which is $4\pi X^2$. The value of this quantity is $4h\pi X^2 V dt$.

To express, as is supposed, that the action of the envelope supplies the place, at every instant, of that which would result from the presence of the medium, it is sufficient to equate the quantity $4h\pi X^2 V dt$ to the value which the expression $-4K\pi X^2 \frac{dv}{dx} dt$

receives when we give to x its complete value X; hence we obtain the equation $\frac{dv}{dx} = -\frac{h}{K}v$, which must hold when in the functions $\frac{dv}{dx}$ and v we put instead of x its value X, which we shall denote by writing it in the form $K\frac{dV}{dx} + hV = 0$.

116. The value of $\frac{dv}{dx}$ taken when $x = X$, must therefore have a constant ratio $-\frac{h}{K}$ to the value of v, which corresponds to the same point. Thus we shall suppose that the external cause of the cooling determines always the state of the very thin envelope, in such a manner that the value of $\frac{dv}{dx}$ which results from this state, is proportional to the value of v, corresponding to $x = X$, and that the constant ratio of these two quantities is $-\frac{h}{K}$. This condition being fulfilled by means of some cause always present, which prevents the extreme value of $\frac{dv}{dx}$ from being anything else but $-\frac{h}{K}v$, the action of the envelope will take the place of that of the air.

It is not necessary to suppose the envelope to be extremely thin, and it will be seen in the sequel that it may have an indefinite thickness. Here the thickness is considered to be indefinitely small, so as to fix the attention on the state of the surface only of the solid.

117. Hence it follows that the three equations which are required to determine the function $\phi(x, t)$ or v are the following,

$$\frac{dv}{dt} = \frac{K}{CD}\left(\frac{d^2v}{dx^2} + \frac{2}{x}\frac{dv}{dx}\right), \quad K\frac{dV}{dx} + hV = 0, \quad \phi(x, 0) = 1.$$

The first applies to all possible values of x and t; the second is satisfied when $x = X$, whatever be the value of t; and the third is satisfied when $t = 0$, whatever be the value of x.

It might be supposed that in the initial state all the spherical layers have not the same temperature: which is what would necessarily happen, if the immersion were imagined not to have lasted for an indefinite time. In this case, which is more general than the foregoing, the given function, which expresses the initial temperature of the molecules situated at distance x from the centre of the sphere, will be represented by $F(x)$; the third equation will then be replaced by the following, $\phi(x, 0) = F(x)$.

Nothing more remains than a purely analytical problem, whose solution will be given in one of the following chapters. It consists in finding the value of v, by means of the general condition, and the two special conditions to which it is subject.

SECTION III.

Equations of the varied movement of heat in a solid cylinder.

118. A solid cylinder of infinite length, whose side is perpendicular to its circular base, having been wholly immersed in a liquid whose temperature is uniform, has been gradually heated, in such a manner that all points equally distant from the axis have acquired the same temperature; it is then exposed to a current of colder air; it is required to determine the temperatures of the different layers, after a given time.

x denotes the radius of a cylindrical surface, all of whose points are equally distant from the axis; X is the radius of the cylinder; v is the temperature which points of the solid, situated at distance x from the axis, must have after the lapse of a time denoted by t, since the beginning of the cooling. Thus v is a function of x and t, and if in it t be made equal to 0, the function of x which arises from this must necessarily satisfy the initial state, which is arbitrary.

119. Consider the movement of heat in an infinitely thin portion of the cylinder, included between the surface whose radius is x, and that whose radius is $x + dx$. The quantity of heat which this portion receives during the instant dt, from the part of the solid which it envelops, that is to say, the quantity which during the same time crosses the cylindrical surface

whose radius is x, and whose length is supposed to be equal to unity, is expressed by

$$-2K\pi x \frac{dv}{dx} dt.$$

To find the quantity of heat which, crossing the second surface whose radius is $x+dx$, passes from the infinitely thin shell into the part of the solid which envelops it, we must, in the foregoing expression, change x into $x+dx$, or, which is the same thing, add to the term

$$-2K\pi x \frac{dv}{dx} dt,$$

the differential of this term, taken with respect to x. Hence the difference of the heat received and the heat lost, or the quantity of heat which accumulating in the infinitely thin shell determines the changes of temperature, is the same differential taken with the opposite sign, or

$$2K\pi \,.\, dt \,.\, d\left(x \frac{dv}{dx}\right);$$

on the other hand, the volume of this intervening shell is $2\pi x dx$, and $2CD\pi x dx$ expresses the quantity of heat required to raise it from the temperature 0 to the temperature 1, C being the specific heat, and D the density. Hence the quotient

$$\frac{2K\pi \,.\, dt \,.\, d\left(x \frac{dv}{dx}\right)}{2CD\pi x dx}$$

is the increment which the temperature receives during the instant dt. Whence we obtain the equation

$$\frac{dv}{dt} = \frac{K}{CD}\left(\frac{d^2v}{dx^2} + \frac{1}{x}\frac{dv}{dx}\right).$$

120. The quantity of heat which, during the instant dt, crosses the cylindrical surface whose radius is x, being expressed in general by $2K\pi x \frac{dv}{dx} dt$, we shall find that quantity which escapes during the same time from the surface of the solid, by making $x = X$ in the foregoing value; on the other hand, the

same quantity, dispersed into the air, is, by the principle of the communication of heat, equal to $2\pi Xhvdt$; we must therefore have at the surface the definite equation $-K\dfrac{dv}{dx}=hv$. The nature of these equations is explained at greater length, either in the articles which refer to the sphere, or in those wherein the general equations have been given for a body of any form whatever. The function v which represents the movement of heat in an infinite cylinder must therefore satisfy, 1st, the general equation $\dfrac{dv}{dt}=\dfrac{K}{CD}\left(\dfrac{d^2v}{dx^2}+\dfrac{1}{x}\dfrac{dv}{dx}\right)$, which applies whatever x and t may be; 2nd, the definite equation $\dfrac{h}{K}v+\dfrac{dv}{dx}=0$, which is true, whatever the variable t may be, when $x=X$; 3rd, the definite equation $v=F(x)$. The last condition must be satisfied by all values of v, when t is made equal to 0, whatever the variable x may be. The arbitrary function $F(x)$ is supposed to be known; it corresponds to the initial state.

SECTION IV.

Equations of the uniform movement of heat in a solid prism of infinite length.

121. A prismatic bar is immersed at one extremity in a constant source of heat which maintains that extremity at the temperature A; the rest of the bar, whose length is infinite, continues to be exposed to a uniform current of atmospheric air maintained at temperature 0; it is required to determine the highest temperature which a given point of the bar can acquire.

The problem differs from that of Article 73, since we now take into consideration all the dimensions of the solid, which is necessary in order to obtain an exact solution.

We are led, indeed, to suppose that in a bar of very small thickness all points of the same section would acquire sensibly equal temperatures; but some uncertainty may rest on the results of this hypothesis. It is therefore preferable to solve the problem rigorously, and then to examine, by analysis, up to what point, and in what cases, we are justified in considering the temperatures of different points of the same section to be equal.

122. The section made at right angles to the length of the bar, is a square whose side is $2l$, the axis of the bar is the axis of x, and the origin is at the extremity A. The three rectangular co-ordinates of a point of the bar are x, y, z, and v denotes the fixed temperature at the same point.

The problem consists in determining the temperatures which must be assigned to different points of the bar, in order that they may continue to exist without any change, so long as the extreme surface A, which communicates with the source of heat, remains subject, at all its points, to the permanent temperature A; thus v is a function of x, y, and z.

123. Consider the movement of heat in a prismatic molecule, enclosed between six planes perpendicular to the three axes of x, y, and z. The first three planes pass through the point m whose co-ordinates are x, y, z, and the others pass through the point m' whose co-ordinates are $x + dx$, $y + dy$, $z + dz$.

To find what quantity of heat enters the molecule during unit of time across the first plane passing through the point m and perpendicular to x, we must remember that the extent of the surface of the molecule on this plane is $dydz$, and that the flow across this area is, according to the theorem of Article 98, equal to $-K\dfrac{dv}{dx}$; thus the molecule receives across the rectangle $dydz$ passing through the point m a quantity of heat expressed by $-K\,dydz\dfrac{dv}{dx}$. To find the quantity of heat which crosses the opposite face, and escapes from the molecule, we must substitute, in the preceding expression, $x + dx$ for x, or, which is the same thing, add to this expression its differential taken with respect to x only; whence we conclude that the molecule loses, at its second face perpendicular to x, a quantity of heat equal to

$$-K\,dydz\frac{dv}{dx} - K\,dydz\,d\left(\frac{dv}{dx}\right);$$

we must therefore subtract this from that which enters at the opposite face; the differences of these two quantities is

$$K\,dydz\,d\left(\frac{dv}{dx}\right), \text{ or, } K\,dxdydz\frac{d^2v}{dx^2};$$

this expresses the quantity of heat accumulated in the molecule in consequence of the propagation in direction of x; which accumulated heat would make the temperature of the molecule vary, if it were not balanced by that which is lost in some other direction.

It is found in the same manner that a quantity of heat equal to $-K\,dz\,dx\,\frac{dv}{dy}$ enters the molecule across the plane passing through the point m perpendicular to y, and that the quantity which escapes at the opposite face is

$$-K\,dz\,dx\,\frac{dv}{dy}-K\,dz\,dx\,d\left(\frac{dv}{dy}\right),$$

the last differential being taken with respect to y only. Hence the difference of the two quantities, or $K\,dx\,dy\,dz\,\frac{d^2v}{dy^2}$, expresses the quantity of heat which the molecule acquires, in consequence of the propagation in direction of y.

Lastly, it is proved in the same manner that the molecule acquires, in consequence of the propagation in direction of z, a quantity of heat equal to $K\,dx\,dy\,dz\,\frac{d^2v}{dz^2}$. Now, in order that there may be no change of temperature, it is necessary for the molecule to retain as much heat as it contained at first, so that the heat it acquires in one direction must balance that which it loses in another. Hence the sum of the three quantities of heat acquired must be nothing; thus we form the equation

$$\frac{d^2v}{dx^2}+\frac{d^2v}{dy^2}+\frac{d^2v}{dz^2}=0.$$

124. It remains now to express the conditions relative to the surface. If we suppose the point m to belong to one of the faces of the prismatic bar, and the face to be perpendicular to z, we see that the rectangle $dxdy$, during unit of time, permits a quantity of heat equal to $Vh\,dx\,dy$ to escape into the air, V denoting the temperature of the point m of the surface, namely what $\phi(x, y, z)$ the function sought becomes when z is made equal to l, half the dimension of the prism. On the other hand, the quantity of heat which, by virtue of the action of the

molecules, during unit of time, traverses an infinitely small surface ω, situated within the prism, perpendicular to z, is equal to $-K\omega\frac{dv}{dz}$, according to the theorems quoted above. This expression is general, and applying it to points for which the co-ordinate z has its complete value l, we conclude from it that the quantity of heat which traverses the rectangle $dx\,dy$ taken at the surface is $-K\,dx\,dy\frac{dv}{dz}$, giving to z in the function $\frac{dv}{dz}$ its complete value l. Hence the two quantities $-K\,dx\,dy\frac{dv}{dz}$, and $h\,dx\,dy\,v$, must be equal, in order that the action of the molecules may agree with that of the medium. This equality must also exist when we give to z in the functions $\frac{dv}{dz}$ and v the value $-l$, which it has at the face opposite to that first considered. Further, the quantity of heat which crosses an infinitely small surface ω, perpendicular to the axis of y, being $-K\omega\frac{dv}{dy}$, it follows that that which flows across a rectangle $dz\,dx$ taken on a face of the prism perpendicular to y is $-K\,dz\,dx\frac{dv}{dy}$, giving to y in the function $\frac{dv}{dy}$ its complete value l. Now this rectangle $dz\,dx$ permits a quantity of heat expressed by $hv\,dx\,dy$ to escape into the air; the equation $hv = -K\frac{dv}{dy}$ becomes therefore necessary, when y is made equal to l or $-l$ in the functions v and $\frac{dv}{dy}$.

125. The value of the function v must by hypothesis be equal to A, when we suppose $x=0$, whatever be the values of y and z. Thus the required function v is determined by the following conditions: 1st, for all values of x, y, z, it satisfies the general equation

$$\frac{d^2v}{dx^2}+\frac{d^2v}{dy^2}+\frac{d^2v}{dz^2}=0;$$

2nd, it satisfies the equation $\frac{h}{K}v+\frac{dv}{dy}=0$, when y is equal to

l or $-l$, whatever x and z may be, or satisfies the equation $\frac{h}{K}v+\frac{dv}{dz}=0$, when z is equal to l or $-l$, whatever x and y may be; 3rd, it satisfies the equation $v=A$, when $x=0$, whatever y and z may be.

SECTION V.

Equations of the varied movement of heat in a solid cube.

126. A solid in the form of a cube, all of whose points have acquired the same temperature, is placed in a uniform current of atmospheric air, maintained at temperature 0. It is required to determine the successive states of the body during the whole time of the cooling.

The centre of the cube is taken as the origin of rectangular coordinates; the three perpendiculars dropped from this point on the faces, are the axes of x, y, and z; $2l$ is the side of the cube, v is the temperature to which a point whose coordinates are x, y, z, is lowered after the time t has elapsed since the commencement of the cooling: the problem consists in determining the function v, which depends on x, y, z and t.

127. To form the general equation which v must satisfy, we must ascertain what change of temperature an infinitely small portion of the solid must experience during the instant dt, by virtue of the action of the molecules which are extremely near to it. We consider then a prismatic molecule enclosed between six planes at right angles; the first three pass through the point m, whose co-ordinates are x, y, z, and the three others, through the point m', whose co-ordinates are

$$x+dx,\ y+dy,\ z+dz.$$

The quantity of heat which during the instant dt passes into the molecule across the first rectangle $dy\,dz$ perpendicular to x, is $-K\,dy\,dz\frac{dv}{dx}\,dt$, and that which escapes in the same time from the molecule, through the opposite face, is found by writing $x+dx$ in place of x in the preceding expression, it is

$$-K\,dy\,dz\left(\frac{dv}{dx}\right)dt-K\,dy\,dz\,d\left(\frac{dv}{dx}\right)dt,$$

the differential being taken with respect to x only. The quantity of heat which during the instant dt enters the molecule, across the first rectangle $dz\,dx$ perpendicular to the axis of y, is $-K\,dz\,dx\,\frac{dv}{dy}\,dt$, and that which escapes from the molecule during the same instant, by the opposite face, is

$$-K\,dz\,dx\,\frac{dv}{dy}\,dt - K\,dz\,dx\,d\left(\frac{dv}{dy}\right)dt,$$

the differential being taken with respect to y only. The quantity of heat which the molecule receives during the instant dt, through its lower face, perpendicular to the axis of z, is $-K\,dx\,dy\,\frac{dv}{dz}\,dt$, and that which it loses through the opposite face is

$$-K\,dx\,dy\,\frac{dv}{dz}\,dt - K\,dx\,dy\,d\left(\frac{dv}{dz}\right)dt,$$

the differential being taken with respect to z only.

The sum of all the quantities of heat which escape from the molecule must now be deducted from the sum of the quantities which it receives, and the difference is that which determines its increase of temperature during the instant: this difference is

$$K\,dy\,dz\,d\left(\frac{dv}{dx}\right)dt + K\,dz\,dx\,d\left(\frac{dv}{dy}\right)dt + K\,dx\,dy\,d\left(\frac{dv}{dz}\right)dt;$$

$$\text{or } K\,dx\,dy\,dz\left\{\frac{d^2v}{dx^2} + \frac{d^2v}{dy^2} + \frac{d^2v}{dz^2}\right\}dt.$$

128. If the quantity which has just been found be divided by that which is necessary to raise the molecule from the temperature 0 to the temperature 1, the increase of temperature which is effected during the instant dt will become known. Now, the latter quantity is $CD\,dx\,dy\,dz$: for C denotes the capacity of the substance for heat; D its density, and $dx\,dy\,dz$ the volume of the molecule. The movement of heat in the interior of the solid is therefore expressed by the equation

$$\frac{dv}{dt} = \frac{K}{CD}\left(\frac{d^2v}{dx^2} + \frac{d^2v}{dy^2} + \frac{d^2v}{dz^2}\right) \ldots\ldots\ldots\ldots (d).$$

129. It remains to form the equations which relate to the state of the surface, which presents no difficulty, in accordance with the principles which we have established. In fact, the quantity of heat which, during the instant dt, crosses the rectangle $dz\,dy$, traced on a plane perpendicular to x, is $-K\,dy\,dz\,\frac{dv}{dx}\,dt$. This result, which applies to all points of the solid, ought to hold when the value of x is equal to l, half the thickness of the prism. In this case, the rectangle $dy\,dz$ being situated at the surface, the quantity of heat which crosses it, and is dispersed into the air during the instant dt, is expressed by $hv\,dy\,dz\,dt$, we ought therefore to have, when $x=l$, the equation $hv=-K\frac{dv}{dx}$. This condition must also be satisfied when $x=-l$.

It will be found also that, the quantity of heat which crosses the rectangle $dz\,dx$ situated on a plane perpendicular to the axis of y being in general $-K\,dz\,dx\,\frac{dv}{dz}$, and that which escapes at the surface into the air across the same rectangle being $hv\,dz\,dx\,dt$, we must have the equation $hv+K\frac{dv}{dy}=0$, when $y=l$ or $-l$. Lastly, we obtain in like manner the definite equation

$$kv+K\frac{dv}{dz}=0,$$

which is satisfied when $z=l$ or $-l$.

130. The function sought, which expresses the varied movement of heat in the interior of a solid of cubic form, must therefore be determined by the following conditions:

1st. It satisfies the general equation

$$\frac{dv}{dt}=\frac{K}{C.D}\left(\frac{d^2v}{dx^2}+\frac{d^2v}{dy^2}+\frac{d^2v}{dz^2}\right);$$

2nd. It satisfies the three definite equations

$$hv+K\frac{dv}{dx}=0,\ hv+K\frac{dv}{dy}=0,\ hv+K\frac{dv}{dz}=0,$$

which hold when $x=\pm l$, $y=\pm l$, $z=\pm l$;

3rd. If in the function v which contains x, y, z, t, we make $t=0$, whatever be the values of x, y, and z, we ought to have, according to hypothesis, $v=A$, which is the initial and common value of the temperature.

131. The equation arrived at in the preceding problem represents the movement of heat in the interior of all solids. Whatever, in fact, the form of the body may be, it is evident that, by decomposing it into prismatic molecules, we shall obtain this result. We may therefore limit ourselves to demonstrating in this manner the equation of the propagation of heat. But in order to make the exhibition of principles more complete, and that we may collect into a small number of consecutive articles the theorems which serve to establish the general equation of the propagation of heat in the interior of solids, and the equations which relate to the state of the surface, we shall proceed, in the two following sections, to the investigation of these equations, independently of any particular problem, and without reverting to the elementary propositions which we have explained in the introduction.

SECTION VI.

General equation of the propagation of heat in the interior of solids.

132. THEOREM I. *If the different points of a homogeneous solid mass, enclosed between six planes at right angles, have actual temperatures determined by the linear equation*

$$v = A - ax - by - cz, \ldots\ldots (a),$$

and if the molecules situated at the external surface on the six planes which bound the prism are maintained, by any cause whatever, at the temperature expressed by the equation (a): all the molecules situated in the interior of the mass will of themselves retain their actual temperatures, so that there will be no change in the state of the prism.

v denotes the actual temperature of the point whose co-ordinates are x, y, z; A, a, b, c, are constant coefficients.

To prove this proposition, consider in the solid any three points whatever $mM\mu$, situated on the same straight line $m\mu$, which the point M divides into two equal parts; denote by x, y, z the co-ordinates of the point M, and its temperature by v, the co-ordinates of the point μ by $x+\alpha$, $y+\beta$, $z+\gamma$, and its temperature by w, the co-ordinates of the point m by $x-\alpha$, $y-\beta$, $z-\gamma$, and its temperature by u, we shall have

$$v = A - ax - by - cz,$$
$$w = A - a(x+\alpha) - b(y+\beta) - c(z+\gamma),$$
$$u = A - a(x-\alpha) - b(y-\beta) - c(z-\gamma),$$

whence we conclude that,

$$v - w = a\alpha + b\beta + c\gamma, \quad \text{and} \quad u - v = a\alpha + b\beta + c\gamma;$$

therefore $$v - w = u - v.$$

Now the quantity of heat which one point receives from another depends on the distance between the two points and on the difference of their temperatures. Hence the action of the point M on the point μ is equal to the action of m on M; thus the point M receives as much heat from m as it gives up to the point μ.

We obtain the same result, whatever be the direction and magnitude of the line which passes through the point M, and is divided into two equal parts. Hence it is impossible for this point to change its temperature, for it receives from all parts as much heat as it gives up.

The same reasoning applies to all other points; hence no change can happen in the state of the solid.

133. COROLLARY I. A solid being enclosed between two infinite parallel planes A and B, if the actual temperature of its different points is supposed to be expressed by the equation $v = 1 - z$, and the two planes which bound it are maintained by any cause whatever, A at the temperature 1, and B at the temperature 0; this particular case will then be included in the preceding lemma, if we make $A = 1$, $a = 0$, $b = 0$, $c = 1$.

134. COROLLARY II. If in the interior of the same solid we imagine a plane M parallel to those which bound it, we see that a certain quantity of heat flows across this plane during unit of time; for two very near points, such as m and n, one

of which is below the plane and the other above it, are unequally heated; the first, whose temperature is highest, must therefore send to the second, during each instant, a certain quantity of heat which, in some cases, may be very small, and even insensible, according to the nature of the body and the distance of the two molecules.

The same is true for any two other points whatever separated by the plane. That which is most heated sends to the other a certain quantity of heat, and the sum of these partial actions, or of all the quantities of heat sent across the plane, composes a continual flow whose value does not change, since all the molecules preserve their temperatures. It is easy to prove *that this flow, or the quantity of heat which crosses the plane* M *during the unit of time, is equivalent to that which crosses, during the same time, another plane* N *parallel to the first.* In fact, the part of the mass which is enclosed between the two surfaces M and N will receive continually, across the plane M, as much heat as it loses across the plane N. If the quantity of heat, which in passing the plane M enters the part of the mass which is considered, were not equal to that which escapes by the opposite surface N, the solid enclosed between the two surfaces would acquire fresh heat, or would lose a part of that which it has, and its temperatures would not be constant; which is contrary to the preceding lemma.

135. The measure of the specific conducibility of a given substance is taken to be the quantity of heat which, in an infinite solid, formed of this substance, and enclosed between two parallel planes, flows during unit of time across unit of surface, taken on any intermediate plane whatever, parallel to the external planes, the distance between which is equal to unit of length, one of them being maintained at temperature 1, and the other at temperature 0. This constant flow of the heat which crosses the whole extent of the prism is denoted by the coefficient K, and is the measure of the conducibility.

136. LEMMA. *If we suppose all the temperatures of the solid in question under the preceding article, to be multiplied by any number whatever* g, *so that the equation of temperatures is* $v = g - gz$, *instead of being* $v = 1 - z$, *and if the two external planes are main-*

tained, one at the temperature g, *and the other at temperature* 0, *the constant flow of heat, in this second hypothesis, or the quantity which during unit of time crosses unit of surface taken on an intermediate plane parallel to the bases, is equal to the product of the first flow multiplied by* g.

In fact, since all the temperatures have been increased in the ratio of 1 to g, the differences of the temperatures of any two points whatever m and μ, are increased in the same ratio. Hence, according to the principle of the communication of heat, in order to ascertain the quantity of heat which m sends to μ on the second hypothesis, we must multiply by g the quantity which the same point m sends to μ on the first hypothesis. The same would be true for any two other points whatever. Now, the quantity of heat which crosses a plane M results from the sum of all the actions which the points m, m', m'', m''', etc., situated on the same side of the plane, exert on the points μ, μ', μ'', μ''', etc., situated on the other side. Hence, if in the first hypothesis the constant flow is denoted by K, it will be equal to gK, when we have multiplied all the temperatures by g.

137. THEOREM II. *In a prism whose constant temperatures are expressed by the equation* $v = A - ax - by - cz$, *and which is bounded by six planes at right angles all of whose points are maintained at constant temperatures determined by the preceding equation, the quantity of heat which, during unit of time, crosses unit of surface taken on any intermediate plane whatever perpendicular to* z, *is the same as the constant flow in a solid of the same substance would be, if enclosed between two infinite parallel planes, and for which the equation of constant temperatures is* $v = c - cz$.

To prove this, let us consider in the prism, and also in the infinite solid, two extremely near points m and μ, separated

Fig. 4.

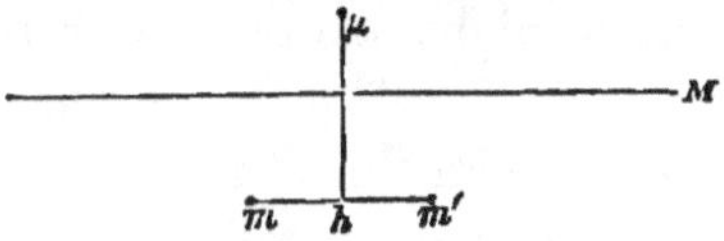

by the plane M perpendicular to the axis of z; μ being above the plane, and m below it (see fig. 4), and above the same plane let us take a point m such that the perpendicular dropped from the point μ on the plane may also be perpendicular to the distance mm' at its middle point h. Denote by x, y, $z+h$, the co-ordinates of the point μ, whose temperature is w, by $x-\alpha$, $y-\beta$, z, the co-ordinates of m, whose temperature is v, and by $x+\alpha$, $y+\beta$, z, the co-ordinates of m', whose temperature is v'.

The action of m on μ, or the quantity of heat which m sends to μ during a certain time, may be expressed by $q(v-w)$. The factor q depends on the distance $m\mu$, and on the nature of the mass. The action of m' on μ will therefore be expressed by $q(v'-w)$; and the factor q is the same as in the preceding expression; hence the sum of the two actions of m on μ, and of m' on μ, or the quantity of heat which μ receives from m and from m', is expressed by

$$q\,(v-w+v'-w).$$

Now, if the points m, μ, m' belong to the prism, we have

$$w=A-ax-by-c\,(z+h),\quad v=A-a\,(x-\alpha)-b\,(y-\beta)-cz,$$

and $$v'=A-a\,(x+\alpha)-b\,(y+\beta)-cz;$$

and if the same points belonged to an infinite solid, we should have, by hypothesis,

$$w=c-c\,(z+h),\quad v=c-cz,\quad \text{and}\quad v'=c-cz.$$

In the first case, we find

$$q\,(v-w+v'-w)=2qch,$$

and, in the second case, we still have the same result. Hence the quantity of heat which μ receives from m and from m' on the first hypothesis, when the equation of constant temperatures is $v=A-ax-by-cz$, is equivalent to the quantity of heat which μ receives from m and from m' when the equation of constant temperatures is $v=c-cz$.

The same conclusion might be drawn with respect to any three other points whatever m', μ', m'', provided that the second μ' be placed at equal distances from the other two, and the altitude of the isosceles triangle $m'\mu'm''$ be parallel to z. Now, the quantity of heat which crosses any plane whatever M, results from the sum of the actions which all the points m, m', m'', m''' etc., situated on one side of this plane, exert on all the points μ, μ', μ'', μ''', etc situated on the other side: hence the constant flow, which, during unit of time, crosses a definite part of the plane M in the infinite solid, is equal to the quantity of heat which flows in the same time across the same portion of the plane M in the prism, all of whose temperatures are expressed by the equation

$$v = A - ax - by - cz.$$

138. COROLLARY. The flow has the value cK in the infinite solid, when the part of the plane which it crosses has unit of surface. *In the prism also it has the same value* cK *or* $-K\frac{dv}{dz}$. It is proved in the same manner, *that the constant flow which takes place, during unit of time, in the same prism across unit of surface, on any plane whatever perpendicular to* y, *is equal to*

$$bK \text{ or } -K\frac{dv}{dy};$$

and that which crosses a plane perpendicular to x *has the value*

$$aK \text{ or } -K\frac{dv}{dx}.$$

139. The propositions which we have proved in the preceding articles apply also to the case in which the instantaneous action of a molecule is exerted in the interior of the mass up to an appreciable distance. In this case, we must suppose that the cause which maintains the external layers of the body in the state expressed by the linear equation, affects the mass up to a finite depth. All observation concurs to prove that in solids and liquids the distance in question is extremely small.

140. THEOREM III. If the temperatures at the points of a solid are expressed by the equation $v = f(x, y, z, t)$, in which x, y, z are the co-ordinates of a molecule whose temperature is equal to v after the lapse of a time t; the flow of heat which crosses part of a plane traced in the solid, perpendicular to one of the three axes, is no longer constant; its value is different for different parts of the plane, and it varies also with the time. This variable quantity may be determined by analysis.

Let ω be an infinitely small circle whose centre coincides with the point m of the solid, and whose plane is perpendicular to the vertical co-ordinate z; during the instant dt there will flow across this circle a certain quantity of heat which will pass from the part of the circle below the plane of the circle into the upper part. This flow is composed of all the rays of heat which depart from a lower point and arrive at an upper point, by crossing a point of the small surface ω. We proceed to shew *that the expression of the value of the flow is* $-K\frac{dv}{dz}\omega dt$.

Let us denote by x', y', z' the coordinates of the point m whose temperature is v'; and suppose all the other molecules to be referred to this point m chosen as the origin of new axes parallel to the former axes: let ξ, η, ζ, be the three co-ordinates of a point referred to the origin m; in order to express the actual temperature w of a molecule infinitely near to m, we shall have the linear equation

$$w = v' + \xi\frac{dv'}{dx} + \eta\frac{dv'}{dy} + \zeta\frac{dv'}{dz}.$$

The coefficients v', $\frac{dv'}{dx}, \frac{dv'}{dy}, \frac{dv'}{dz}$ are the values which are found by substituting in the functions $v, \frac{dv}{dx}, \frac{dv}{dy}, \frac{dv}{dz}$, for the variables x, y z, the constant quantities x', y', z', which measure the distances of the point m from the first three axes of x, y, and z.

Suppose now that the point m is also an internal molecule of a rectangular prism, enclosed between six planes perpendicular to the three axes whose origin is m; that w the actual temperature of each molecule of this prism, whose dimensions are finite, is expressed by the linear equation $w = A + a\xi + b\eta + c\zeta$, and that the six faces which bound the prism are maintained at the fixed temperatures which the last equation assigns to them. The state of the internal molecules will also be permanent, and a quantity of heat measured by the expression $-Kc\,\omega dt$ will flow during the instant dt across the circle ω.

This arranged, if we take as the values of the constants A, a, b, c, the quantities v', $\frac{dv'}{dx}, \frac{dv'}{dy}, \frac{dv'}{dz}$, the fixed state of the prism will be expressed by the equation

$$w = v' + \frac{dv'}{dx}\xi + \frac{dv'}{dy}\eta + \frac{dv'}{dz}\zeta.$$

Thus the molecules infinitely near to the point m will have, during the instant dt, the same actual temperature in the solid whose state is variable, and in the prism whose state is constant. Hence the flow which exists at the point m, during the instant dt, across the infinitely small circle ω, is the same in either solid; it is therefore expressed by $-K\frac{dv'}{dz}\omega dt$.

From this we derive the following proposition

If in a solid whose internal temperatures vary with the time, by virtue of the action of the molecules, we trace any straight line whatever, and erect (see fig. 5), at the different points of this line, the ordinates pm *of a plane curve equal to the temperatures of these points taken at the same moment; the flow of heat, at each point* p *of the straight line, will be proportional to the tangent of the angle* α *which the element of the curve makes with the parallel to the abscissæ;* that is to say, if at the point p we place the centre of an

Fig. 5.

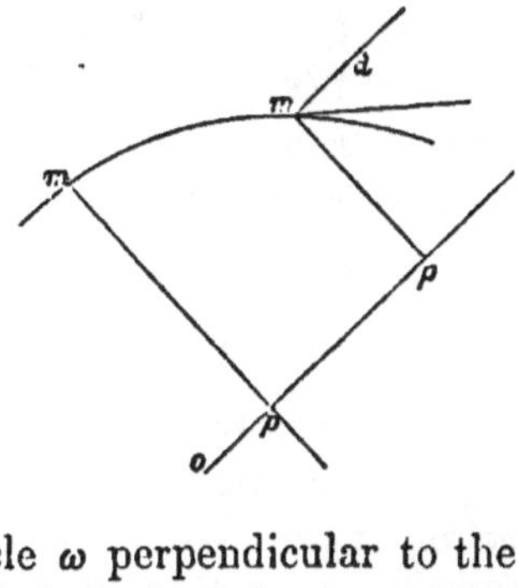

infinitely small circle ω perpendicular to the line, the quantity of heat which has flowed during the instant dt, across this circle, in the direction in which the abscissæ op increase, will be measured by the product of four factors, which are, the tangent of the angle α, a constant coefficient K, the area ω of the circle, and the duration dt of the instant.

141. COROLLARY. If we represent by ϵ the abscissa of this curve or the distance of a point p of the straight line from a

fixed point o, and by v the ordinate which represents the temperature of the point p, v will vary with the distance ϵ and will be a certain function $f(\epsilon)$ of that distance; the quantity of heat which would flow across the circle ω, placed at the point p perpendicular to the line, will be $-K\frac{dv}{d\epsilon}\omega dt$, or

$$-Kf'(\epsilon)\,\omega dt,$$

denoting the function $\frac{df(\epsilon)}{d\epsilon}$ by $f'(\epsilon)$.

We may express this result in the following manner, which facilitates its application.

To obtain the actual flow of heat at a point p *of a straight line drawn in a solid, whose temperatures vary by action of the molecules, we must divide the difference of the temperatures at two points infinitely near to the point* p *by the distance between these points. The flow is proportional to the quotient.*

142. THEOREM IV. From the preceding Theorems it is easy to deduce the general equations of the propagation of heat.

Suppose the different points of a homogeneous solid of any form whatever, to have received initial temperatures which vary successively by the effect of the mutual action of the molecules, and suppose the equation $v = f(x, y, z, t)$ *to represent the successive states of the solid, it may now be shewn that* v *a function of four variables necessarily satisfies the equation*

$$\frac{dv}{dt} = \frac{K}{CD}\left(\frac{d^2v}{dx^2} + \frac{d^2v}{dy^2} + \frac{d^2v}{dz^2}\right).$$

In fact, let us consider the movement of heat in a molecule enclosed between six planes at right angles to the axes of x, y, and z; the first three of these planes pass through the point m whose coordinates are x, y, z, the other three pass through the point m', whose coordinates are $x + dx$, $y + dy$, $z + dz$.

During the instant dt, the molecule receives, across the lower rectangle $dxdy$, which passes through the point m, a quantity of heat equal to $-K\,dx\,dy\,\frac{dv}{dz}\,dt$. To obtain the quantity which escapes from the molecule by the opposite face, it is sufficient to change z into $z + dz$ in the preceding expression, that is to say, to add to this expression its own differential taken with respect to z only; we then have

$$-K dx\, dy \frac{dv}{dz} dt - K dx\, dy \frac{d\left(\frac{dv}{dz}\right)}{dz} dz\, dt$$

as the value of the quantity which escapes across the upper rectangle. The same molecule receives also across the first rectangle $dz\, dx$ which passes through the point m, a quantity of heat equal to $-K\frac{dv}{dy} dz\, dx\, dt$; and if we add to this expression its own differential taken with respect to y only, we find that the quantity which escapes across the opposite face $dz\, dx$ is expressed by

$$-K\frac{dv}{dy} dz\, dx\, dt - K\frac{d\left(\frac{dv}{dy}\right)}{dy} dy\, dz\, dx\, dt.$$

Lastly, the molecule receives through the first rectangle $dy\, dz$ a quantity of heat equal to $-K\frac{dv}{dx} dy\, dz\, dt$, and that which it loses across the opposite rectangle which passes through m' is expressed by

$$-K\frac{dv}{dx} dy\, dz\, dt - K\frac{d\left(\frac{dv}{dx}\right)}{dx} dx\, dy\, dz\, dt.$$

We must now take the sum of the quantities of heat which the molecule receives and subtract from it the sum of those which it loses. Hence it appears that during the instant dt, a total quantity of heat equal to

$$K\left(\frac{d^2v}{dx^2} + \frac{d^2v}{dy^2} + \frac{d^2v}{dz^2}\right) dx\, dy\, dz\, dt$$

accumulates in the interior of the molecule. It remains only to obtain the increase of temperature which must result from this addition of heat.

D being the density of the solid, or the weight of unit of volume, and C the specific capacity, or the quantity of heat which raises the unit of weight from the temperature 0 to the temperature 1; the product $CD dx dy dz$ expresses the quantity of heat required to raise from 0 to 1 the molecule whose volume is $dx\,dy\,dz$. Hence dividing by this product the quantity of heat which the molecule has just acquired, we shall have its increase of temperature. Thus we obtain the general equation

$$\frac{dv}{dt} = \frac{K}{CD}\left(\frac{d^2v}{dx^2} + \frac{d^2v}{dy^2} + \frac{d^2v}{dz^2}\right) \ldots\ldots\ldots\ldots\ldots (A),$$

which is the equation of the propagation of heat in the interior of all solid bodies.

143. Independently of this equation the system of temperatures is often subject to several definite conditions, of which no general expression can be given, since they depend on the nature of the problem.

If the dimensions of the mass in which heat is propagated are finite, and if the surface is maintained by some special cause in a given state; for example, if all its points retain, by virtue of that cause, the constant temperature 0, we shall have, denoting the unknown function v by $\phi\,(x, y, z, t)$, the equation of condition $\phi\,(x, y, z, t) = 0$; which must be satisfied by all values of x, y, z which belong to points of the external surface, whatever be the value of t. Further, if we suppose the initial temperatures of the body to be expressed by the known function $F(x, y, z)$, we have also the equation $\phi\,(x, y, z, 0) = F(x, y, z)$; the condition expressed by this equation must be fulfilled by all values of the co-ordinates x, y, z which belong to any point whatever of the solid.

144. Instead of submitting the surface of the body to a constant temperature, we may suppose the temperature not to be the same at different points of the surface, and that it varies with the time according to a given law; which is what takes place in the problem of terrestrial temperature. In this case the equation relative to the surface contains the variable t.

145. In order to examine by itself, and from a very general point of view, the problem of the propagation of heat, the solid whose initial state is given must be supposed to have all its dimensions infinite; no special condition disturbs then the diffusion of heat, and the law to which this principle is submitted becomes more manifest; it is expressed by the general equation

$$\frac{dv}{dt} = \frac{K}{CD}\left(\frac{d^2v}{dx^2} + \frac{d^2v}{dy^2} + \frac{dv^2}{dz^2}\right),$$

to which must be added that which relates to the initial arbitrary state of the solid.

Suppose the initial temperature of a molecule, whose co-ordinates are x, y, z, to be a known function $F(x, y, z)$, and denote the unknown value v by $\phi(x, y, z, t)$, we shall have the definite equation $\phi(x, y, z, 0) = F(x, y, z)$; thus the problem is reduced to the integration of the general equation (A) in such a manner that it may agree, when the time is zero, with the equation which contains the arbitrary function F.

SECTION VII.

General equation relative to the surface.

146. If the solid has a definite form, and if its original heat is dispersed gradually into atmospheric air maintained at a constant temperature, a third condition relative to the state of the surface must be added to the general equation (A) and to that which represents the initial state.

We proceed to examine, in the following articles, the nature of the equation which expresses this third condition.

Consider the variable state of a solid whose heat is dispersed into air, maintained at the fixed temperature 0. Let ω be an infinitely small part of the external surface, and μ a point of ω, through which a normal to the surface is drawn; different points of this line have at the same instant different temperatures.

Let v be the actual temperature of the point μ, taken at a definite instant, and w the corresponding temperature of a point ν of the solid taken on the normal, and distant from μ by an infinitely small quantity α. Denote by x, y, z the co-ordinates of the point μ, and those of the point ν by $x + \delta x$, $y + \delta y$, $z + \delta z$; let $f(x, y, z) = 0$ be the known equation to the surface of the solid, and $v = \phi(x, y, z, t)$ the general equation which ought to give the value of v as a function of the four variables x, y, z, t. Differentiating the equation $f(x, y, z) = 0$, we shall have

$$mdx + ndy + pdz = 0;$$

m, n, p being functions of x, y, z.

It follows from the corollary enunciated in Article 141, that the flow in direction of the normal, or the quantity of heat which during the instant dt would cross the surface ω, if it were placed at any point whatever of this line, at right angles to its direction, is proportional to the quotient which is obtained by dividing the difference of temperature of two points infinitely near by their distance. Hence the expression for the flow at the end of the normal is

$$-K\frac{w-v}{\alpha}\omega dt;$$

K denoting the specific conducibility of the mass. On the other hand, the surface ω permits a quantity of heat to escape into the air, during the time dt, equal to $hv\omega dt$; h being the conducibility relative to atmospheric air. Thus the flow of heat at the end of the normal has two different expressions, that is to say:

$$hv\omega dt \text{ and } -K\frac{w-v}{\alpha}\omega dt;$$

hence these two quantities are equal; and it is by the expression of this equality that the condition relative to the surface is introduced into the analysis.

147. We have

$$w = v + \delta v = v + \frac{dv}{dx}\delta x + \frac{dv}{dy}\delta y + \frac{dv}{dz}\delta z.$$

Now, it follows from the principles of geometry, that the coordinates $\delta x, \delta y, \delta z$, which fix the position of the point ν of the normal relative to the point μ, satisfy the following conditions:

$$p\delta x = m\delta z, \quad p\delta y = n\delta z.$$

We have therefore

$$w - v = \frac{1}{p}\left(m\frac{dv}{dx} + n\frac{dv}{dy} + p\frac{dv}{dz}\right)\delta z:$$

we have also

$$\alpha = \sqrt{\delta x^2 + \delta y^2 + \delta z^2} = \frac{1}{p}(m^2 + n^2 + p^2)^{\frac{1}{2}}\,\delta z,$$

or $$\alpha = \frac{q}{p}\,\delta z,\ \text{denoting by } q \text{ the quantity } (m^2 + n^2 + p^2)^{\frac{1}{2}},$$

hence $$\frac{w - v}{\alpha} = \left(m\frac{dv}{dx} + n\frac{dv}{dy} + p\frac{dv}{dz}\right)\frac{1}{q};$$

consequently the equation

$$hv\omega dt = -k\left(\frac{w - v}{\alpha}\right)\omega dt$$

becomes the following[1]:

$$m\frac{dv}{dx} + n\frac{dv}{dy} + p\frac{dv}{dx} + \frac{h}{K}vq = 0 \ldots\ldots\ldots\ldots\ldots(B).$$

This equation is definite and applies only to points at the surface; it is that which must be added to the general equation of the propagation of heat (A), and to the condition which determines the initial state of the solid; m, n, p, q, are known functions of the co-ordinates of the points on the surface.

148. The equation (B) signifies in general that the decrease of the temperature, in the direction of the normal, at the boundary of the solid, is such that the quantity of heat which tends to escape by virtue of the action of the molecules, is equivalent always to that which the body must lose in the medium.

The mass of the solid might be imagined to be prolonged, in such a manner that the surface, instead of being exposed to the air, belonged at the same time to the body which it bounds, and to the mass of a solid envelope which contained it. If, on this hypothesis, any cause whatever regulated at every instant the decrease of the temperatures in the solid envelope, and determined it in such a manner that the condition expressed by the equation (B) was always satisfied, the action of the envelope would take the

[1] Let N be the normal,

$$K\frac{dv}{dN} + hv = 0,$$

$$\frac{dv}{dN} = \frac{m}{q}\frac{dv}{dx} + \&c.;$$

the rest as in the text. [R. L. E.]

place of that of the air, and the movement of heat would be the same in either case: we can suppose then that this cause exists, and determine on this hypothesis the variable state of the solid; which is what is done in the employment of the two equations (A) and (B).

By this it is seen how the interruption of the mass and the action of the medium, disturb the diffusion of heat by submitting it to an accidental condition.

149. We may also consider the equation (B), which relates to the state of the surface under another point of view: but we must first derive a remarkable consequence from Theorem III. (Art. 140). We retain the construction referred to in the corollary of the same theorem (Art. 141). Let x, y, z be the co-ordinates of the point p, and

$$x+\delta x, \quad y+\delta y, \quad z+\delta z$$

those of a point q infinitely near to p, and taken on the straight line in question: if we denote by v and w the temperatures of the two points p and q taken at the same instant, we have

$$w = v + \delta v = v + \frac{dv}{dx}\delta x + \frac{dv}{dy}\delta y + \frac{dv}{dz}\delta z;$$

hence the quotient

$$\frac{\delta v}{\delta \epsilon} = \frac{dv}{dx}\frac{\delta x}{\delta \epsilon} + \frac{dv}{dx}\frac{dy}{\delta \epsilon} + \frac{dv}{dz}\frac{\delta z}{\delta \epsilon}, \text{ and } \delta\epsilon = \sqrt{\delta x^2 + \delta y^2 + \delta z^2};$$

thus the quantity of heat which flows across the surface ω placed at the point m, perpendicular to the straight line, is

$$-K\omega dt\left\{\frac{dv}{dx}\frac{\delta x}{\delta \epsilon} + \frac{dv}{dy}\frac{\delta y}{\delta \epsilon} + \frac{dv}{dz}\frac{\delta z}{\delta \epsilon}\right\}.$$

The first term is the product of $-K\dfrac{dv}{dx}$ by dt and by $\omega\dfrac{\delta x}{\delta \epsilon}$. The latter quantity is, according to the principles of geometry, the area of the projection of ω on the plane of y and z; thus the product represents the quantity of heat which would flow across the area of the projection, if it were placed at the point p perpendicular to the axis of x.

The second term $-K\dfrac{dv}{dy}\,\omega\,\dfrac{\delta y}{\delta\epsilon}\,dt$ represents the quantity of heat which would cross the projection of ω, made on the plane of x and z, if this projection were placed parallel to itself at the point p.

Lastly, the third term $-K\dfrac{dv}{dz}\,\omega\,\dfrac{\delta z}{\delta\epsilon}\,dt$ represents the quantity of heat which would flow during the instant dt, across the projection of ω on the plane of x and y, if this projection were placed at the point p, perpendicular to the co-ordinate z.

By this it is seen *that the quantity of heat which flows across every infinitely small part of a surface drawn in the interior of the solid, can always be decomposed into three other quantities of flow, which penetrate the three orthogonal projections of the surface, along the directions perpendicular to the planes of the projections.* The result gives rise to properties analogous to those which have been noticed in the theory of forces.

150. The quantity of heat which flows across a plane surface ω, infinitely small, given in form and position, being equivalent to that which would cross its three orthogonal projections, it follows that, if in the interior of the solid an element be imagined of any form whatever, the quantities of heat which pass into this polyhedron by its different faces, compensate each other reciprocally: or more exactly, the sum of the terms of the first order, which enter into the expression of the quantities of heat received by the molecule, is zero; so that the heat which is in fact accumulated in it, and makes its temperature vary, cannot be expressed except by terms infinitely smaller than those of the first order.

This result is distinctly seen when the general equation (A) has been established, by considering the movement of heat in a prismatic molecule (Articles 127 and 142); the demonstration may be extended to a molecule of any form whatever, by substituting for the heat received through each face, that which its three projections would receive.

In other respects it is necessary that this should be so: for, if one of the molecules of the solid acquired during each instant a quantity of heat expressed by a term of the first order, the variation of its temperature would be infinitely greater than that of other molecules, that is to say, during each infinitely small instant its temperature would increase or decrease by a finite quantity, which is contrary to experience.

151. We proceed to apply this remark to a molecule situated at the external surface of the solid.

Fig. 6.

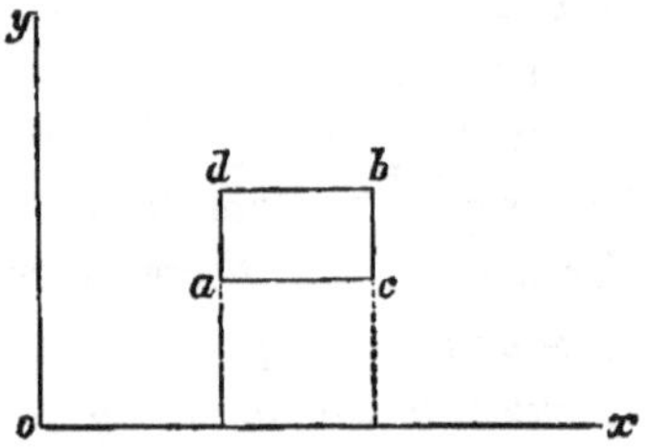

Through a point a (see fig. 6), taken on the plane of x and y, draw two planes perpendicular, one to the axis of x the other to the axis of y. Through a point b of the same plane, infinitely near to a, draw two other planes parallel to the two preceding planes; the ordinates z, raised at the points a, b, c, d, up to the external surface of the solid, will mark on this surface four points a', b', c', d', and will be the edges of a truncated prism, whose base is the rectangle $abcd$. If through the point a' which denotes the least elevated of the four points a', b', c', d', a plane be drawn parallel to that of x and y, it will cut off from the truncated prism a molecule, one of whose faces, that is to say $a'b'c'd'$, coincides with the surface of the solid. The values of the four ordinates aa', cc', dd', bb' are the following:

$$aa' = z,$$

$$cc' = z + \frac{dz}{dx}\,dx,$$

$$dd' = z + \frac{dz}{dy}\,dy,$$

$$bb' = z + \frac{dz}{dx}\,dx + \frac{dz}{dy}\,dy.$$

152. One of the faces perpendicular to x is a triangle, and the opposite face is a trapezium. The area of the triangle is

$$\frac{1}{2}\, dy\, \frac{dz}{dy}\, dy,$$

and the flow of heat in the direction perpendicular to this surface being $-K\frac{dv}{dx}$ we have, omitting the factor dt,

$$-K\frac{dv}{dx}\,\frac{1}{2}\, dy\, \frac{dz}{dy}\, dy,$$

as the expression of the quantity of heat which in one instant passes into the molecule, across the triangle in question.

The area of the opposite face is

$$\frac{1}{2}\, dy \left(\frac{dz}{dx}\, dx + \frac{dz}{dx}\, dx + \frac{dz}{dy}\, dy\right),$$

and the flow perpendicular to this face is also $-K\frac{dv}{dx}$, suppressing terms of the second order infinitely smaller than those of the first; subtracting the quantity of heat which escapes by the second face from that which enters by the first we find

$$K\frac{dv}{dx}\,\frac{dz}{dx}\, dx\, dy.$$

This term expresses the quantity of heat the molecule receives through the faces perpendicular to x.

It will be found, by a similar process, that the same molecule receives, through the faces perpendicular to y, a quantity of heat equal to $K\frac{dv}{dy}\,\frac{dz}{dy}\, dx\, dy$.

The quantity of heat which the molecule receives through the rectangular base is $-K\frac{dv}{dz}\, dx\, dy$. Lastly, across the upper surface $a'b'c'd'$, a certain quantity of heat is permitted to escape, equal to the product of hv into the extent ω of that surface. The value of ω is, according to known principles, the same as that of $dx\, dy$ multiplied by the ratio $\frac{\epsilon}{z}$; ϵ denoting the length of the normal between the external surface and the plane of x and y, and

$$\epsilon = z\left\{1 + \left(\frac{dz}{dx}\right)^2 + \left(\frac{dz}{dy}\right)^2\right\}^{\frac{1}{2}},$$

hence the molecule loses across its surface $a'b'c'd'$ a quantity of heat equal to $hv\,dx\,dy\,\frac{\epsilon}{z}$.

Now, the terms of the first order which enter into the expression of the total quantity of heat acquired by the molecule, must cancel each other, in order that the variation of temperature may not be at each instant a finite quantity; we must then have the equation

$$K\left(\frac{dv}{dx}\frac{dz}{dx}\,dx\,dy + \frac{dv}{dy}\frac{dz}{dy}\,dx\,dy - \frac{dv}{dz}\,dx\,dy\right) - hv\frac{\epsilon}{z}\,dx\,dy = 0,$$

$$\text{or}\quad \frac{h}{K}v\frac{\epsilon}{z} = \frac{dv}{dx}\frac{dz}{dx} + \frac{dv}{dy}\frac{dz}{dy} - \frac{dv}{dz}.$$

153. Substituting for $\frac{dz}{dx}$ and $\frac{dz}{dy}$ their values derived from the equation

$$m\,dx + n\,dy + p\,dz = 0,$$

and denoting by q the quantity

$$(m^2 + n^2 + p^2)\,,$$

we have

$$K\left(m\frac{dv}{dx} + n\frac{dv}{dy} + p\frac{dv}{dz}\right) + hvq = 0 \ldots\ldots\ldots\ldots (B),$$

thus we know distinctly what is represented by each of the terms of this equation.

Taking them all with contrary signs and multiplying them by $dx\,dy$, the first expresses how much heat the molecule receives through the two faces perpendicular to x, the second how much it receives through its two faces perpendicular to y, the third how much it receives through the face perpendicular to z, and the fourth how much it receives from the medium. The equation therefore expresses that the sum of all the terms of the first order is zero, and that the heat acquired cannot be represented except by terms of the second order.

154. To arrive at equation (B), we in fact consider one of the molecules whose base is in the surface of the solid, as a vessel which receives or loses heat through its different faces. The equation signifies that all the terms of the first order which

enter into the expression of the heat acquired cancel each other; so that the gain of heat cannot be expressed except by terms of the second order. We may give to the molecule the form, either of a right prism whose axis is normal to the surface of the solid, or that of a truncated prism, or any form whatever.

The general equation (A), (Art. 142) supposes that all the terms of the first order cancel each other in the interior of the mass, which is evident for prismatic molecules enclosed in the solid. The equation (B), (Art. 147) expresses the same result for molecules situated at the boundaries of bodies.

Such are the general points of view from which we may look at this part of the theory of heat.

The equation $\frac{dv}{dt} = \frac{K}{CD}\left(\frac{d^2v}{dx^2} + \frac{d^2v}{dy^2} + \frac{d^2v}{dz^2}\right)$ represents the movement of heat in the interior of bodies. It enables us to ascertain the distribution from instant to instant in all substances solid or liquid; from it we may derive the equation which belongs to each particular case.

In the two following articles we shall make this application to the problem of the cylinder, and to that of the sphere.

SECTION VIII.

Application of the general equations.

155. Let us denote the variable radius of any cylindrical envelope by r, and suppose, as formerly, in Article 118, that all the molecules equally distant from the axis have at each instant a common temperature; v will be a function of r and t; r is a function of y, z, given by the equation $r^2 = y^2 + z^2$. It is evident in the first place that the variation of v with respect to x is nul; thus the term $\frac{d^2v}{dx^2}$ must be omitted. We shall have then, according to the principles of the differential calculus, the equations

$$\frac{dv}{dy} = \frac{dv}{dr}\frac{dr}{dy} \quad \text{and} \quad \frac{d^2v}{dy^2} = \frac{d^2v}{dr^2}\left(\frac{dr}{dy}\right)^2 + \frac{dv}{dr}\left(\frac{d^2r}{dy^2}\right),$$

$$\frac{dv}{dz} = \frac{dv}{dr}\frac{dr}{dz} \quad \text{and} \quad \frac{d^2v}{dz^2} = \frac{d^2v}{dr^2}\left(\frac{dr}{dz}\right)^2 + \frac{dv}{dr}\left(\frac{d^2r}{dz^2}\right);$$

whence

$$\frac{d^2v}{dy^2}+\frac{d^2v}{dz^2}=\frac{d^2v}{dr^2}\left\{\left(\frac{dr}{dy}\right)^2+\left(\frac{dr}{dz}\right)^2+\frac{dv}{dr}\left(\frac{d^2r}{dy^2}+\frac{d^2r}{dz^2}\right)\right\}\ldots\ldots(a).$$

In the second member of the equation, the quantities

$$\frac{dr}{dy},\ \frac{dr}{dz},\ \frac{d^2r}{dy^2},\ \frac{d^2r}{dz^2},$$

must be replaced by their respective values; for which purpose we derive from the equation $y^2+z^2=r^2$,

$$y=r\frac{dr}{dy}\ \text{and}\ 1=\left(\frac{dr}{dy}\right)^2+r\frac{d^2r}{dy^2},$$

$$z=r\frac{dr}{dz}\ \text{and}\ 1=\left(\frac{dr}{dz}\right)^2+r\frac{d^2r}{dz^2},$$

and consequently

$$y^2+z^2=r^2\left\{\left(\frac{dr}{dy}\right)^2+\left(\frac{dr}{dz}\right)^2\right\},$$

$$2=\left(\frac{dr}{dy}\right)^2+\left(\frac{dr}{dz}\right)^2+r\left\{\frac{d^2r}{dy^2}+\frac{d^2r}{dz^2}\right\}.$$

The first equation, whose first member is equal to r^2, gives

$$\left(\frac{dr}{dy}\right)^2+\left(\frac{dr}{dz}\right)^2=1\ldots\ldots\ldots\ldots\ldots\ldots(b);$$

the second gives, when we substitute for

$$\left(\frac{dr}{dy}\right)^2+\left(\frac{dr}{dz}\right)^2$$

its value 1,

$$\frac{d^2r}{dy^2}+\frac{d^2r}{dz^2}=\frac{1}{r}\ldots\ldots\ldots\ldots\ldots\ldots\ldots\ldots(c).$$

If the values given by equations (b) and (c) be now substituted in (a), we have

$$\frac{d^2v}{dy^2}+\frac{d^2v}{dz^2}=\frac{d^2v}{dr^2}+\frac{1}{r}\frac{dv}{dr}.$$

Hence the equation which expresses the movement of heat in the cylinder, is

$$\frac{dv}{dt}=\frac{K}{CD}\left(\frac{d^2v}{dr^2}+\frac{1}{r}\frac{dv}{dr}\right),$$

as was found formerly, Art. 119.

We might also suppose that particles equally distant from the centre have not received a common initial temperature; in this case we should arrive at a much more general equation.

156. To determine, by means of equation (A), the movement of heat in a sphere which has been immersed in a liquid, we shall regard v as a function of r and t; r is a function of x, y, z, given by the equation

$$r^2 = x^2 + y^2 + z^2,$$

r being the variable radius of an envelope. We have then

$$\frac{dv}{dx} = \frac{dv}{dr}\frac{dr}{dx} \quad\text{and}\quad \frac{d^2v}{dx^2} = \frac{d^2v}{dr^2}\left(\frac{dr}{dx}\right)^2 + \frac{dv}{dr}\frac{d^2r}{dx^2},$$

$$\frac{dv}{dy} = \frac{dv}{dr}\frac{dr}{dy} \quad\text{and}\quad \frac{d^2v}{dy^2} = \frac{d^2v}{dr^2}\left(\frac{dr}{dy}\right)^2 + \frac{dv}{dr}\frac{d^2r}{dy^2},$$

$$\frac{dv}{dz} = \frac{dv}{dr}\frac{dr}{dz} \quad\text{and}\quad \frac{d^2v}{dz^2} = \frac{d^2v}{dr^2}\left(\frac{dr}{dz}\right)^2 + \frac{dv}{dr}\frac{d^2r}{dz^2}.$$

Making these substitutions in the equation

$$\frac{dv}{dt} = \frac{K}{CD}\left\{\frac{d^2v}{dx^2} + \frac{d^2v}{dy^2} + \frac{d^2v}{dz^2}\right\},$$

we shall have

$$\frac{dv}{dt} = \frac{K}{CD}\left[\frac{d^2v}{dr^2}\left\{\left(\frac{dr}{dx}\right)^2 + \left(\frac{dr}{dy}\right)^2 + \left(\frac{dz}{dr}\right)^2\right\} + \frac{dv}{dr}\left\{\frac{d^2r}{dx^2} + \frac{d^2r}{dy^2} + \frac{d^2r}{dz^2}\right\}\right] \quad (a).$$

The equation $x^2 + y^2 + z^2 = r^2$ gives the following results;

$$x = r\frac{dr}{dx} \quad\text{and}\quad 1 = \left(\frac{dr}{dx}\right)^2 + r\frac{d^2r}{dx^2},$$

$$y = r\frac{dr}{dy} \quad\text{and}\quad 1 = \left(\frac{dr}{dy}\right)^2 + r\frac{d^2r}{dy^2},$$

$$z = r\frac{dr}{dz} \quad\text{and}\quad 1 = \left(\frac{dr}{dz}\right)^2 + r\frac{d^2r}{dz^2}.$$

The three equations of the first order give:

$$x^2 + y^2 + z^2 = r^2\left\{\left(\frac{dr}{dx}\right)^2 + \left(\frac{dr}{dy}\right)^2 + \left(\frac{dr}{dz}\right)^2\right\}.$$

The three equations of the second order give:

$$3=\left(\frac{dr}{dx}\right)^2+\left(\frac{dr}{dy}\right)^2+\left(\frac{dr}{dz}\right)^2+r\left\{\frac{d^2r}{dx^2}+\frac{d^2r}{dy^2}+\frac{d^2r}{dz^2}\right\}:$$

and substituting for

$$\left(\frac{dr}{dx}\right)^2+\left(\frac{dr}{dy}\right)^2+\left(\frac{dr}{dz}\right)^2$$

its value 1, we have

$$\frac{d^2r}{dx^2}+\frac{d^2r}{dy^2}+\frac{d^2r}{dz^2}=\frac{2}{r}.$$

Making these substitutions in the equation (*a*) we have the equation

$$\frac{dv}{dt}=\frac{K}{CD}\left\{\frac{d^2v}{dr^2}+\frac{2}{r}\frac{dv}{dr}\right\},$$

which is the same as that of Art. 114.

The equation would contain a greater number of terms, if we supposed molecules equally distant from the centre not to have received the same initial temperature.

We might also deduce from the definite equation (B), the equations which express the state of the surface in particular cases, in which we suppose solids of given form to communicate their heat to the atmospheric air; but in most cases these equations present themselves at once, and their form is very simple, when the co-ordinates are suitably chosen.

SECTION IX.

General Remarks.

157. The investigation of the laws of movement of heat in solids now consists in the integration of the equations which we have constructed; this is the object of the following chapters. We conclude this chapter with general remarks on the nature of the quantities which enter into our analysis.

In order to measure these quantities and express them numerically, they must be compared with different kinds of units, five in number, namely, the unit of length, the unit of time, that of temperature, that of weight, and finally the unit which serves to measure quantities of heat. For the last unit, we might have chosen the quantity of heat which raises a given volume of a certain substance from the temperature 0 to the temperature 1. The choice of this unit would have been preferable in many respects to that of the quantity of heat required to convert a mass of ice of a given weight, into an equal mass of water at 0, without raising its temperature. We have adopted the last unit only because it had been in a manner fixed beforehand in several works on physics; besides, this supposition would introduce no change into the results of analysis.

158. The specific elements which in every body determine the measurable effects of heat are three in number, namely, the conducibility proper to the body, the conducibility relative to the atmospheric air, and the capacity for heat. The numbers which express these quantities are, like the specific gravity, so many natural characters proper to different substances.

We have already remarked, Art. 36, that the conducibility of the surface would be measured in a more exact manner, if we had sufficient observations on the effects of radiant heat in spaces deprived of air.

It may be seen, as has been mentioned in the first section of Chapter I., Art. 11, that only three specific coefficients, K, h, C, enter into the investigation; they must be determined by observation; and we shall point out in the sequel the experiments adapted to make them known with precision.

159. The number C which enters into the analysis, is always multiplied by the density D, that is to say, by the number of units of weight which are equivalent to the weight of unit of volume; thus the product CD may be replaced by the coefficient c. In this case we must understand by the specific capacity for heat, the quantity required to raise from temperature 0 to temperature 1 unit of volume of a given substance, and not unit of weight of that substance.

With the view of not departing from the common definition, we have referred the capacity for heat to the weight and not to the volume; but it would be preferable to employ the coefficient c which we have just defined; magnitudes measured by the unit of weight would not then enter into the analytical expressions: we should have to consider only, 1st, the linear dimension x, the temperature v, and the time t; 2nd, the coefficients c, h, and K. The three first quantities are undetermined, and the three others are, for each substance, constant elements which experiment determines. As to the unit of surface and the unit of volume, they are not absolute, but depend on the unit of length.

160. It must now be remarked that every undetermined magnitude or constant has one *dimension* proper to itself, and that the terms of one and the same equation could not be compared, if they had not the same *exponent of dimension*. We have introduced this consideration into the theory of heat, in order to make our definitions more exact, and to serve to verify the analysis; it is derived from primary notions on quantities; for which reason, in geometry and mechanics, it is the equivalent of the fundamental lemmas which the Greeks have left us without proof.

161. In the analytical theory of heat, every equation (E) expresses a necessary relation between the existing magnitudes x, t, v, c, h, K. This relation depends in no respect on the choice of the unit of length, which from its very nature is contingent, that is to say, if we took a different unit to measure the linear dimensions, the equation (E) would still be the same. Suppose then the unit of length to be changed, and its second value to be equal to the first divided by m. Any quantity whatever x which in the equation (E) represents a certain line ab, and which, consequently, denotes a certain number of times the unit of length, becomes mx, corresponding to the same length ab; the value t of the time, and the value v of the temperature will not be changed; the same is not the case with the specific elements h, K, c: the first, h, becomes $\frac{h}{m^2}$; for it expresses the quantity of heat which escapes, during the unit of time, from the unit of surface at the temperature 1. If we examine attentively the nature of the coefficient K, as we have defined it in Articles 68 and 135,

we perceive that it becomes $\frac{K}{m}$; for the flow of heat varies directly as the area of the surface, and inversely as the distance between two infinite planes (Art. 72). As to the coefficient c which represents the product CD, it also depends on the unit of length and becomes $\frac{c}{m^3}$; hence equation (E) must undergo no change when we write mx instead of x, and at the same time $\frac{K}{m}$, $\frac{h}{m^2}$, $\frac{c}{m^3}$, instead of K, h, c; the number m disappears after these substitutions: thus the dimension of x with respect to the unit of length is 1, that of K is -1, that of h is -2, and that of c is -3. If we attribute to each quantity its own *exponent of dimension*, the equation will be homogeneous, since every term will have the same total exponent. Numbers such as S, which represent surfaces or solids, are of two dimensions in the first case, and of three dimensions in the second. Angles, sines, and other trigonometrical functions, logarithms or exponents of powers, are, according to the principles of analysis, *absolute* numbers which do not change with the unit of length; their dimensions must therefore be taken equal to 0, which is the dimension of all abstract numbers.

If the unit of time, which was at first 1, becomes $\frac{1}{n}$, the number t will become nt, and the numbers x and v will not change. The coefficients K, h, c will become $\frac{K}{n}$, $\frac{h}{n}$, c. Thus the dimensions of x, t, v with respect to the unit of time are 0, 1, 0, and those of K, h, c are -1, -1, 0.

If the unit of temperature be changed, so that the temperature 1 becomes that which corresponds to an effect other than the boiling of water; and if that effect requires a less temperature, which is to that of boiling water in the ratio of 1 to the number p; v will become vp, x and t will keep their values, and the coefficients K, h, c will become $\frac{K}{p}$, $\frac{h}{p}$, $\frac{c}{p}$.

The following table indicates the dimensions of the three undetermined quantities and the three constants, with respect to each kind of unit.

Quantity or Constant.	Length.	Duration.	Temperature.
Exponent of dimension of x ...	1	0	0
" " t ...	0	1	0
" " v ...	0	0	1
The specific conducibility, K ...	-1	-1	-1
The surface conducibility, h ...	-2	-1	-1
The capacity for heat, c ...	-3	0	1

162. If we retained the coefficients C and D, whose product has been represented by c, we should have to consider the unit of weight, and we should find that the exponent of dimension, with respect to the unit of length, is -3 for the density D, and 0 for C.

On applying the preceding rule to the different equations and their transformations, it will be found that they are homogeneous with respect to each kind of unit, and that the dimension of every angular or exponential quantity is nothing. If this were not the case, some error must have been committed in the analysis, or abridged expressions must have been introduced.

If, for example, we take equation (b) of Art. 105,

$$\frac{dv}{dt} = \frac{K}{CD}\frac{d^2v}{dx^2} - \frac{hl}{CDS}v,$$

we find that, with respect to the unit of length, the dimension of each of the three terms is 0; it is 1 for the unit of temperature, and -1 for the unit of time.

In the equation $v = Ae^{-x\sqrt{\frac{2h}{Kl}}}$ of Art. 76, the linear dimension of each term is 0, and it is evident that the dimension of the exponent $x\sqrt{\frac{2h}{Kl}}$ is always nothing, whatever be the units of length, time, or temperature.

CHAPTER III.

PROPAGATION OF HEAT IN AN INFINITE RECTANGULAR SOLID.

SECTION I.

Statement of the problem.

163. PROBLEMS relative to the uniform propagation, or to the varied movement of heat in the interior of solids, are reduced, by the foregoing methods, to problems of pure analysis, and the progress of this part of physics will depend in consequence upon the advance which may be made in the art of analysis. The differential equations which we have proved contain the chief results of the theory; they express, in the most general and most concise manner, the necessary relations of numerical analysis to a very extensive class of phenomena; and they connect for ever with mathematical science one of the most important branches of natural philosophy.

It remains now to discover the proper treatment of these equations in order to derive their complete solutions and an easy application of them. The following problem offers the first example of analysis which leads to such solutions; it appeared to us better adapted than any other to indicate the elements of the method which we have followed.

164. Suppose a homogeneous solid mass to be contained between two planes B and C vertical, parallel, and infinite, and to be divided into two parts by a plane A perpendicular to the other two (fig. 7); we proceed to consider the temperatures of the mass BAC bounded by the three infinite planes A, B, C. The other part $B'AC'$ of the infinite solid is supposed to be a constant source of heat, that is to say, all its points are maintained at the temperature 1, which cannot alter. The two

lateral solids bounded, one by the plane C and the plane A produced, the other by the plane B and the plane A pro-

Fig. 7.

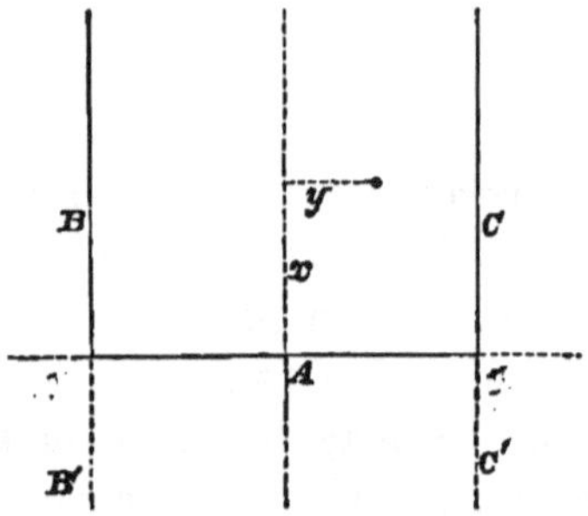

duced, have at all points the constant temperature 0, some external cause maintaining them always at that temperature; lastly, the molecules of the solid bounded by A, B and C have the initial temperature 0. Heat will pass continually from the source A into the solid BAC, and will be propagated there in the longitudinal direction, which is infinite, and at the same time will turn towards the cool masses B and C, which will absorb great part of it. The temperatures of the solid BAC will be raised gradually: but will not be able to surpass nor even to attain a maximum of temperature, which is different for different points of the mass. It is required to determine the final and constant state to which the variable state continually approaches.

If this final state were known, and were then formed, it would subsist of itself, and this is the property which distinguishes it from all other states. Thus the actual problem consists in determining the permanent temperatures of an infinite rectangular solid, bounded by two masses of ice B and C, and a mass of boiling water A; the consideration of such simple and primary problems is one of the surest modes of discovering the laws of natural phenomena, and we see, by the history of the sciences, that every theory has been formed in this manner.

165. To express more briefly the same problem, suppose a rectangular plate BAC, of infinite length, to be heated at its base A, and to preserve at all points of the base a constant

temperature 1, whilst each of the two infinite sides B and C, perpendicular to the base A, is submitted also at every point to a constant temperature 0; it is required to determine what must be the stationary temperature at any point of the plate.

It is supposed that there is no loss of heat at the surface of the plate, or, which is the same thing, we consider a solid formed by superposing an infinite number of plates similar to the preceding: the straight line Ax which divides the plate into two equal parts is taken as the axis of x, and the co-ordinates of any point m are x and y; lastly, the width A of the plate is represented by $2l$, or, to abridge the calculation, by π, the value of the ratio of the diameter to the circumference of a circle.

Imagine a point m of the solid plate BAC, whose co-ordinates are x and y, to have the actual temperature v, and that the quantities v, which correspond to different points, are such that no change can happen in the temperatures, provided that the temperature of every point of the base A is always 1, and that the sides B and C retain at all their points the temperature 0.

If at each point m a vertical co-ordinate be raised, equal to the temperature v, a curved surface would be formed which would extend above the plate and be prolonged to infinity. We shall endeavour to find the nature of this surface, which passes through a line drawn above the axis of y at a distance equal to unity, and which cuts the horizontal plane of xy along two infinite straight lines parallel to x.

166. To apply the general equation

$$\frac{dv}{dt} = \frac{K}{CD}\left(\frac{d^2v}{dx^2} + \frac{d^2v}{dy^2} + \frac{d^2v}{dz^2}\right),$$

we must consider that, in the case in question, abstraction is made of the co-ordinate z, so that the term $\frac{d^2v}{dz^2}$ must be omitted; with respect to the first member $\frac{dv}{dt}$, it vanishes, since we wish to determine the stationary temperatures; thus the equation which belongs to the actual problem, and determines the properties of the required curved surface, is the following:

$$\frac{d^2v}{dx^2} + \frac{d^2v}{dy^2} = 0 \ldots\ldots\ldots\ldots\ldots\ldots\ldots\ldots (a).$$

The function of x and y, $\phi(x, y)$, which represents the permanent state of the solid BAC, must, 1st, satisfy the equation (a); 2nd, become nothing when we substitute $-\frac{1}{2}\pi$ or $+\frac{1}{2}\pi$ for y, whatever the value of x may be; 3rd, must be equal to unity when we suppose $x = 0$ and y to have any value included between $-\frac{1}{2}\pi$ and $+\frac{1}{2}\pi$.

Further, this function $\phi(x, y)$ ought to become extremely small when we give to x a very large value, since all the heat proceeds from the source A.

167. In order to consider the problem in its elements, we shall in the first place seek for the simplest functions of x and y, which satisfy equation (a); we shall then generalise the value of v in order to satisfy all the stated conditions. By this method the solution will receive all possible extension, and we shall prove that the problem proposed admits of no other solution.

Functions of two variables often reduce to less complex expressions, when we attribute to one of the variables or to both of them infinite values; this is what may be remarked in algebraic functions which, in this particular case, take the form of the product of a function of x by a function of y.

We shall examine first if the value of v can be represented by such a product; for the function v must represent the state of the plate throughout its whole extent, and consequently that of the points whose co-ordinate x is infinite. We shall then write $v = F(x) f(y)$; substituting in equation (a) and denoting $\frac{d^2F(x)}{dx^2}$ by $F''(x)$ and $\frac{d^2f(y)}{dy^2}$ by $f''(y)$, we shall have

$$\frac{F''(x)}{F(x)} + \frac{f''(y)}{f(y)} = 0;$$

we then suppose $\frac{F''(x)}{F(x)} = m$ and $\frac{f''(y)}{f(y)} = -m$, m being any constant quantity, and as it is proposed only to find a particular value of v, we deduce from the preceding equations $F(x)=e^{-mx}$, $f(y)=\cos my$.

168. We could not suppose m to be a negative number, and we must necessarily exclude all particular values of v, into which terms such as e^{mx} might enter, m being a positive number, since the temperature v cannot become infinite when x is infinitely great. In fact, no heat being supplied except from the constant source A, only an extremely small portion can arrive at those parts of space which are very far removed from the source. The remainder is diverted more and more towards the infinite edges B and C, and is lost in the cold masses which bound them.

The exponent m which enters into the function $e^{-mx}\cos my$ is unknown, and we may choose for this exponent any positive number: but, in order that v may become nul on making $y=-\frac{1}{2}\pi$ or $y=+\frac{1}{2}\pi$, whatever x may be, m must be taken to be one of the terms of the series, 1, 3, 5, 7, &c.; by this means the second condition will be fulfilled.

169. A more general value of v is easily formed by adding together several terms similar to the preceding, and we have

$$v=ae^{-x}\cos y+be^{-3x}\cos 3y+ce^{-5x}\cos 5y+de^{-7x}\cos 7y+\&c. \ldots\ldots(b).$$

It is evident that the function v denoted by $\phi(x, y)$ satisfies the equation $\frac{d^2v}{dx^2}+\frac{d^2v}{dy^2}=0$, and the condition $\phi(x, \pm\frac{1}{2}\pi)=0$. A third condition remains to be fulfilled, which is expressed thus, $\phi(0, y)=1$, and it is essential to remark that this result must exist when we give to y any value whatever included between $-\frac{1}{2}\pi$ and $+\frac{1}{2}\pi$. Nothing can be inferred as to the values which the function $\phi(0, y)$ would take, if we substituted in place of y a quantity not included between the limits $-\frac{1}{2}\pi$ and $+\frac{1}{2}\pi$. Equation (b) must therefore be subject to the following condition:

$$1=a\cos y+b\cos 3y+c\cos 5y+d\cos 7y+\&c.$$

The coefficients, a, b, c, d, &c., whose number is infinite, are determined by means of this equation.

The second member is a function of y, which is equal to 1

so long as the variable y is included between the limits $-\frac{1}{2}\pi$ and $+\frac{1}{2}\pi$. It may be doubted whether such a function exists, but this difficulty will be fully cleared up by the sequel.

170. Before giving the calculation of the coefficients, we may notice the effect represented by each one of the terms of the series in equation (b).

Suppose the fixed temperature of the base A, instead of being equal to unity at every point, to diminish as the point of the line A becomes more remote from the middle point, being proportional to the cosine of that distance; in this case it will easily be seen what is the nature of the curved surface, whose vertical ordinate expresses the temperature v or $\phi(x, y)$. If this surface be cut at the origin by a plane perpendicular to the axis of x, the curve which bounds the section will have for its equation $v = a \cos y$; the values of the coefficients will be the following:

$$a = a,\ b = 0,\ c = 0,\ d = 0,$$

and so on, and the equation of the curved surface will be

$$v = ae^{-x} \cos y.$$

If this surface be cut at right angles to the axis of y, the section will be a logarithmic spiral whose convexity is turned towards the axis; if it be cut at right angles to the axis of x, the section will be a trigonometric curve whose concavity is turned towards the axis.

It follows from this that the function $\frac{d^2v}{dx^2}$ is always positive, and $\frac{d^2v}{dy^2}$ is always negative. Now the quantity of heat which a molecule acquires in consequence of its position between two others in the direction of x is proportional to the value of $\frac{d^2v}{dx^2}$ (Art. 123): it follows then that the intermediate molecule receives from that which precedes it, in the direction of x, more heat than it communicates to that which follows it. But, if the same molecule be considered as situated between two others in the direction of y, the function $\frac{d^2v}{dy^2}$ being negative, it appears that the intermediate molecule communicates to that which follows it more heat than it receives from that which precedes it. Thus it follows that the excess of the heat which it acquires in the direction of x, is exactly compensated by that which it loses in the direction of y, as the equation $\frac{d^2v}{dx^2}+\frac{d^2v}{dy^2}=0$ denotes. Thus then the route followed by the heat which escapes from the source A becomes known. It is propagated in the direction of x, and at the same time it is decomposed into two parts, one of which is directed towards one of the edges, whilst the other part continues to separate from the origin, to be decomposed like the preceding, and so on to infinity. The surface which we are considering is generated by the trigonometric curve which corresponds to the base A, moved with its plane at right angles to the axis of x along that axis, each one of its ordinates decreasing indefinitely in proportion to successive powers of the same fraction.

Analogous inferences might be drawn, if the fixed temperatures of the base A were expressed by the term

$$b \cos 3y \text{ or } c \cos 5y, \&c.;$$

and in this manner an exact idea might be formed of the movement of heat in the most general case; for it will be seen by the sequel that the movement is always compounded of a multitude of elementary movements, each of which is accomplished as if it alone existed.

SECTION II.

First example of the use of trigonometric series in the theory of heat.

171. Take now the equation

$$1 = a \cos y + b \cos 3y + c \cos 5y + d \cos 7y + \&c.,$$

in which the coefficients a, b, c, d, &c. are to be determined. In order that this equation may exist, the constants must necessarily satisfy the equations which are obtained by successive differentiations; whence the following results,

$$1 = a\cos y + b\cos 3y + c\cos 5y + d\cos 7y + \&c.,$$
$$0 = a\sin y + 3b\sin 3y + 5c\sin 5y + 7d\sin 7y + \&c.,$$
$$0 = a\cos y + 3^2b\cos 3y + 5^2c\cos 5y + 7^2d\cos 7y + \&c.,$$
$$0 = a\sin y + 3^3b\sin 3y + 5^3c\sin 5y + 7^3d\sin 7y + \&c.,$$

and so on to infinity.

These equations necessarily hold when $y=0$, thus we have

$$1 = a + b + c + d + e + f + g + \ldots \&c.,$$
$$0 = a + 3^2b + 5^2c + 7^2d + 9^2e + 11^2f + \ldots \&c.,$$
$$0 = a + 3^4b + 5^4c + 7^4d + 9^4e + \ldots \&c.,$$
$$0 = a + 3^6b + 5^6c + 7^6d + \ldots \&c.,$$
$$0 = a + 3^8b + 5^8c + \ldots \&c.,$$
&c.

The number of these equations is infinite like that of the unknowns $a, b, c, d, e, \ldots$ &c. The problem consists in eliminating all the unknowns, except one only.

172. In order to form a distinct idea of the result of these eliminations, the number of the unknowns $a, b, c, d, \ldots$ &c., will be supposed at first definite and equal to m. We shall employ the first m equations only, suppressing all the terms containing the unknowns which follow the m first. If in succession m be made equal to 2, 3, 4, 5, and so on, the values of the unknowns will be found on each one of these hypotheses. The quantity a, for example, will receive one value for the case of two unknowns, others for the cases of three, four, or successively a greater number of unknowns. It will be the same with the unknown b, which will receive as many different values as there have been cases of elimination; each one of the other unknowns is in like manner susceptible of an infinity of different values. Now the value of one of the unknowns, for the case in which their number is infinite, is the limit towards which the values which it receives by means of the successive eliminations tend. It is required then to examine whether, according as the number of unknowns increases, the value of each one of $a, b, c, d \ldots$ &c. does not converge to a finite limit which it continually approaches.

Suppose the six following equations to be employed:

$$1 = a + b + c + d + e + f + \&c.,$$
$$0 = a + 3^2 b + 5^2 c + 7^2 d + 9^2 e + 11^2 f + \&c.,$$
$$0 = a + 3^4 b + 5^4 c + 7^4 d + 9^4 e + 11^4 f + \&c.,$$
$$0 = a + 3^6 b + 5^6 c + 7^6 d + 9^6 e + 11^6 f + \&c.,$$
$$0 = a + 3^8 b + 5^8 c + 7^8 d + 9^8 e + 11^8 f + \&c.,$$
$$0 = a + 3^{10} b + 5^{10} c + 7^{10} d + 9^{10} e + 11^{10} f + \&c.$$

The five equations which do not contain f are:

$$11^2 = a(11^2-1^2) + b(11^2-3^2) + c(11^2-5^2) + d(11^2-7^2) + e(11^2-9^2),$$
$$0 = a(11^2-1^2) + 3^2 b(11^2-3^2) + 5^2 c(11^2-5^2) + 7^2 d(11^2-7^2) + 9^2 e(11^2-9^2),$$
$$0 = a(11^2-1^2) + 3^4 b(11^2-3^2) + 5^4 c(11^2-5^2) + 7^4 d(11^2-7^2) + 9^4 e(11^2-9^2),$$
$$0 = a(11^2-1^2) + 3^6 b(11^2-3^2) + 5^6 c(11^2-5^2) + 7^6 d(11^2-7^2) + 9^6 e(11^2-9^2),$$
$$0 = a(11^2-1^2) + 3^8 b(11^2-3^2) + 5^8 c(11^2-5^2) + 7^8 d(11^2-7^2) + 9^8 e(11^2-9^2).$$

Continuing the elimination we shall obtain the final equation in a, which is:

$$a\,(11^2-1^2)\,(9^2-1^2)\,(7^2-1^2)\,(5^2-1^2)\,(3^2-1^2) = 11^2 . 9^2 . 7^2 . 5^2 . 3^2 . 1^2.$$

173. If we had employed a number of equations greater by unity, we should have found, to determine a, an equation analogous to the preceding, having in the first member one factor more, namely, $13^2 - 1^2$, and in the second member 13^2 for the new factor. The law to which these different values of a are subject is evident, and it follows that the value of a which corresponds to an infinite number of equations is expressed thus:

$$a = \frac{3^2}{3^2-1^2} \cdot \frac{5^2}{5^2-1^2} \cdot \frac{7^2}{7^2-1^2} \cdot \frac{9^2}{9^2-1^2} \cdot \frac{11^2}{11^2-1^2} \cdot \&c.,$$

or
$$a = \frac{3.3}{2.4} \cdot \frac{5.5}{4.6} \cdot \frac{7.7}{6.8} \cdot \frac{9.9}{8.10} \cdot \frac{11.11}{10.12} \cdot \&c.$$

Now the last expression is known and, in accordance with Wallis' Theorem, (*) we conclude that $a = \frac{4}{\pi}$. It is required then only to ascertain the values of the other unknowns.

174. The five equations which remain after the elimination of f may be compared with the five simpler equations which would have been employed if there had been only five unknowns.

The last equations differ from the equations of Art. 172, in that in them e, d, c, b, a are found to be multiplied respectively by the factors

$$\frac{11^2-9^2}{11^2}, \quad \frac{11^2-7^2}{11^2}, \quad \frac{11^2-5^2}{11^2}, \quad \frac{11^2-3^2}{11^2}, \quad \frac{11^2-1^2}{11^2}.$$

It follows from this that if we had solved the five linear equations which must have been employed in the case of five unknowns, and had calculated the value of each unknown, it would have been easy to derive from them the value of the unknowns of the same name corresponding to the case in which six equations should have been employed. It would suffice to multiply the values of e, d, c, b, a, found in the first case, by the known factors. It will be easy in general to pass from the value of one of these quantities, taken on the supposition of a certain number of equations and unknowns, to the value of the same quantity, taken in the case in which there should have been one unknown and one equation more. For example, if the value of e, found on the hypothesis of five equations and five unknowns, is represented by E, that of the same quantity, taken in the case of one unknown more, will be $E\frac{11^2}{11^2-9^2}$. The same value, taken in the case of seven unknowns, will be, for the same reason,

$$E\frac{11^2}{11^2-9^2}\cdot\frac{13^2}{13^2-9^2},$$

and in the case of eight unknowns it will be

$$E\frac{11^2}{11^2-9^2}\cdot\frac{13^2}{13^2-9^2}\cdot\frac{15^2}{15^2-9^2},$$

and so on. In the same manner it will suffice to know the value of b, corresponding to the case of two unknowns, to derive from it that of the same letter which corresponds to the cases of three, four, five unknowns, &c. We shall only have to multiply this first value of b by

$$\frac{5^2}{5^2-3^2}\cdot\frac{7^2}{7^2-3^2}\cdot\frac{9^2}{9^2-3^2}\cdots \&c.$$

Similarly if we knew the value of c for the case of three unknowns, we should multiply this value by the successive factors

$$\frac{7^2}{7^2-5^2}\cdot\frac{9^2}{9^2-5^2}\cdot\frac{11^2}{11^2-5^2}\ldots \text{\&c.}$$

We should calculate the value of d for the case of four unknowns only, and multiply this value by

$$\frac{9^2}{9^2-7^2}\cdot\frac{11^2}{11^2-7^2}\cdot\frac{13^2}{13^2-7^2}\ldots \text{\&c.}$$

The calculation of the value of a is subject to the same rule, for if its value be taken for the case of one unknown, and multiplied successively by

$$\frac{3^2}{3^2-1^2},\ \frac{5^2}{5^2-1^2},\ \frac{7^2}{7^2-1^2},\ \frac{9^2}{9^2-1^2},$$

the final value of this quantity will be found.

175. The problem is therefore reduced to determining the value of a in the case of one unknown, the value of b in the case of two unknowns, that of c in the case of three unknowns, and so on for the other unknowns.

It is easy to conclude, by inspection only of the equations and without any calculation, that the results of these successive eliminations must be

$$a=1,$$

$$b=\frac{1^2}{1^2-3^2},$$

$$c=\frac{1^2}{1^2-5^2}\cdot\frac{3^2}{3^2-5^2},$$

$$d=\frac{1^2}{1^2-7^2}\cdot\frac{3^2}{3^2-7^2}\cdot\frac{5^2}{5^2-7^2},$$

$$e=\frac{1^2}{1^2-9^2}\cdot\frac{3^2}{3^2-9^2}\cdot\frac{5^2}{5^2-9^2}\cdot\frac{7^2}{7^2-9^2}.$$

176. It remains only to multiply the preceding quantities by the series of products which ought to complete them, and which we have given (Art. 174). We shall have consequently, for the

final values of the unknowns a, b, c, d, e, f, &c., the following expressions:

$$a=1\cdot\frac{3^2}{3^2-1^2}\cdot\frac{5^2}{5^2-1^2}\cdot\frac{7^2}{7^2-1^2}\cdot\frac{9^2}{9^2-1^2}\cdot\frac{11^2}{11^2-1^2}\text{ \&c.,}$$

$$b=\frac{1^2}{1^2-3^2}\cdot\frac{5^2}{5^2-3^2}\cdot\frac{7^2}{7^2-3^2}\cdot\frac{9^2}{9^2-3^2}\cdot\frac{11^2}{11^2-3^2}\text{ \&c.,}$$

$$c=\frac{1^2}{1^2-5^2}\cdot\frac{3^2}{3^2-5^2}\cdot\frac{7^2}{7^2-5^2}\cdot\frac{9^2}{9^2-5^2}\cdot\frac{11^2}{11^2-5^2}\text{ \&c.,}$$

$$d=\frac{1^2}{1^2-7^2}\cdot\frac{3^2}{3^2-7^2}\cdot\frac{5^2}{5^2-7^2}\cdot\frac{9^2}{9^2-7^2}\cdot\frac{11^2}{11^2-7^2}\text{ \&c.,}$$

$$e=\frac{1^2}{1^2-9^2}\cdot\frac{3^2}{3^2-9^2}\cdot\frac{5^2}{5^2-9^2}\cdot\frac{7^2}{7^2-9^2}\cdot\frac{11^2}{11^2-9^2}\cdot\frac{13^2}{13^2-9^2}\text{ \&c.,}$$

$$f=\frac{1^2}{1^2-11^2}\cdot\frac{3^2}{3^2-11^2}\cdot\frac{5^2}{5^2-11^2}\cdot\frac{7^2}{7^2-11^2}\cdot\frac{9^2}{9^2-11^2}\cdot\frac{13^2}{13^2-11^2}\text{ \&c.,}$$

or,

$$a=+1\cdot\frac{3.3}{2.4}\cdot\frac{5.5}{4.6}\cdot\frac{7.7}{6.8}\text{ \&c.,}$$

$$b=-\frac{1.1}{2.4}\cdot\frac{5.5}{2.8}\cdot\frac{7.7}{4.10}\cdot\frac{9.9}{6.12}\text{ \&c.,}$$

$$c=+\frac{1.1}{4.6}\cdot\frac{3.3}{2.8}\cdot\frac{7.7}{2.12}\cdot\frac{9.9}{4.14}\cdot\frac{11.11}{6.16}\text{ \&c.,}$$

$$d=-\frac{1.1}{6.8}\cdot\frac{3.3}{4.10}\cdot\frac{5.5}{2.12}\cdot\frac{9.9}{2.16}\cdot\frac{11.11}{4.18}\text{ \&c.,}$$

$$e=+\frac{1.1}{8.10}\cdot\frac{3.3}{6.12}\cdot\frac{5.5}{4.14}\cdot\frac{7.7}{2.16}\cdot\frac{11.11}{2.20}\cdot\frac{13.13}{4.22}\text{ \&c.,}$$

$$f=-\frac{1.1}{10.12}\cdot\frac{3.3}{8.14}\cdot\frac{5.5}{6.16}\cdot\frac{7.7}{4.18}\cdot\frac{9.9}{2.20}\cdot\frac{13.13}{2.24}\cdot\frac{15.15}{4.26}\text{ \&c.}$$

The quantity $\frac{1}{2}\pi$ or a quarter of the circumference is equivalent, according to Wallis' Theorem, to

$$\frac{2.2}{1.3}\cdot\frac{4.4}{3.5}\cdot\frac{6.6}{5.7}\cdot\frac{8.8}{7.9}\cdot\frac{10.10}{9.11}\cdot\frac{12.12}{11.13}\cdot\frac{14.14}{13.15}\text{ \&c.}$$

If now in the values of a, b, c, d, &c., we notice what are the factors which must be joined on to numerators and denominators to complete the double series of odd and even numbers, we find that the factors to be supplied are:

$$\left.\begin{array}{l} \text{for } b\ \dfrac{3.3}{6}, \\ \text{for } c\ \dfrac{5.5}{10}, \\ \text{for } d\ \dfrac{7.7}{14}, \\ \text{for } e\ \dfrac{9.9}{18}, \\ \text{for } f\ \dfrac{11.11}{22}, \end{array}\right\} \text{ whence we conclude } \left\{\begin{array}{l} a = \quad 2\,.\,\dfrac{2}{\pi}, \\ b = -2\,.\,\dfrac{2}{3\pi}, \\ c = \quad 2\,.\,\dfrac{2}{5\pi}, \\ d = -2\,.\,\dfrac{2}{7\pi}, \\ e = \quad 2\,.\,\dfrac{2}{9\pi}, \\ f = -2\,.\,\dfrac{2}{11\pi}.\text{[1]} \end{array}\right.$$

177. Thus the eliminations have been completely effected, and the coefficients a, b, c, d, &c., determined in the equation

$$1 = a\cos y + b\cos 3y + c\cos 5y + d\cos 7y + e\cos 9y + \text{\&c.}$$

The substitution of these coefficients gives the following equation:

$$\frac{\pi}{4} = \cos y - \frac{1}{3}\cos 3y + \frac{1}{5}\cos 5y - \frac{1}{7}\cos 7y + \frac{1}{9}\cos 9y - \text{\&c.}\text{[2]}$$

The second member is a function of y, which does not change in value when we give to the variable y a value included between $-\frac{1}{2}\pi$ and $+\frac{1}{2}\pi$. It would be easy to prove that this series is always convergent, that is to say that writing instead of y any number whatever, and following the calculation of the coefficients, we approach more and more to a fixed value, so that the difference of this value from the sum of the calculated terms becomes less than any assignable magnitude. Without stopping for a proof,

[1] It is a little better to deduce the value of b in a, of c in b, &c. [R. L. E.]

[2] The coefficients a, b, c, &c., might be determined, according to the methods of Section VI., by multiplying both sides of the first equation by $\cos y$, $\cos 3y$, $\cos 5y$, &c., respectively, and integrating from $-\frac{1}{2}\pi$ to $+\frac{1}{2}\pi$, as was done by D. F. Gregory, *Cambridge Mathematical Journal*, Vol. I. p. 106. [A. F.]

which the reader may supply, we remark that the fixed value which is continually approached is $\frac{1}{4}\pi$, if the value attributed to y is included between 0 and $\frac{1}{2}\pi$, but that it is $-\frac{1}{4}\pi$, if y is included between $\frac{1}{2}\pi$ and $\frac{3}{2}\pi$; for, in this second interval, each term of the series changes in sign. In general the limit of the series is alternately positive and negative; in other respects, the convergence is not sufficiently rapid to produce an easy approximation, but it suffices for the truth of the equation.

178. The equation

$$y = \cos x - \frac{1}{3}\cos 3x + \frac{1}{5}\cos 5x - \frac{1}{7}\cos 7x + \&c.$$

belongs to a line which, having x for abscissa and y for ordinate, is composed of separated straight lines, each of which is parallel to the axis, and equal to the circumference. These parallels are situated alternately above and below the axis, at the distance $\frac{1}{4}\pi$, and joined by perpendiculars which themselves make part of the line. To form an exact idea of the nature of this line, it must be supposed that the number of terms of the function

$$\cos x - \frac{1}{3}\cos 3x + \frac{1}{5}\cos 5x - \&c.$$

has first a definite value. In the latter case the equation

$$y = \cos x - \frac{1}{3}\cos 3x + \frac{1}{5}\cos 5x - \&c.$$

belongs to a curved line which passes alternately above and below the axis, cutting it every time that the abscissa x becomes equal to one of the quantities

$$0, \ \pm\frac{1}{2}\pi, \ \pm\frac{3}{2}\pi, \ \pm\frac{5}{2}\pi, \ \&c.$$

According as the number of terms of the equation increases, the curve in question tends more and more to coincidence with the preceding line, composed of parallel straight lines and of perpendicular lines; so that this line is the limit of the different curves which would be obtained by increasing successively the number of terms.

SECTION III.

Remarks on these series.

179. We may look at the same equations from another point of view, and prove directly the equation

$$\frac{\pi}{4} = \cos x - \frac{1}{3}\cos 3x + \frac{1}{5}\cos 5x - \frac{1}{7}\cos 7x + \frac{1}{9}\cos 9x - \&c.$$

The case where x is nothing is verified by Leibnitz' series,

$$\frac{\pi}{4} = 1 - \frac{1}{3} + \frac{1}{5} - \frac{1}{7} + \frac{1}{9} - \&c.$$

We shall next assume that the number of terms of the series

$$\cos x - \frac{1}{3}\cos 3x + \frac{1}{5}\cos 5x - \frac{1}{7}\cos 7x + \&c.$$

instead of being infinite is finite and equal to m. We shall consider the value of the finite series to be a function of x and m. We shall express this function by a series arranged according to negative powers of m; and it will be found that the value of the function approaches more nearly to being constant and independent of x, as the number m becomes greater.

Let y be the function required, which is given by the equation

$$y = \cos x - \frac{1}{3}\cos 3x + \frac{1}{5}\cos 5x - \frac{1}{7}\cos 7x + \ldots - \frac{1}{2m-1}\cos(2m-1)x,$$

m, the number of terms, being supposed even. This equation differentiated with respect to x gives

$$-\frac{dy}{dx} = \sin x - \sin 3x + \sin 5x - \sin 7x + \ldots$$
$$+ \sin(2m-3)x - \sin(2m-1)x;$$

multiplying by $2\sin 2x$, we have

$$-2\frac{dy}{dx}\sin 2x = 2\sin x\sin 2x - 2\sin 3x\sin 2x + 2\sin 5x\sin 2x \ldots$$
$$+ 2\sin(2m-3)x\sin 2x - 2\sin(2m-1)x\sin 2x.$$

Each term of the second member being replaced by the difference of two cosines, we conclude that

$$-2\frac{dy}{dx}\sin 2x = \cos(-x) - \cos 3x$$
$$- \cos x + \cos 5x$$
$$+ \cos 3x - \cos 7x$$
$$- \cos 5x + \cos 9x$$
$$\dots\dots\dots\dots\dots\dots$$
$$+ \cos(2m-5)x - \cos(2m-1)x$$
$$- \cos(2m-3x) + \cos(2m+1)x.$$

The second member reduces to

$$\cos(2m+1)x - \cos(2m-1)x, \quad \text{or} \quad -2\sin 2mx \sin x;$$

hence
$$y = \frac{1}{2}\int\left(dx\frac{\sin 2mx}{\cos x}\right).$$

180. We shall integrate the second member by parts, distinguishing in the integral between the factor $\sin 2mx\, dx$ which must be integrated successively, and the factor $\frac{1}{\cos x}$ or $\sec x$ which must be differentiated successively; denoting the results of these differentiations by $\sec' x$, $\sec'' x$, $\sec''' x$, ... &c., we shall have

$$2y = \text{const.} - \frac{1}{2m}\cos 2mx \sec x + \frac{1}{2^2m^2}\sin 2mx \sec' x$$
$$+ \frac{1}{2^3m^3}\cos 2mx \sec'' x + \&c.;$$

thus the value of y or

$$\cos x - \frac{1}{3}\cos 3x + \frac{1}{5}\cos 5x - \frac{1}{7}\cos 7x + \dots - \frac{1}{2m-1}\cos(2m-1)x,$$

which is a function of x and m, becomes expressed by an infinite series; and it is evident that the more the number m increases, the more the value of y tends to become constant. For this reason, when the number m is infinite, the function y has a definite value which is always the same, whatever be the positive value of x, less than $\frac{1}{2}\pi$. Now, if the arc x be supposed nothing, we have

$$y = 1 - \frac{1}{3} + \frac{1}{5} - \frac{1}{7} + \frac{1}{9} - \&c.,$$

which is equal to $\frac{1}{4}\pi$. Hence generally we shall have

$$\frac{1}{4}\pi = \cos x - \frac{1}{3}\cos 3x + \frac{1}{5}\cos 5x - \frac{1}{7}\cos 7x + \frac{1}{9}\cos 9x - \&c. \ldots (b).$$

181. If in this equation we assume $x = \frac{1}{2}\frac{\pi}{2}$, we find

$$\frac{\pi}{2\sqrt{2}} = 1 + \frac{1}{3} - \frac{1}{5} - \frac{1}{7} + \frac{1}{9} + \frac{1}{11} - \frac{1}{13} - \frac{1}{15} + \ldots \&c.;$$

by giving to the arc x other particular values, we should find other series, which it is useless to set down, several of which have been already published in the works of Euler. If we multiply equation (b) by dx, and integrate it, we have

$$\frac{\pi x}{4} = \sin x - \frac{1}{3^2}\sin 3x + \frac{1}{5^2}\sin 5x - \frac{1}{7^2}\sin 7x + \&c.$$

Making in the last equation $x = \frac{1}{2}\pi$, we find

$$\frac{\pi^2}{8} = 1 + \frac{1}{3^2} + \frac{1}{5^2} + \frac{1}{7^2} + \frac{1}{9^2} + \&c.,$$

a series already known. Particular cases might be enumerated to infinity; but it agrees better with the object of this work to determine, by following the same process, the values of the different series formed of the sines or cosines of multiple arcs.

182. Let

$$y = \sin x - \frac{1}{2}\sin 2x + \frac{1}{3}\sin 3x - \frac{1}{4}\sin 4x \ldots$$
$$+ \frac{1}{m-1}\sin(m-1)x - \frac{1}{m}\sin mx,$$

m being any even number. We derive from this equation

$$\frac{dy}{dx} = \cos x - \cos 2x + \cos 3x - \cos 4x \ldots + \cos(m-1)x - \cos mx;$$

multiplying by $2\sin x$, and replacing each term of the second member by the difference of two sines, we shall have

$$
\begin{aligned}
2\sin x\frac{dy}{dx} = {} & \sin(x+x) - \sin(x-x) \\
& -\sin(2x+x) + \sin(2x-x) \\
& +\sin(3x+x) - \sin(3x-x) \\
& \cdots\cdots\cdots\cdots\cdots\cdots \\
& +\sin\{(m-1)x-x\} - \sin\{(m+1)x-x\} \\
& -\sin(mx+x) + \sin(mx-x);
\end{aligned}
$$

and, on reduction,

$$2\sin x\frac{dy}{dx} = \sin x + \sin mx - \sin(mx+x):$$

the quantity $\sin mx - \sin(mx+x)$,

or $\sin(mx + \tfrac{1}{2}x - \tfrac{1}{2}x) - \sin(mx + \tfrac{1}{2}x + \tfrac{1}{2}x)$,

is equal to $-2\sin\tfrac{1}{2}x\cos(mx+\tfrac{1}{2}x)$;

we have therefore

$$\frac{dy}{dx} = \frac{1}{2} - \frac{\sin\frac{1}{2}x}{\sin x}\cos(mx+\tfrac{1}{2}x),$$

or

$$\frac{dy}{dx} = \frac{1}{2} - \frac{\cos(mx+\frac{1}{2}x)}{2\cos\frac{1}{2}x};$$

whence we conclude

$$y = \tfrac{1}{2}x - \int dx\,\frac{\cos(mx+\frac{1}{2}x)}{2\cos\frac{1}{2}x}.$$

If we integrate this by parts, distinguishing between the factor $\dfrac{1}{\cos\frac{1}{2}x}$ or $\sec\frac{1}{2}x$, which must be successively differentiated, and the factor $\cos(mx+\frac{1}{2}x)$, which is to be integrated several times in succession, we shall form a series in which the powers of $m+\dfrac{1}{2}$ enter into the denominators. As to the constant it is nothing, since the value of y begins with that of x.

It follows from this that the value of the finite series

$$\sin x - \frac{1}{2}\sin 2x + \frac{1}{3}\sin 3x - \frac{1}{5}\sin 5x + \frac{1}{7}\sin 7x - \ldots - \frac{1}{m}\sin mx$$

differs very little from that of $\frac{1}{2}x$, when the number of terms is very great; and if this number is infinite, we have the known equation

$$\frac{1}{2}x = \sin x - \frac{1}{2}\sin 2x + \frac{1}{3}\sin 3x - \frac{1}{4}\sin 4x + \frac{1}{5}\sin 5x - \&c.$$

From the last series, that which has been given above for the value of $\frac{1}{4}\pi$ might also be derived.

183. Let now

$$y = \frac{1}{2}\cos 2x - \frac{1}{4}\cos 4x + \frac{1}{6}\cos 6x - \ldots$$

$$+ \frac{1}{2m-2}\cos(2m-2)x - \frac{1}{2m}\cos 2mx.$$

Differentiating, multiplying by $2\sin 2x$, substituting the differences of cosines, and reducing, we shall have

$$2\frac{dy}{dx} = -\tan x + \frac{\sin(2m+1)x}{\cos x},$$

or $$2y = c - \int dx \tan x + \int dx \frac{\sin(2m+1)x}{\cos x};$$

integrating by parts the last term of the second member, and supposing m infinite, we have $y = c + \frac{1}{2}\log\cos x$. If in the equation

$$y = \frac{1}{2}\cos 2x - \frac{1}{4}\cos 4x + \frac{1}{6}\cos 6x - \frac{1}{8}\cos 8x + \ldots \&c.$$

we suppose x nothing, we find

$$y = \frac{1}{2} - \frac{1}{4} + \frac{1}{6} - \frac{1}{8} + \ldots \&c. = \frac{1}{2}\log 2;$$

therefore $$y = \frac{1}{2}\log 2 + \frac{1}{2}\log\cos x.$$

Thus we meet with the series given by Euler,

$$\log(2\cos\tfrac{1}{2}x) = \cos x - \frac{1}{2}\cos 2x + \frac{1}{3}\cos 3x - \frac{1}{4}\cos 4x + \&c.$$

184. Applying the same process to the equation

$$y = \sin x + \frac{1}{3}\sin 3x + \frac{1}{5}\sin 5x + \frac{1}{7}\sin 7x + \&c.,$$

we find the following series, which has not been noticed,

$$\frac{1}{4}\pi = \sin x + \frac{1}{3}\sin 3x + \frac{1}{5}\sin 5x + \frac{1}{7}\sin 7x + \frac{1}{9}\sin 9x + \&c.$$ [1]

It must be observed with respect to all these series, that the equations which are formed by them do not hold except when the variable x is included between certain limits. Thus the function

$$\cos x - \frac{1}{3}\cos 3x + \frac{1}{5}\cos 5x - \frac{1}{7}\cos 7x + \&c.$$

is not equal to $\frac{1}{4}\pi$, except when the variable x is contained between the limits which we have assigned. It is the same with the series

$$\sin x - \frac{1}{2}\sin 2x + \frac{1}{3}\sin 3x - \frac{1}{4}\sin 4x + \frac{1}{5}\sin 5x - \&c.$$

This infinite series, which is always convergent, has the value $\frac{1}{2}x$ so long as the arc x is greater than 0 and less than π. But it is not equal to $\frac{1}{2}x$, if the arc exceeds π; it has on the contrary values very different from $\frac{1}{2}x$; for it is evident that in the interval from $x = \pi$ to $x = 2\pi$, the function takes with the contrary sign all the values which it had in the preceding interval from $x = 0$ to $x = \pi$. This series has been known for a long time, but the analysis which served to discover it did not indicate why the result ceases to hold when the variable exceeds π.

The method which we are about to employ must therefore be examined attentively, and the origin of the limitation to which each of the trigonometrical series is subject must be sought.

185. To arrive at it, it is sufficient to consider that the values expressed by infinite series are not known with exact certainty except in the case where the limits of the sum of the terms which complete them can be assigned; it must therefore be supposed that we employ only the first terms of these series,

[1] This may be derived by integration from 0 to π as in Art. 222. [R. L. E.]

and the limits between which the remainder is included must be found.

We will apply this remark to the equation

$$y = \cos x - \frac{1}{3}\cos 3x + \frac{1}{5}\cos 5x - \frac{1}{7}\cos 7x \ldots + \frac{\cos(2m-3)x}{2m-3} - \frac{\cos(2m-1)x}{2m-1}.$$

The number of terms is even and is represented by m; from it is derived the equation $\frac{2dy}{dx} = \frac{\sin 2mx}{\cos x}$, whence we may infer the value of y, by integration by parts. Now the integral $\int uvdx$ may be resolved into a series composed of as many terms as may be desired, u and v being functions of x. We may write, for example,

$$\int uvdx = c + u\int vdx - \frac{du}{dx}\int dx\int vdx + \frac{d^2u}{dx^2}\int dx\int dx\int vdx - \int\left\{d\left(\frac{d^2u}{dx^2}\right)\int dx\int dx\int vdx\right\},$$

an equation which is verified by differentiation.

Denoting $\sin 2mx$ by v and $\sec x$ by u, it will be found that

$$2y = c - \frac{1}{2m}\sec x\cos 2mx + \frac{1}{2^2m^2}\sec' x\sin 2mx + \frac{1}{2^3m^3}\sec'' x\cos 2mx - \int\left(d\,\frac{\sec'' x}{2^3m^3}\,.\cos 2mx\right).$$

186. It is required now to ascertain the limits between which the integral $\frac{1}{2^3m^3}\int\{d(\sec'' x)\cos 2mx\}$ which completes the series is included. To form this integral an infinity of values must be given to the arc x, from 0, the limit at which the integral begins, up to x, which is the final value of the arc; for each one of these values of x the value of the differential $d(\sec'' x)$ must be determined, and that of the factor $\cos 2mx$, and all the partial products must be added: now the variable factor $\cos 2mx$ is necessarily a positive or negative fraction; consequently the integral is composed of the sum of the variable values of the differential $d(\sec'' x)$, multiplied respectively by these fractions.

The total value of the integral is then less than the sum of the differentials $d(\sec'' x)$, taken from $x = 0$ up to x, and it is greater than this sum taken negatively: for in the first case we replace the variable factor $\cos 2mx$ by the constant quantity 1, and in the second case we replace this factor by -1: now the sum of the differentials $d(\sec'' x)$, or which is the same thing, the integral $\int d(\sec'' x)$, taken from $x = 0$, is $\sec'' x - \sec'' 0$; $\sec'' x$ is a certain function of x, and $\sec'' 0$ is the value of this function, taken on the supposition that the arc x is nothing.

The integral required is therefore included between

$$+(\sec'' x - \sec'' 0) \text{ and } -(\sec'' x - \sec'' 0);$$

that is to say, representing by k an unknown fraction positive or negative, we have always

$$\int \{d(\sec'' x) \cos 2mx\} = k(\sec'' x - \sec'' 0).$$

Thus we obtain the equation

$$2y = c - \frac{1}{2m} \sec x \cos 2mx + \frac{1}{2^2 m^2} \sec' x \sin 2mx + \frac{1}{2^3 m^3} \sec'' x \cos 2mx \mp \frac{k}{2^3 m^3} (\sec'' x - \sec'' 0),$$

in which the quantity $\frac{k}{2^3 m^3}(\sec'' x - \sec'' 0)$ expresses exactly the sum of all the last terms of the infinite series.

187. If we had investigated two terms only we should have had the equation

$$2y = c - \frac{1}{2m} \sec x \cos 2mx + \frac{1}{2^2 m^2} \sec' x \sin 2mx + \frac{k}{2^2 m^2} (\sec' x - \sec' 0).$$

From this it follows that we can develope the value of y in as many terms as we wish, and express exactly the remainder of the series; we thus find the set of equations

$$2y = c - \frac{1}{2m} \sec x \cos 2mx \pm \frac{k}{2^2 m^2} (\sec x - \sec 0),$$

$$2y = c - \frac{1}{2m} \sec x \cos 2mx + \frac{1}{2^2 m^2} \sec' x \sin 2mx \mp \frac{k}{2^2 m^2} (\sec' x - \sec' 0),$$

$$2y = c - \frac{1}{2m} \sec x \cos 2mx + \frac{1}{2^2 m^2} \sec' x \sin 2mx + \frac{1}{2^3 m^3} \sec'' x \cos 2mx \mp \frac{k}{2^3 m^3} (\sec'' x - \sec'' 0).$$

The number k which enters into these equations is not the same for all, and it represents in each one a certain quantity which is always included between 1 and -1; m is equal to the number of terms of the series

$$\cos x - \frac{1}{3}\cos 3x + \frac{1}{5}\cos 5x - \ldots - \frac{1}{2m-1}\cos(2m-1)x,$$

whose sum is denoted by y.

188. These equations could be employed if the number m were given, and however great that number might be, we could determine as exactly as we pleased the variable part of the value of y. If the number m be infinite, as is supposed, we consider the first equation only; and it is evident that the two terms which follow the constant become smaller and smaller; so that the exact value of $2y$ is in this case the constant c; this constant is determined by assuming $x=0$ in the value of y, whence we conclude

$$\frac{\pi}{4} = \cos x - \frac{1}{3}\cos 3x + \frac{1}{5}\cos 5x - \frac{1}{7}\cos 7x + \frac{1}{9}\cos 9x - \&c.$$

It is easy to see now that the result necessarily holds if the arc x is less than $\frac{1}{2}\pi$. In fact, attributing to this arc a definite value X as near to $\frac{1}{2}\pi$ as we please, we can always give to m a value so great, that the term $\frac{k}{2m}(\sec x - \sec 0)$, which completes the series, becomes less than any quantity whatever; but the exactness of this conclusion is based on the fact that the term $\sec x$ acquires no value which exceeds all possible limits, whence it follows that the same reasoning cannot apply to the case in which the arc x is not less than $\frac{1}{2}\pi$.

The same analysis could be applied to the series which express the values of $\frac{1}{2}x$, $\log\cos x$, and by this means we can assign the limits between which the variable must be included, in order that the result of analysis may be free from all uncertainty; moreover, the same problems may be treated otherwise by a method founded on other principles[1].

189. The expression of the law of fixed temperatures in a solid plate supposed the knowledge of the equation

[1] Cf. De Morgan's *Diff. and Int. Calculus*, pp. 605—609. [A. F.]

$$\frac{\pi}{4} = \cos x - \frac{1}{3}\cos 3x + \frac{1}{5}\cos 5x - \frac{1}{7}\cos 7x + \frac{1}{9}\cos 9x - \&c.$$

A simpler method of obtaining this equation is as follows:

If the sum of two arcs is equal to $\frac{1}{2}\pi$, a quarter of the circumference, the product of their tangent is 1; we have therefore in general

$$\frac{1}{2}\pi = \text{arc tan}\, u + \text{arc tan}\frac{1}{u} \quad\dots\dots\dots\dots\dots (c);$$

the symbol arc tan u denotes the length of the arc whose tangent is u, and the series which gives the value of that arc is well known; whence we have the following result:

$$\frac{1}{2}\pi = u + \frac{1}{u} - \frac{1}{3}\left(u^3 + \frac{1}{u^3}\right) + \frac{1}{5}\left(u^5 + \frac{1}{u^5}\right) - \frac{1}{7}\left(u^7 + \frac{1}{u^7}\right)$$
$$+ \frac{1}{9}\left(u^9 + \frac{1}{u^9}\right) - \&c. \quad\dots\dots\dots\dots\dots\dots (d).$$

If now we write $e^{x\sqrt{-1}}$ instead of u in equation (c), and in equation (d), we shall have

$$\frac{1}{2}\pi = \text{arc tan}\, e^{x\sqrt{-1}} + \text{arc tan}\, e^{-x\sqrt{-1}},$$

and $\frac{1}{4}\pi = \cos x - \frac{1}{3}\cos 3x + \frac{1}{5}\cos 5x - \frac{1}{7}\cos 7x + \frac{1}{9}\cos 9x - \&c.$

The series of equation (d) is always divergent, and that of equation (b) (Art. 180) is always convergent; its value is $\frac{1}{4}\pi$ or $-\frac{1}{4}\pi$.

SECTION IV.

General solution.

190. We can now form the complete solution of the problem which we have proposed; for the coefficients of equation (b) (Art. 169) being determined, nothing remains but to substitute them, and we have

$$\frac{\pi v}{4} = e^{-x}\cos y - \frac{1}{3}e^{-3x}\cos 3y + \frac{1}{5}e^{-5x}\cos 5y - \frac{1}{7}e^{-7x}\cos 7y + \&c. \dots (\alpha).$$

This value of v satisfies the equation $\frac{d^2v}{dx^2}+\frac{d^2v}{dy^2}=0$; it becomes nothing when we give to y a value equal to $\frac{1}{2}\pi$ or $-\frac{1}{2}\pi$; lastly, it is equal to unity when x is nothing and y is included between $-\frac{1}{2}\pi$ and $+\frac{1}{2}\pi$. Thus all the physical conditions of the problem are exactly fulfilled, and it is certain that, if we give to each point of the plate the temperature which equation (α) determines, and if the base A be maintained at the same time at the temperature 1, and the infinite edges B and C at the temperature 0, it would be impossible for any change to occur in the system of temperatures.

191. The second member of equation (α) having the form of an exceedingly convergent series, it is always easy to determine numerically the temperature of a point whose co-ordinates x and y are known. The solution gives rise to various results which it is necessary to remark, since they belong also to the general theory.

If the point m, whose fixed temperature is considered, is very distant from the origin A, the value of the second member of the equation (α) will be very nearly equal to $e^{-x}\cos y$; it reduces to this term if x is infinite.

The equation $v=\frac{4}{\pi}e^{-x}\cos y$ represents also a state of the solid which would be preserved without any change, if it were once formed; the same would be the case with the state represented by the equation $v=\frac{4}{3\pi}e^{-3x}\cos 3y$, and in general each term of the series corresponds to a particular state which enjoys the same property. All these partial systems exist at once in that which equation (α) represents; they are superposed, and the movement of heat takes place with respect to each of them as if it alone existed. In the state which corresponds to any one of these terms, the fixed temperatures of the points of the base A differ from one point to another, and this is the only condition of the problem which is not fulfilled; but the general state which results from the sum of all the terms satisfies this special condition.

According as the point whose temperature is considered is

more distant from the origin, the movement of heat is less complex: for if the distance x is sufficiently great, each term of the series is very small with respect to that which precedes it, so that the state of the heated plate is sensibly represented by the first three terms, or by the first two, or by the first only, for those parts of the plate which are more and more distant from the origin.

The curved surface whose vertical ordinate measures the fixed temperature v, is formed by adding the ordinates of a multitude of particular surfaces whose equations are

$$\frac{\pi v_1}{4} = e^{-x} \cos y, \quad \frac{\pi v_2}{4} = -\tfrac{1}{3} e^{-3x} \cos 3y, \quad \frac{\pi v_3}{4} = \tfrac{1}{5} e^{-5x} \cos 5y, \text{ \&c.}$$

The first of these coincides with the general surface when x is infinite, and they have a common asymptotic sheet.

If the difference $v - v_1$ of their ordinates is considered to be the ordinate of a curved surface, this surface will coincide, when x is infinite, with that whose equation is $\frac{1}{4}\pi v_2 = -\frac{1}{3} e^{-3x} \cos 3y$. All the other terms of the series produce similar results.

The same results would again be found if the section at the origin, instead of being bounded as in the actual hypothesis by a straight line parallel to the axis of y, had any figure whatever formed of two symmetrical parts. It is evident therefore that the particular values

$$ae^{-x} \cos y, \quad be^{-3x} \cos 3y, \quad ce^{-5x} \cos 5y, \text{ \&c.},$$

have their origin in the physical problem itself, and have a necessary relation to the phenomena of heat. Each of them expresses a simple mode according to which heat is established and propagated in a rectangular plate, whose infinite sides retain a constant temperature. The general system of temperatures is compounded always of a multitude of simple systems, and the expression for their sum has nothing arbitrary but the coefficients a, b, c, d, &c.

192. Equation (a) may be employed to determine all the circumstances of the permanent movement of heat in a rectangular plate heated at its origin. If it be asked, for example, what is the expenditure of the source of heat, that is to say,

what is the quantity which, during a given time, passes across the base A and replaces that which flows into the cold masses B and C; we must consider that the flow perpendicular to the axis of y is expressed by $-K\dfrac{dv}{dx}$. The quantity which during the instant dt flows across a part dy of the axis is therefore

$$-K\frac{dv}{dx}\,dy\,dt;$$

and, as the temperatures are permanent, the amount of the flow, during unit of time, is $-K\dfrac{dv}{dx}\,dy$. This expression must be integrated between the limits $y=-\frac{1}{2}\pi$ and $y=+\frac{1}{2}\pi$, in order to ascertain the whole quantity which passes the base, or which is the same thing, must be integrated from $y=0$ to $y=\frac{1}{2}\pi$, and the result doubled. The quantity $\dfrac{dv}{dx}$ is a function of x and y, in which x must be made equal to 0, in order that the calculation may refer to the base A, which coincides with the axis of y. The expression for the expenditure of the source of heat is therefore $2\int\left(-K\dfrac{dv}{dx}\,dy\right)$. The integral must be taken from $y=0$ to $y=\frac{1}{2}\pi$; if, in the function $\dfrac{dv}{dx}$, x is not supposed equal to 0, but $x=x$, the integral will be a function of x which will denote the quantity of heat which flows in unit of time across a transverse edge at a distance x from the origin.

193. If we wish to ascertain the quantity of heat which, during unit of time, passes across a line drawn on the plate parallel to the edges B and C, we employ the expression $-K\dfrac{dv}{dy}$, and, multiplying it by the element dx of the line drawn, integrate with respect to x between the given boundaries of the line; thus the integral $\int\left(-K\dfrac{dv}{dy}\,dx\right)$ shews how much heat flows across the whole length of the line; and if before or after the integration we make $y=\frac{1}{2}\pi$, we determine the quantity of heat which, during unit of time, escapes from the plate across the infinite edge C. We may next compare the latter quantity with the expenditure

of the source of heat; for the source must necessarily supply continually the heat which flows into the masses B and C. If this compensation did not exist at each instant, the system of temperatures would be variable.

194. Equation (α) gives

$$-K\frac{dv}{dx}=\frac{4K}{\pi}(e^{-x}\cos y-e^{-3x}\cos 3y+e^{-5x}\cos 5y-e^{-7x}\cos 7y+\&c.);$$

multiplying by dy, and integrating from $y=0$, we have

$$\frac{4K}{\pi}\left(e^{-x}\sin y-\frac{1}{3}e^{-3x}\sin 3y+\frac{1}{5}e^{-5x}\sin 5y-\frac{1}{7}e^{-7x}\sin 7y+\&c.\right).$$

If y be made $=\frac{1}{2}\pi$, and the integral doubled, we obtain

$$\frac{8K}{\pi}\left(e^{-x}+\frac{1}{3}e^{-3x}+\frac{1}{5}e^{-5x}+\frac{1}{7}e^{-7x}+\&c.\right)$$

as the expression for the quantity of heat which, during unit of time, crosses a line parallel to the base, and at a distance x from that base.

From equation (α) we derive also

$$-K\frac{dv}{dy}=\frac{4K}{\pi}(e^{-x}\sin y-e^{-3x}\sin 3y+e^{-5x}\sin 5y-e^{-7x}\sin 7y+\&c.):$$

hence the integral $\int -K\left(\frac{dv}{dy}\right)dx$, taken from $x=0$, is

$$\frac{4K}{\pi}\{(1-e^{-x})\sin y-(1-e^{-3x})\sin 3y+(1-e^{-5x})\sin 5y$$
$$-(1-e^{-7}x)\sin 7y+\&c.\}.$$

If this quantity be subtracted from the value which it assumes when x is made infinite, we find

$$\frac{4K}{\pi}\left(e^{-x}\sin y-\frac{1}{3}e^{-3x}\sin 3y+\frac{1}{5}e^{-5x}\sin 5y-\&c.\right);$$

and, on making $y=\frac{1}{2}\pi$, we have an expression for the whole quantity of heat which crosses the infinite edge C, from the point whose distance from the origin is x up to the end of the plate; namely,

$$\frac{4K}{\pi}\left(e^{-x}+\frac{1}{3}e^{-3x}+\frac{1}{5}e^{-5x}+\frac{1}{7}e^{-7x}+\&c.\right),$$

which is evidently equal to half the quantity which in the same time passes beyond the transverse line drawn on the plate at a distance x from the origin. We have already remarked that this result is a necessary consequence of the conditions of the problem; if it did not hold, the part of the plate which is situated beyond the transverse line and is prolonged to infinity would not receive through its base a quantity of heat equal to that which it loses through its two edges; it could not therefore preserve its state, which is contrary to hypothesis.

195. As to the expenditure of the source of heat, it is found by supposing $x=0$ in the preceding expression; hence it assumes an infinite value, the reason for which is evident if it be remarked that, according to hypothesis, every point of the line A has and retains the temperature 1: parallel lines which are very near to this base have also a temperature very little different from unity: hence, the extremities of all these lines contiguous to the cold masses B and C communicate to them a quantity of heat incomparably greater than if the decrease of temperature were continuous and imperceptible. In the first part of the plate, at the ends near to B or C, a *cataract* of heat, or an infinite flow, exists. This result ceases to hold when the distance x becomes appreciable.

196. The length of the base has been denoted by π. If we assign to it any value $2l$, we must write $\frac{1}{2}\pi\frac{y}{l}$ instead of y, and multiplying also the values of x by $\frac{\pi}{2l}$, we must write $\frac{1}{2}\pi\frac{x}{l}$ instead of x. Denoting by A the constant temperature of the base, we must replace v by $\frac{v}{A}$. These substitutions being made in the equation (α), we have

$$v = \frac{4A}{\pi}\left(e^{-\frac{\pi x}{2l}}\cos\frac{\pi y}{2l} - \frac{1}{3}e^{-\frac{3\pi x}{2l}}\cos 3\frac{\pi y}{2l} + \frac{1}{5}e^{-\frac{5\pi x}{2l}}\cos 5\frac{\pi y}{2l} - \frac{1}{7}e^{-\frac{7\pi x}{2l}}\cos 7\frac{\pi y}{2l} + \&c.\right) \quad\ldots\ldots(\beta).$$

This equation represents exactly the system of permanent temperature in an infinite rectangular prism, included between two masses of ice B and C, and a constant source of heat.

197. It is easy to see either by means of this equation, or from Art. 171, that heat is propagated in this solid, by separating more and more from the origin, at the same time that it is directed towards the infinite faces B and C. Each section parallel to that of the base is traversed by a wave of heat which is renewed at each instant with the same intensity: the intensity diminishes as the section becomes more distant from the origin. Similar movements are effected with respect to any plane parallel to the infinite faces; each of these planes is traversed by a constant wave which conveys its heat to the lateral masses.

The developments contained in the preceding articles would be unnecessary, if we had not to explain an entirely new theory, whose principles it is requisite to fix. With that view we add the following remarks.

198. Each of the terms of equation (α) corresponds to only one particular system of temperatures, which might exist in a rectangular plate heated at its end, and whose infinite edges are maintained at a constant temperature. Thus the equation $v = e^{-x} \cos y$ represents the permanent temperatures, when the points of the base A are subject to a fixed temperature, denoted by $\cos y$. We may now imagine the heated plate to be part of a plane which is prolonged to infinity in all directions, and denoting the co-ordinates of any point of this plane by x and y, and the temperature of the same point by v, we may apply to the entire plane the equation $v = e^{-x} \cos y$; by this means, the edges B and C receive the constant temperature 0; but it is not the same with contiguous parts BB and CC; they receive and keep lower temperatures. The base A has at every point the permanent temperature denoted by $\cos y$, and the contiguous parts AA have higher temperatures. If we construct the curved surface whose vertical ordinate is equal to the permanent temperature at each point of the plane, and if it be cut by a vertical plane passing through the line A or parallel to that line, the form of the section will be that of a trigonometrical line whose ordinate represents the infinite and periodic series of cosines. If the same curved surface be cut by a vertical plane parallel to the axis of x, the form of the section will through its whole length be that of a logarithmic curve.

199. By this it may be seen how the analysis satisfies the two conditions of the hypothesis, which subjected the base to a temperature equal to $\cos y$, and the two sides B and C to the temperature 0. When we express these two conditions we solve in fact the following problem: If the heated plate formed part of an infinite plane, what must be the temperatures at all the points of the plane, in order that the system may be self-permanent, and that the fixed temperatures of the infinite rectangle may be those which are given by the hypothesis?

We have supposed in the foregoing part that some external causes maintained the faces of the rectangular solid, one at the temperature 1, and the two others at the temperature 0. This effect may be represented in different manners; but the hypothesis proper to the investigation consists in regarding the prism as part of a solid all of whose dimensions are infinite, and in determining the temperatures of the mass which surrounds it, so that the conditions relative to the surface may be always observed.

200. To ascertain the system of permanent temperatures in a rectangular plate whose extremity A is maintained at the temperature 1, and the two infinite edges at the temperature 0, we might consider the changes which the temperatures undergo, from the initial state which is given, to the fixed state which is the object of the problem. Thus the variable state of the solid would be determined for all values of the time, and it might then be supposed that the value was infinite.

The method which we have followed is different, and conducts more directly to the expression of the final state, since it is founded on a distinctive property of that state. We now proceed to shew that the problem admits of no other solution than that which we have stated. The proof follows from the following propositions.

201. If we give to all the points of an infinite rectangular plate temperatures expressed by equation (α), and if at the two edges B and C we maintain the fixed temperature 0, whilst the end A is exposed to a source of heat which keeps all points of the line A at the fixed temperature 1; no change can happen in the state of the solid. In fact, the equation $\frac{d^2v}{dx^2}+\frac{d^2v}{dy^2}=0$ being

satisfied, it is evident (Art. 170) that the quantity of heat which determines the temperature of each molecule can be neither increased nor diminished.

The different points of the same solid having received the temperatures expressed by equation (a) or $v = \phi(x, y)$, suppose that instead of maintaining the edge A at the temperature 1, the fixed temperature 0 be given to it as to the two lines B and C; the heat contained in the plate BAC will flow across the three edges A, B, C, and by hypothesis it will not be replaced, so that the temperatures will diminish continually, and their final and common value will be zero. This result is evident since the points infinitely distant from the origin A have a temperature infinitely small from the manner in which equation (a) was formed.

The same effect would take place in the opposite direction, if the system of temperatures were $v = -\phi(x, y)$, instead of being $v = \phi(x, y)$; that is to say, all the initial negative temperatures would vary continually, and would tend more and more towards their final value 0, whilst the three edges A, B, C preserved the temperature 0.

202. Let $v = \phi(x, y)$ be a given equation which expresses the initial temperature of points in the plate BAC, whose base A is maintained at the temperature 1, whilst the edges B and C preserve the temperature 0.

Let $v = F(x, y)$ be another given equation which expresses the initial temperature of each point of a solid plate BAC exactly the same as the preceding, but whose three edges B, A, C are maintained at the temperature 0.

Suppose that in the first solid the variable state which succeeds to the final state is determined by the equation $v = \phi(x, y, t)$, t denoting the time elapsed, and that the equation $v = \Phi(x, y, t)$ determines the variable state of the second solid, for which the initial temperatures are $F(x, y)$.

Lastly, suppose a third solid like each of the two preceding: let $v = f(x, y) + F(x, y)$ be the equation which represents its initial state, and let 1 be the constant temperature of the base A, 0 and 0 those of the two edges B and C.

We proceed to shew that the variable state of the third solid is determined by the equation $v = \phi(x, y, t) + \Phi(x, y, t)$.

In fact, the temperature of a point m of the third solid varies, because that molecule, whose volume is denoted by M, acquires or loses a certain quantity of heat Δ. The increase of temperature during the instant dt is

$$\frac{\Delta}{cM}\,dt,$$

the coefficient c denoting the specific capacity with respect to volume. The variation of the temperature of the same point in the first solid is $\frac{d}{cM}\,dt$, and $\frac{D}{cM}\,dt$ in the second, the letters d and D representing the quantity of heat positive or negative which the molecule acquires by virtue of the action of all the neighbouring molecules. Now it is easy to perceive that Δ is equal to $d + D$. For proof it is sufficient to consider the quantity of heat which the point m receives from another point m' belonging to the interior of the plate, or to the edges which bound it.

The point m_1, whose initial temperature is denoted by f_1, transmits, during the instant dt, to the molecule m, a quantity of heat expressed by $q_1(f_1 - f)dt$, the factor q_1 representing a certain function of the distance between the two molecules. Thus the whole quantity of heat acquired by m is $\Sigma q_1(f_1 - f)dt$, the sign Σ expressing the sum of all the terms which would be found by considering the other points m_2, m_3, m_4 &c. which act on m; that is to say, writing q_2, f_2 or q_3, f_3, or q_4, f_4 and so on, instead of q_1, f_1. In the same manner $\Sigma q_1(F_1 - F)dt$ will be found to be the expression of the whole quantity of heat acquired by the same point m of the second solid; and the factor q_1 is the same as in the term $\Sigma q_1(f_1 - f)dt$, since the two solids are formed of the same matter, and the position of the points is the same; we have then

$$d = \Sigma q_1(f_1 - f)dt \text{ and } D = \Sigma q_1(F_1 - F)dt.$$

For the same reason it will be found that

$$\Delta = \Sigma q_1\{f_1 + F_1 - (f + F)\}dt;$$

hence $$\Delta = d + D \text{ and } \frac{\Delta}{cM} = \frac{d}{cM} + \frac{D}{cM}.$$

It follows from this that the molecule m of the third solid acquires, during the instant dt, an increase of temperature equal to the sum of the two increments which the same point would have gained in the two first solids. Hence at the end of the first instant, the original hypothesis will again hold, since any molecule whatever of the third solid has a temperature equal to the sum of those which it has in the two others. Thus the same relation exists at the beginning of each instant, that is to say, the variable state of the third solid can always be represented by the equation

$$v = \phi(x, y, t) + \Phi(x, y, t).$$

203. The preceding proposition is applicable to all problems relative to the uniform or varied movement of heat. It shews that the movement can always be decomposed into several others, each of which is effected separately as if it alone existed. This superposition of simple effects is one of the fundamental elements in the theory of heat. It is expressed in the investigation, by the very nature of the general equations, and derives its origin from the principle of the communication of heat.

Let now $v = \phi(x, y)$ be the equation (α) which expresses the permanent state of the solid plate BAC, heated at its end A, and whose edges B and C preserve the temperature 1; the initial state of the plate is such, according to hypothesis, that all its points have a nul temperature, except those of the base A, whose temperature is 1. The initial state can then be considered as formed of two others, namely: a first, in which the initial temperatures are $-\phi(x, y)$, the three edges being maintained at the temperature 0, and a second state, in which the initial temperatures are $+\phi(x, y)$, the two edges B and C preserving the temperature 0, and the base A the temperature 1; the superposition of these two states produces the initial state which results from the hypothesis. It remains then only to examine the movement of heat in each one of the two partial states. Now, in the second, the system of temperatures can undergo no change; and in the first, it has been remarked in Article 201 that the temperatures vary continually, and end with being nul. Hence the final state, properly so called, is that which is represented by $v = \phi(x, y)$ or equation (α).

If this state were formed at first it would be self-existent, and it is this property which has served to determine it for us. If the solid plate be supposed to be in another initial state, the difference between the latter state and the fixed state forms a partial state, which imperceptibly disappears. After a considerable time, the difference has nearly vanished, and the system of fixed temperatures has undergone no change. Thus the variable temperatures converge more and more to a final state, independent of the primitive heating.

204. We perceive by this that the final state is unique; for, if a second state were conceived, the difference between the second and the first would form a partial state, which ought to be self-existent, although the edges A, B, C were maintained at the temperature 0. Now the last effect cannot occur; similarly if we supposed another source of heat independent of that which flows from the origin A; besides, this hypothesis is not that of the problem we have treated, in which the initial temperatures are nul. It is evident that parts very distant from the origin can only acquire an exceedingly small temperature.

Since the final state which must be determined is unique, it follows that the problem proposed admits no other solution than that which results from equation (α). Another form may be given to this result, but the solution can be neither extended nor restricted without rendering it inexact.

The method which we have explained in this chapter consists in forming first very simple particular values, which agree with the problem, and in rendering the solution more general, to the intent that v or $\phi(x, y)$ may satisfy three conditions, namely:

$$\frac{d^2v}{dx^2} + \frac{d^2v}{dy^2} = 0, \quad \phi(x, 0) = 1, \quad \phi(x, \pm \tfrac{1}{2}\pi) = 0.$$

It is clear that the contrary order might be followed, and the solution obtained would necessarily be the same as the foregoing. We shall not stop over the details, which are easily supplied, when once the solution is known. We shall only give in the following section a remarkable expression for the function $\phi(x, y)$ whose value was developed in a convergent series in equation (α).

SECTION V.

Finite expression of the result of the solution.

205. The preceding solution might be deduced from the integral of the equation $\dfrac{d^2v}{dx^2} + \dfrac{d^2v}{dy^2} = 0$,[1] which contains imaginary quantities, under the sign of the arbitrary functions. We shall confine ourselves here to the remark that the integral

$$v = \phi\,(x + y\sqrt{-1}) + \psi\,(x - y\sqrt{-1}),$$

has a manifest relation to the value of v given by the equation

$$\frac{\pi v}{4} = e^{-x}\cos y - \frac{1}{3}\,e^{-3x}\cos 3y + \frac{1}{5}\,e^{-5x}\cos 5y - \&\text{c.}$$

In fact, replacing the cosines by their imaginary expressions, we have

$$\frac{\pi v}{2} = e^{-(x-y\sqrt{-1})} - \frac{1}{3}\,e^{-3(x-y\sqrt{-1})} + \frac{1}{5}\,e^{-5(x-y\sqrt{-1})} - \&\text{c.}$$

$$+ e^{-(x+y\sqrt{-1})} - \frac{1}{3}\,e^{-3(x+y\sqrt{-1})} + \frac{1}{5}\,e^{-5(x+y\sqrt{-1})} - \&\text{c.}$$

The first series is a function of $x - y\sqrt{-1}$, and the second series is the same function of $x + y\sqrt{-1}$.

Comparing these series with the known development of arc tan z in functions of z its tangent, it is immediately seen that the first is arc tan $e^{-(x-y\sqrt{-1})}$, and the second is arc tan $e^{-(x+y\sqrt{-1})}$; thus equation (α) takes the finite form

$$\frac{\pi v}{2} = \text{arc tan } e^{-(x+y\sqrt{-1})} + \text{arc tan } e^{-(x-y\sqrt{-1})}\ldots\ldots\ldots(B).$$

In this mode it conforms to the general integral

$$v = \phi\,(x + y\sqrt{-1}) + \psi\,(x - y\sqrt{-1})\ldots\ldots\ldots(A),$$

the function $\phi\,(z)$ is arc tan e^{-z}, and similarly the function $\psi\,(z)$.

[1] D. F. Gregory derived the solution from the form

$$v = \cos\left(y\,\frac{d}{dx}\right)\phi\,(x) + \sin\left(y\,\frac{d}{dx}\right)\psi\,(x).$$

Camb. Math. Journal, Vol. I. p. 105. [A. F.]

If in equation (B) we denote the first term of the second member by p and the second by q, we have

$$\frac{1}{2}\pi v = p + q, \quad \tan p = e^{-(x+y\sqrt{-1})}, \quad \tan q = e^{-(x-y\sqrt{-1})};$$

whence $\quad \tan(p+q)$ or $\dfrac{\tan p + \tan q}{1 - \tan p \tan q} = \dfrac{2e^{-x}\cos y}{1 - e^{-2x}} = \dfrac{2\cos y}{e^{x} - e^{-x}}$;

whence we deduce the equation $\dfrac{1}{2}\pi v = \text{arc tan}\left(\dfrac{2\cos y}{e^{x} - e^{-x}}\right) \ldots\ldots(C)$.

This is the simplest form under which the solution of the problem can be presented.

206. This value of v or $\phi(x, y)$ satisfies the conditions relative to the ends of the solid, namely, $\phi(x, \pm\frac{1}{2}\pi) = 0$, and $\phi(0, y) = 1$; it satisfies also the general equation $\dfrac{d^2v}{dx^2} + \dfrac{d^2v}{dy^2} = 0$, since equation (C) is a transformation of equation (B). Hence it represents exactly the system of permanent temperatures; and since that state is unique, it is impossible that there should be any other solution, either more general or more restricted.

The equation (C) furnishes, by means of tables, the value of one of the three unknowns v, x, y, when two of them are given; it very clearly indicates the nature of the surface whose vertical ordinate is the permanent temperature of a given point of the solid plate. Finally, we deduce from the same equation the values of the differential coefficients $\dfrac{dv}{dx}$ and $\dfrac{dv}{dy}$ which measure the velocity with which heat flows in the two orthogonal directions; and we consequently know the value of the flow in any other direction.

These coefficients are expressed thus,

$$\frac{dv}{dx} = -2\cos y\left(\frac{e^{x} + e^{-x}}{e^{2x} + 2\cos 2y + e^{-2x}}\right),$$

$$\frac{dv}{dy} = -2\sin y\left(\frac{e^{x} - e^{-x}}{e^{2x} + 2\cos 2y + e^{-2x}}\right).$$

It may be remarked that in Article 194 the value of $\dfrac{dv}{dx}$, and that of $\dfrac{dv}{dy}$ are given by infinite series, whose sums may be easily found, by replacing the trigonometrical quantities by imaginary exponentials. We thus obtain the values of $\frac{dv}{dx}$ and $\frac{dv}{dy}$ which we have just stated.

The problem which we have now dealt with is the first which we have solved in the theory of heat, or rather in that part of the theory which requires the employment of analysis. It furnishes very easy numerical applications, whether we make use of the trigonometrical tables or convergent series, and it represents exactly all the circumstances of the movement of heat. We pass on now to more general considerations.

SECTION VI.

Development of an arbitrary function in trigonometric series.

207. The problem of the propagation of heat in a rectangular solid has led to the equation $\frac{d^2v}{dx^2}+\frac{d^2v}{dy^2}=0$; and if it be supposed that all the points of one of the faces of the solid have a common temperature, the coefficients a, b, c, d, etc. of the series

$$a\cos x+b\cos 3x+c\cos 5x+d\cos 7x+\ldots \&c.,$$

must be determined so that the value of this function may be equal to a constant whenever the arc x is included between $-\frac{1}{2}\pi$ and $+\frac{1}{2}\pi$. The value of these coefficients has just been assigned; but herein we have dealt with a single case only of a more general problem, which consists in developing any function whatever in an infinite series of sines or cosines of multiple arcs. This problem is connected with the theory of partial differential equations, and has been attacked since the origin of that analysis. It was necessary to solve it, in order to integrate suitably the equations of the propagation of heat; we proceed to explain the solution.

We shall examine, in the first place, the case in which it is required to reduce into a series of sines of multiple arcs, a function whose development contains only odd powers of the variable. Denoting such a function by $\phi(x)$, we arrange the equation

$$\phi(x) = a \sin x + b \sin 2x + c \sin 3x + d \sin 4x + \dots \text{\&c.},$$

in which it is required to determine the value of the coefficients a, b, c, d, &c. First we write the equation

$$\phi(x) = x\phi'(0) + \frac{x^2}{\lfloor 2}\phi''(0) + \frac{x^3}{\lfloor 3}\phi'''(0) + \frac{x^4}{\lfloor 4}\phi^{\text{iv}}(0) + \frac{x^5}{\lfloor 5}\phi^{\text{v}}(0) + \dots \text{\&c.},$$

in which $\phi'(0)$, $\phi''(0)$, $\phi'''(0)$, $\phi^{\text{iv}}(0)$, &c. denote the values taken by the coefficients

$$\frac{d\phi(x)}{dx}, \quad \frac{d^2\phi(x)}{dx^2}, \quad \frac{d^3\phi(x)}{dx^3}, \quad \frac{d^4\phi(x)}{dx^4}, \text{ \&c.},$$

when we suppose $x = 0$ in them. Thus, representing the development according to powers of x by the equation

$$\phi(x) = Ax - B\frac{x^3}{\lfloor 3} + C\frac{x^5}{\lfloor 5} - D\frac{x^7}{\lfloor 7} + E\frac{x^9}{\lfloor 9} - \text{\&c.},$$

we have

$$\begin{array}{ll} \phi(0) = 0, \text{ and} & \phi'(0) = A, \\ \phi''(0) = 0, & \phi'''(0) = B, \\ \phi^{\text{iv}}(0) = 0, & \phi^{\text{v}}(0) = C, \\ \phi^{\text{vi}}(0) = 0, & \phi^{\text{vii}}(0) = D, \\ \text{\&c.} & \text{\&c.} \end{array}$$

If now we compare the preceding equation with the equation

$$\phi(x) = a \sin x + b \sin 2x + c \sin 3x + d \sin 4x + e \sin 5x + \text{\&c.},$$

developing the second member with respect to powers of x, we have the equations

$$\begin{aligned} A &= a + 2b + 3c + 4d + 5e + \text{\&c.}, \\ B &= a + 2^3b + 3^3c + 4^3d + 5^3e + \text{\&c.}, \\ C &= a + 2^5b + 3^5c + 4^5d + 5^5e + \text{\&c.}, \\ D &= a + 2^7b + 3^7c + 4^7d + 5^7e + \text{\&c.}, \\ E &= a + 2^9b + 3^9c + 4^9d + 5^9e + \text{\&c.} \; \dots\dots\dots\dots (a). \end{aligned}$$

These equations serve to find the coefficients a, b, c, d, e, &c., whose number is infinite. To determine them, we first regard the number of unknowns as finite and equal to m; thus we suppress all the equations which follow the first m equations, and we omit from each equation all the terms of the second member which follow the first m terms which we retain. The whole number m being given, the coefficients a, b, c, d, e, &c. have fixed values which may be found by elimination. Different values would be obtained for the same quantities, if the number of the equations and that of the unknowns were greater by one unit. Thus the value of the coefficients varies as we increase the number of the coefficients and of the equations which ought to determine them. It is required to find what the limits are towards which the values of the unknowns converge continually as the number of equations increases. These limits are the true values of the unknowns which satisfy the preceding equations when their number is infinite.

208. We consider then in succession the cases in which we should have to determine one unknown by one equation, two unknowns by two equations, three unknowns by three equations, and so on to infinity.

Suppose that we denote as follows different systems of equations analogous to those from which the values of the coefficients must be derived:

$$a_1 = A_1, \qquad \begin{aligned} a_2 + 2b_2 &= A_2, \\ a_2 + 2^3 b_2 &= B_2, \end{aligned} \qquad \begin{aligned} a_3 + 2b_3 + 3c_3 &= A_3, \\ a_3 + 2^3 b_3 + 3^3 c_3 &= B_3, \\ a_3 + 2^5 b_3 + 3^5 c_3 &= C_3, \end{aligned}$$

$$\begin{aligned} a_4 + 2b_4 + 3c_4 + 4d_4 &= A_4, \\ a_4 + 2^3 b_4 + 3^3 c_4 + 4^3 d_4 &= B_4, \\ a_4 + 2^5 b_4 + 3^5 c_4 + 4^5 d_4 &= C_4, \\ a_4 + 2^7 b_4 + 3^7 c_4 + 4^7 d_4 &= D_4, \end{aligned}$$

$$\begin{aligned} a_5 + 2b_5 + 3c_5 + 4d_5 + 5e_5 &= A_5, \\ a_5 + 2^3 b_5 + 3^3 c_5 + 4^3 d_5 + 5^3 e_5 &= B_5, \\ a_5 + 2^5 b_5 + 3^5 c_5 + 4^5 d_5 + 5^5 e_5 &= C_5, \\ a_5 + 2^7 b_5 + 3^7 c_5 + 4^7 d_5 + 5^7 e_5 &= D_5, \\ a_5 + 2^9 b_5 + 3^9 c_5 + 4^9 d_5 + 5^9 e_5 &= E_5, \end{aligned}$$

&c. &c. .... (b).

If now we eliminate the last unknown e_5 by means of the five equations which contain A_5, B_5, C_5, D_5, E_5, &c., we find

$$a_5(5^2-1^2)+2b_5(5^2-2^2)+3c_5(5^2-3^2)+4d_5(5^2-4^2)=5^2A_5-B_5,$$

$$a_5(5^2-1^2)+2^3b_5(5^2-2^2)+3^3c_5(5^2-3^2)+4^3d_5(5^2-4^2)=5^2B_5-C_5,$$

$$a_5(5^2-1^2)+2^5b_5(5^2-2^2)+3^5c_5(5^2-3^2)+4^5d_5(5^2-4^2)=5^2C_5-D_5,$$

$$a_5(5^2-1^2)+2^7b_5(5^2-2^2)+3^7c_5(5^2-3^2)+4^7d_5(5^2-4^2)=5^2D_5-E_5.$$

We could have deduced these four equations from the four which form the preceding system, by substituting in the latter instead of

$$a_4, \quad (5^2-1^2)\,a_5,$$
$$b_4, \quad (5^2-2^2)\,b_5,$$
$$c_4, \quad (5^2-3^2)\,c_5,$$
$$d_4, \quad (5^2-4^2)\,d_5;$$

and instead of

$$A_4, \quad 5^2A_5-B_5,$$
$$B_4, \quad 5^2B_5-C_5,$$
$$C_4, \quad 5^2C_5-D_5,$$
$$D_4, \quad 5^2D_5-E_5.$$

By similar substitutions we could always pass from the case which corresponds to a number m of unknowns to that which corresponds to the number $m+1$. Writing in order all the relations between the quantities which correspond to one of the cases and those which correspond to the following case, we shall have

$$a_1=a_2(2^2-1),$$
$$a_2=a_3(3^2-1),\quad b_2=b_3(3^2-2^2),$$
$$a_3=a_4(4^2-1),\quad b_3=b_4(4^2-2^2),\quad c_3=c_4(4^2-3^2),$$
$$a_4=a_5(5^2-1),\quad b_4=b_5(5^2-2^2),\quad c_4=c_5(5^2-3^2),\quad d_4=d_5(5^2-4^2),$$
$$a_5=a_6(6^2-1),\quad b_5=b_6(6^2-2^2),\quad c_5=c_6(6^2-3^2),\quad d_5=d_6(6^2-4^2),$$
$$e_5=e_6(6^2-5^2),$$

&c. &c. ....(c).

We have also

$$A_1 = 2^2 A_2 - B_2,$$
$$A_2 = 3^2 A_3 - B_3, \quad B_2 = 3^2 B_3 - C_3,$$
$$A_3 = 4^2 A_4 - B_4, \quad B_3 = 4^2 B_4 - C_4, \quad C_3 = 4^2 C_4 - D_4,$$
$$A_4 = 5^2 A_5 - B_5, \quad B_4 = 5^2 B_5 - C_5, \quad C_4 = 5^2 C_5 - D_5, \quad D_4 = 5^2 D_5 - E_5,$$
$$\&\text{c.} \qquad\qquad \&\text{c.} \quad \ldots\ldots\ldots\ldots\ldots\ldots\ldots\ldots (d).$$

From equations (c) we conclude that on representing the unknowns, whose number is infinite, by a, b, c, d, e, &c., we must have

$$a = \frac{a_1}{(2^2-1)(3^2-1)(4^2-1)(6^2-1)\ldots},$$

$$b = \frac{b_2}{(3^2-2^2)(4^2-2^2)(5^2-2^2)(6^2-2^2)\ldots},$$

$$c = \frac{c_3}{(4^2-3^2)(5^2-3^2)(6^2-3^2)(7^2-3^2)\ldots},$$

$$d = \frac{d_4}{(5^2-4^2)(6^2-4^2)(7^2-4^2)(8^2-4^2)\ldots},$$

$$\&\text{c.} \qquad\qquad \&\text{c.} \quad \ldots\ldots\ldots\ldots\ldots\ldots (e).$$

209. It remains then to determine the values of a_1, b_2, c_3, d_4, e_5, &c.; the first is given by one equation, in which A_1 enters; the second is given by two equations into which $A_2 B_2$ enter; the third is given by three equations, into which $A_3 B_3 C_3$ enter; and so on. It follows from this that if we knew the values of

$$A_1, \quad A_2 B_2, \quad A_3 B_3 C_3, \quad A_4 B_4 C_4 D_4 \ldots, \ \&\text{c.},$$

we could easily find a_1 by solving one equation, $a_2 b_2$ by solving two equations, $a_3 b_3 c_3$ by solving three equations, and so on: after which we could determine a, b, c, d, e, &c. It is required then to calculate the values of

$$A_1, \quad A_2 B_2, \quad A_3 B_3 C_3, \quad A_4 B_4 C_4 D_4, \quad A_5 B_5 C_5 D_5 E_5 \ldots, \ \&\text{c.},$$

by means of equations (d). 1st, we find the value of A_2 in terms of A_2 and B_2; 2nd, by two substitutions we find this value of A_1 in terms of $A_3 B_3 C_3$; 3rd, by three substitutions we find the same value of A_1 in terms of $A_4B_4C_4D_4$, and so on. The successive values of A_1 are

$$A_1 = A_2 2^2 - B_2,$$

$$A_1 = A_3 2^2 . 3^2 - B_3 (2^2 + 3^2) + C_3,$$

$$A_1 = A_4 2^2 . 3^2 . 4^2 - B_4 (2^2 . 3^2 + 2^2 . 4^2 + 3^2 . 4^2) + C_4 (2^2 + 3^2 + 4^2) - D_4,$$

$$A_1 = A_5 2^2 . 3^2 . 4^2 . 5^2 - B_5 (2^2 . 3^2 . 4^2 + 2^2 . 3^2 . 5^2 + 2^2 . 4^2 . 5^2 + 3^2 . 4^2 . 5^2)$$
$$+ C_5 (2^2 . 3^2 + 2^2 . 4^2 + 2^2 . 5^2 + 3^2 . 4^2 + 3^2 . 5^2 + 4^2 . 5^2)$$
$$- D_5 (2^2 + 3^2 + 4^2 + 5^2) + E_5, \ \&c.,$$

the law of which is readily noticed. The last of these values, which is that which we wish to determine, contains the quantities A, B, C, D, E, &c., with an infinite index, and these quantities are known; they are the same as those which enter into equations (a).

Dividing the ultimate value of A_1 by the infinite product

$$2^2 . 3^2 . 4^2 . 5^2 . 6^2 \ldots \&c.,$$

we have

$$A - B\left(\frac{1}{2^2} + \frac{1}{3^2} + \frac{1}{4^2} + \frac{1}{5^2} + \&c.\right) + C\left(\frac{1}{2^2 . 3^2} + \frac{1}{2^2 . 4^2} + \frac{1}{3^2 . 4^2} + \&c.\right)$$
$$- D\left(\frac{1}{2^2 . 3^2 . 4^2} + \frac{1}{2^2 . 3^2 . 5^2} + \frac{1}{3^2 . 4^2 . 5^2} + \&c.\right)$$
$$+ E\left(\frac{1}{2^2 . 3^2 . 4^2 . 5^2} + \frac{1}{2^2 . 3^2 . 4^2 . 6^2} + \&c.\right) + \&c.$$

The numerical coefficients are the sums of the products which could be formed by different combinations of the fractions

$$\frac{1}{1^2}, \ \frac{1}{2^2}, \ \frac{1}{3^2}, \ \frac{1}{5^2}, \ \frac{1}{6^2} \ldots \&c.,$$

after having removed the first fraction $\frac{1}{1^2}$. If we represent the respective sums of products by $P_1, Q_1, R_1, S_1, T_1, \ldots$ &c., and if we employ the first of equations (e) and the first of equations (b), we have, to express the value of the first coefficient a, the equation

$$\frac{a (2^2 - 1)(3^2 - 1)(4^2 - 1)(5^2 - 1) \ldots}{2^2 . 3^2 . 4^2 . 5^2 \ldots}$$
$$= A - BP_1 + CQ_1 - DR_1 + ES_1 - \&c.,$$

now the quantities P_1, Q_1, R_1, S_1, T_1... &c. may be easily determined, as we shall see lower down; hence the first coefficient a becomes entirely known.

210. We must pass on now to the investigation of the following coefficients b, c, d, e, &c., which from equations (e) depend on the quantities b_2, c_3, d_4, e_5, &c. For this purpose we take up equations (b), the first has already been employed to find the value of a_1, the two following give the value of b_2, the three following the value of c_3, the four following the value of d_4, and so on.

On completing the calculation, we find by simple inspection of the equations the following results for the values of b_2, c_3, d_4, &c.

$$2b_2(1^2-2^2)=A_2 1^2-B_2,$$

$$3c_3(1^2-3^2)(2^2-3^2)=A_3 1^2.2^2-B_3(1^2+2^2)+C_3,$$

$$4d_4(1^2-4^2)(2^2-4^2)(3^2-4^2)$$
$$=A_4 1^2.2^2.3^2-B_4(1^2.2^2+1^2.3^2+2^2.3^2)+C_4(1^2+2^2+3^2)-D_4,$$

&c.

It is easy to perceive the law which these equations follow; it remains only to determine the quantities A_2B_2, $A_3B_3C_3$, $A_4B_4C_4$, &c.

Now the quantities A_2B_2 can be expressed in terms of $A_3B_3C_3$, the latter in terms of $A_4B_4C_4D_4$. For this purpose it suffices to effect the substitutions indicated by equations (d); the successive changes reduce the second members of the preceding equations so as to contain only the $ABCD$, &c. with an infinite suffix, that is to say, the known quantities $ABCD$, &c. which enter into equations (a); the coefficients become the different products which can be made by combining the squares of the numbers $1^2 2^2 3^2 4^2 5^2$ to infinity. It need only be remarked that the first of these squares 1^2 will not enter into the coefficients of the value of a_1; that the second 2^2 will not enter into the coefficients of the value of b_2; that the third square 3^2 will be omitted only from those which serve to form the coefficients of the value of c_3; and so of the rest to infinity. We have then for the values of

$b_2 c_3 d_4 e_5$, &c., and consequently for those of $bcde$, &c., results entirely analogous to that which we have found above for the value of the first coefficient a_1.

211. If now we represent by P_2, Q_2, R_2, S_2, &c., the quantities

$$\frac{1}{1^2}+\frac{1}{3^2}+\frac{1}{4^2}+\frac{1}{5^2}+\dots,$$

$$\frac{1}{1^2.3^2}+\frac{1}{1^2.4^2}+\frac{1}{1^2.5^2}+\frac{1}{3^2.4^2}+\frac{1}{3^2.5^2}+\dots,$$

$$\frac{1}{1^2.3^2.4^2}+\frac{1}{1^2.3^2.5^2}+\frac{1}{3^2.4^2.5^2}+\dots$$

$$\frac{1}{1^2.3^2.4^2.5^2}+\frac{1}{1^2.4^2.5^2.6^2}+\dots,$$

&c.,

which are formed by combinations of the fractions $\frac{1}{1^2}$, $\frac{1}{2^2}$, $\frac{1}{3^2}$, $\frac{1}{4^2}$, $\frac{1}{5^2}$ ... &c. to infinity, omitting $\frac{1}{2^2}$ the second of these fractions we have, to determine the value of b_2, the equation

$$2b_2\,\frac{1^2-2^2}{1^2.3^2.4^2.5^2\dots}=A-BP_2+CQ_2-DR_2+ES_2-\&c.$$

Representing in general by $P_nQ_nR_nS_n$ ... the sums of the products which can be made by combining all the fractions $\frac{1}{1^2}$, $\frac{1}{2^2}$, $\frac{1}{3^2}$, $\frac{1}{4^2}$, $\frac{1}{5^2}$ ... to infinity, after omitting the fraction $\frac{1}{n^2}$ only; we have in general to determine the quantities a_1, b_2, c_3, d_4, e_5..., &c., the following equations:

$$A-BP_1+CQ_1-DR_1+ES_1-\&c.=\ a_1\,\frac{1}{2^2.3^2.4^2.5^2\dots},$$

$$A-BP_2+CQ_2-DR_2+ES_2-\&c.=2b_2\,\frac{(1^2-2^2)}{1^2.3^2.4^2.5^2\dots},$$

$$A-BP_3+CQ_3-DR_3+ES_3-\&c.=3c_3\,\frac{(1^2-3^2)\,(2^2-3^2)}{1^2.2^2.4^2.5^2.6^2\dots},$$

$$A-BP_4+CQ_4-DR_4+ES_4-\&c.=4d_4\,\frac{(1^2-4^2)\,(2^2-4^2)\,(3^2-4^2)}{1^2.2^2.3^2.5^2.6^2\dots},$$

&c.

212. If we consider now equations (e) which give the values of the coefficients a, b, c, d, &c., we have the following results:

$$a\,\frac{(2^2-1^2)\,(3^2-1^2)\,(4^2-1^2)\,(5^2-1^2)\ldots}{2^2.\,3^2.\,4^2.\,5^2\ldots}$$
$$= A - BP_1 + CQ_1 - DR_1 + ES_1 - \&c.,$$

$$2b\,\frac{(1^2-2^2)\,(3^2-2^2)\,(4^2-2^2)\,(5^2-2^2)\ldots}{1^2.\,3^2.\,4^2.\,5^2\ldots}$$
$$= A - BP_2 + CQ_2 - DR_2 + ES_2 - \&c.,$$

$$3c\,\frac{(1^2-3^2)\,(2^2-3^2)\,(4^2-3^2)\,(5^2-3^2)\ldots}{1^2.\,2^2.\,4^2.\,5^2\ldots}$$
$$= A - BP_3 - CQ_3 - DR_3 + ES_3 - \&c.,$$

$$4d\,\frac{(1^2-4^2)\,(2^2-4^2)\,(3^2-4^2)\,(5^2-4^2)\ldots}{1^2.\,2^2.\,3^2.\,5^2\ldots}$$
$$= A - BP_4 + CQ_4 - DR_4 + ES_4 - \&c.,$$

&c.

Remarking the factors which are wanting to the numerators and denominators to complete the double series of natural numbers, we see that the fraction is reduced, in the first equation to $\frac{1}{1}\cdot\frac{1}{2}$; in the second to $-\frac{2}{2}\cdot\frac{2}{4}$; in the third to $\frac{3}{3}\cdot\frac{3}{6}$; in the fourth to $-\frac{4}{4}\cdot\frac{4}{8}$; so that the products which multiply a, $2b$, $3c$, $4d$, &c., are alternately $\frac{1}{2}$ and $-\frac{1}{2}$. It is only required then to find the values of $P_1Q_1R_1S_1$, $P_2Q_2R_2S_2$, $P_3Q_3R_3S_3$, &c.

To obtain them we may remark that we can make these values depend upon the values of the quantities $PQRST$, &c., which represent the different products which may be formed with the fractions $\frac{1}{1^2}$, $\frac{1}{2^2}$, $\frac{1}{3^2}$, $\frac{1}{4^2}$, $\frac{1}{5^2}$, $\frac{1}{6^2}$, &c., without omitting any.

With respect to the latter products, their values are given by the series for the developments of the sine. We represent then the series

$$\frac{1}{1^2}+\frac{1}{2^2}+\frac{1}{3^2}+\frac{1}{4^2}+\frac{1}{5^2}+\&c.$$

$$\frac{1}{1^2.2^2}+\frac{1}{1^2.3^2}+\frac{1}{1^2.4^2}+\frac{1}{2^2.3^2}+\frac{1}{2^2.4^2}+\frac{1}{3^2.4^2}+\&c.$$

$$\frac{1}{1^2.2^2.3^2}+\frac{1}{1^2.2^2.4^2}+\frac{1}{1^2.3^2.4^2}+\frac{1}{2^2.3^2.4^2}+\&c.$$

$$\frac{1}{1^2.2^2.3^2.4^2}+\frac{1}{2^2.3^2.4^2.5^2}+\frac{1}{1^2.2^2.3^2.5^2}+\&c.$$

by P, Q, R, S, &c.

The series $\sin x = x - \frac{x^3}{\lfloor 3} + \frac{x^5}{\lfloor 5} - \frac{x^7}{\lfloor 7} + \&c.$

furnishes the values of the quantities P, Q, R, S, &c. In fact, the value of the sine being expressed by the equation

$$\sin x = x\left(1-\frac{x^2}{1^2\pi^2}\right)\left(1-\frac{x^2}{2^2\pi^2}\right)\left(1-\frac{x^2}{3^2\pi^2}\right)\left(1-\frac{x^2}{4^2\pi^2}\right)\left(1-\frac{x^2}{5^2\pi^2}\right)\&c.$$

we have

$$1-\frac{x^2}{\lfloor 3}+\frac{x^4}{\lfloor 5}-\frac{x^6}{\lfloor 7}+\&c.$$

$$=\left(1-\frac{x^2}{1^2\pi^2}\right)\left(1-\frac{x^2}{2^2\pi^2}\right)\left(1-\frac{x^2}{3^2\pi^2}\right)\left(1-\frac{x^2}{4^2\pi^2}\right)\&c.$$

Whence we conclude at once that

$$P=\frac{\pi^2}{\lfloor 3},\quad Q=\frac{\pi^4}{\lfloor 5},\quad R=\frac{\pi^6}{\lfloor 7},\quad S=\frac{\pi^8}{\lfloor 9},\ \&c.$$

213. Suppose now that P_n, Q_n, R_n, S_n, &c., represent the sums of the different products which can be made with the fractions $\frac{1}{1^2}, \frac{1}{2^2}, \frac{1}{3^2}, \frac{1}{4^2}, \frac{1}{5^2}$, &c., from which the fraction $\frac{1}{n^2}$ has been removed, n being any integer whatever; it is required to determine P_n, Q_n, R_n, S_n, &c., by means of P, Q, R, S, &c. If we denote by

$$1-qP_n+q^2Q_n-q^3R_n+q^4S_n-\&c.,$$

the products of the factors

$$\left(1-\frac{q}{1^2}\right)\left(1-\frac{q}{2^2}\right)\left(1-\frac{q}{3^2}\right)\left(1-\frac{q}{4^2}\right)\&c.,$$

among which the factor $\left(1-\frac{q}{n^2}\right)$ only has been omitted; it follows that on multiplying by $\left(1-\frac{q}{n^2}\right)$ the quantity

$$1 - qP_n + q^2Q_n - q^3R_n + q^4S_n - \&c.,$$

we obtain
$$1 - qP + q^2Q - q^3R + q^4S - \&c.$$

This comparison gives the following relations:

$$P_n + \frac{1}{n^2} = P,$$

$$Q_n + \frac{1}{n^2} P_n = Q,$$

$$R_n + \frac{1}{n^2} Q_n = R,$$

$$S_n + \frac{1}{n^2} R_n = S,$$

$$\&c.;$$

or
$$P_n = P - \frac{1}{n^2},$$

$$Q_n = Q - \frac{1}{n^2} P + \frac{1}{n^4},$$

$$R_n = R - \frac{1}{n^2} Q + \frac{1}{n^4} P - \frac{1}{n^6},$$

$$S_n = S - \frac{1}{n^2} R + \frac{1}{n^4} Q - \frac{1}{n^6} P + \frac{1}{n^8},$$

$$\&c.$$

Employing the known values of P, Q, R, S, and making n equal to 1, 2, 3, 4, 5, &c. successively, we shall have the values of $P_1 Q_1 R_1 S_1$, &c.; those of $P_2 Q_2 R_2 S_2$, &c.; those of $P_3 Q_3 R_3 S_3$, &c.

214. From the foregoing theory it follows that the values of a, b, c, d, e, &c., derived from the equations

$$a + 2b + 3c + 4d + 5e + \&c. = A,$$

$$a + 2^3b + 3^3c + 4^3d + 5^3e + \&c. = B,$$

$$a + 2^5b + 3^5c + 4^5d + 5^5e + \&c. = C,$$

$$a + 2^7b + 3^7c + 4^7d + 5^7e + \&c. = D,$$

$$a + 2^9b + 3^9c + 4^9d + 5^9e + \&c. = E,$$

$$\&c.,$$

are thus expressed,

$$\frac{1}{2}\,a = A - B\left(\frac{\pi^2}{\lfloor 3} - \frac{1}{1^2}\right) + C\left(\frac{\pi^4}{\lfloor 5} - \frac{1}{1^2}\frac{\pi^2}{\lfloor 3} + \frac{1}{1^4}\right)$$
$$- D\left(\frac{\pi^6}{\lfloor 7} - \frac{1}{1^2}\frac{\pi^4}{\lfloor 5} + \frac{1}{1^4}\frac{\pi^2}{\lfloor 3} - \frac{1}{1^6}\right)$$
$$+ E\left(\frac{\pi^8}{\lfloor 9} - \frac{1}{1^2}\frac{\pi^6}{\lfloor 7} + \frac{1}{1^4}\frac{\pi^4}{\lfloor 5} - \frac{1}{1^6}\frac{\pi^2}{\lfloor 3} + \frac{1}{1^8}\right) - \&c.;$$

$$-\frac{1}{2}\,2b = A - B\left(\frac{\pi^2}{\lfloor 3} - \frac{1}{2^2}\right) + C\left(\frac{\pi^4}{\lfloor 5} - \frac{1}{2^2}\frac{\pi^4}{\lfloor 3} + \frac{1}{2^4}\right)$$
$$- D\left(\frac{\pi^6}{\lfloor 7} - \frac{1}{2^2}\frac{\pi^4}{\lfloor 5} + \frac{1}{2^4}\frac{\pi^2}{\lfloor 3} - \frac{1}{2^6}\right)$$
$$+ E\left(\frac{\pi^8}{\lfloor 9} - \frac{1}{2^2}\frac{\pi^6}{\lfloor 7} + \frac{1}{2^4}\frac{\pi^4}{\lfloor 5} - \frac{1}{2^6}\frac{\pi^2}{\lfloor 3} + \frac{1}{2^8}\right) - \&c.;$$

$$\frac{1}{2}\,3c = A - B\left(\frac{\pi^2}{\lfloor 3} - \frac{1}{3^2}\right) + C\left(\frac{\pi^4}{\lfloor 5} - \frac{1}{3^2}\frac{\pi^2}{\lfloor 3} + \frac{1}{3^4}\right)$$
$$- D\left(\frac{\pi^6}{\lfloor 7} - \frac{1}{3^2}\frac{\pi^4}{\lfloor 5} + \frac{1}{3^4}\frac{\pi^2}{\lfloor 3} - \frac{1}{3^6}\right)$$
$$+ E\left(\frac{\pi^8}{\lfloor 9} - \frac{1}{3^2}\frac{\pi^6}{\lfloor 7} + \frac{1}{3^4}\frac{\pi^4}{\lfloor 5} - \frac{1}{3^6}\frac{\pi^2}{\lfloor 3} + \frac{1}{3^8}\right) - \&c.;$$

$$-\frac{1}{2}\,4d = A - B\left(\frac{\pi^2}{\lfloor 3} - \frac{1}{4^2}\right) + C\left(\frac{\pi^4}{\lfloor 5} - \frac{1}{4^2}\frac{\pi^4}{\lfloor 3} + \frac{1}{4^4}\right)$$
$$- D\left(\frac{\pi^6}{\lfloor 7} - \frac{1}{4^2}\frac{\pi^4}{\lfloor 5} + \frac{1}{4^4}\frac{\pi^2}{\lfloor 3} - \frac{1}{4^6}\right)$$
$$+ E\left(\frac{\pi^8}{\lfloor 9} - \frac{1}{4^2}\frac{\pi^6}{\lfloor 7} + \frac{1}{4^4}\frac{\pi^4}{\lfloor 5} - \frac{1}{4^6}\frac{\pi^2}{\lfloor 3} + \frac{1}{4^8}\right) - \&c.;$$

&c.

215. Knowing the values of a, b, c, d, e, &c., we can substitute them in the proposed equation

$$\phi(x) = a\sin x + b\sin 2x + c\sin 3x + d\sin 4x + e\sin 5x + \&c.,$$

and writing also instead of the quantities A, B, C, D, E, &c., their values $\phi'(0)$, $\phi'''(0)$, $\phi^{v}(0)$, $\phi^{vii}(0)$, $\phi^{ix}(0)$, &c., we have the general equation

$$\frac{1}{2}\phi(x) = \sin x\left\{\phi'(0)+\phi'''(0)\left(\frac{\pi^2}{\lfloor 3}-\frac{1}{1^2}\right)+\phi^{v}(0)\left(\frac{\pi^4}{\lfloor 5}-\frac{1}{1^2}\frac{\pi^2}{\lfloor 3}+\frac{1}{1^4}\right)\right.$$
$$\left.+\phi^{vii}(0)\left(\frac{\pi^6}{\lfloor 7}-\frac{1}{1^2}\frac{\pi^4}{\lfloor 5}+\frac{1}{1^4}\frac{\pi^2}{\lfloor 3}-\frac{1}{1^6}\right)+\text{\&c.}\right\};$$
$$-\frac{1}{2}\sin 2x\left\{\phi'(0)+\phi'''(0)\left(\frac{\pi^2}{\lfloor 3}-\frac{1}{2^2}\right)+\phi^{v}(0)\left(\frac{\pi^4}{\lfloor 5}-\frac{1}{2^2}\frac{\pi^2}{\lfloor 3}+\frac{1}{2^4}\right)\right.$$
$$\left.+\phi^{vii}(0)\left(\frac{\pi^6}{\lfloor 7}-\frac{1}{2^2}\frac{\pi^4}{\lfloor 5}+\frac{1}{2^4}\frac{\pi^2}{\lfloor 3}-\frac{1}{2^6}\right)+\text{\&c.}\right\};$$
$$+\frac{1}{3}\sin 3x\left\{\phi'(0)+\phi'''(0)\left(\frac{\pi^2}{\lfloor 3}-\frac{1}{3^2}\right)+\phi^{v}(0)\left(\frac{\pi^4}{\lfloor 5}-\frac{1}{3^2}\frac{\pi^2}{\lfloor 3}+\frac{1}{3^4}\right)\right.$$
$$\left.+\phi^{vii}(0)\left(\frac{\pi^6}{\lfloor 7}-\frac{1}{3^2}\frac{\pi^4}{\lfloor 5}+\frac{1}{3^4}\frac{\pi^2}{\lfloor 3}-\frac{1}{3^6}\right)+\text{\&c.}\right\};$$
$$-\frac{1}{4}\sin 4x\left\{\phi'(0)+\phi'''(0)\left(\frac{\pi^2}{\lfloor 3}-\frac{1}{4^2}\right)+\phi^{v}(0)\left(\frac{\pi^4}{\lfloor 5}-\frac{1}{4^2}\frac{\pi^2}{\lfloor 3}+\frac{1}{4^4}\right)\right.$$
$$\left.+\phi^{vii}(0)\left(\frac{\pi^6}{\lfloor 7}-\frac{1}{4^2}\frac{\pi^4}{\lfloor 5}+\frac{1}{4^4}\frac{\pi^2}{\lfloor 3}-\frac{1}{4^6}\right)+\text{\&c.}\right\};$$
$$+\text{\&c.} \qquad (A),$$

We may make use of the preceding series to reduce into a series of sines of multiple arcs any proposed function whose development contains only odd powers of the variable.

216. The first case which presents itself is that in which $\phi(x) = x$; we find then $\phi'(0) = 1$, $\phi'''(0) = 0$, $\phi^{v}(0) = 0$, &c., and so for the rest. We have therefore the series

$$\frac{1}{2}x = \sin x - \frac{1}{2}\sin 2x + \frac{1}{3}\sin 3x - \frac{1}{4}\sin 4x + \text{\&c.},$$

which has been given by Euler.

If we suppose the proposed function to be x^3, we shall have

$$\phi'(0) = 0,\ \phi'''(0) = \lfloor 3,\ \ \phi^{v}(0) = 0,\ \phi^{vii}(0) = 0,\ \text{\&c.},$$

which gives the equation

$$\frac{1}{2}x^3 = \left(\pi^2 - \frac{\lfloor 3}{1^2}\right)\sin x - \left(\pi^2 - \frac{\lfloor 3}{2^2}\right)\frac{1}{2}\sin 2x + \left(\pi^2 - \frac{\lfloor 3}{3^2}\right)\frac{1}{3}\sin 3x + \text{\&c.}$$

~~(A)~~.

We should arrive at the same result, starting from the preceding equation,

$$\frac{1}{2}x = \sin x - \frac{1}{2}\sin 2x + \frac{1}{3}\sin 3x - \frac{1}{4}\sin 4x + \&c.$$

In fact, multiplying each member by dx, and integrating, we have

$$C - \frac{x^2}{4} = \cos x - \frac{1}{2^2}\cos 2x + \frac{1}{3^2}\cos 3x - \frac{1}{4^2}\cos 4x + \&c.\,;$$

the value of the constant C is

$$1 - \frac{1}{2^2} + \frac{1}{3^2} - \frac{1}{4^2} + \frac{1}{5^2} - \&c.\,;$$

a series whose sum is known to be $\frac{1}{2}\frac{\pi^2}{\underline{|3}}$. Multiplying by dx the two members of the equation

$$\frac{1}{2}\frac{\pi^2}{\underline{|3}} - \frac{x^2}{4} = \cos x - \frac{1}{2^2}\cos 2x + \frac{1}{3^2}\cos 3x - \&c.,$$

and integrating we have

$$\frac{1}{2}\frac{\pi^2 x}{\underline{|3}} - \frac{1}{2}\frac{x^3}{\underline{|3}} = \sin x - \frac{1}{2^3}\sin 2x + \frac{1}{3^3}\sin 3x - \&c.$$

If now we write instead of x its value derived from the equation

$$\frac{1}{2}x = \sin x - \frac{1}{2}\sin 2x + \frac{1}{3}\sin 3x - \frac{1}{4}\sin 4x + \&c.,$$

we shall obtain the same equation as above, namely,

$$\frac{1}{2}\frac{x^3}{\underline{|3}} = \sin x\left(\frac{\pi^2}{\underline{|3}} - \frac{1}{1^2}\right) - \frac{1}{2}\sin 2x\left(\frac{\pi^2}{\underline{|3}} - \frac{1}{2^2}\right) + \frac{1}{3}\sin 3x\left(\frac{\pi^2}{\underline{|3}} - \frac{1}{3^2}\right) - \&c.$$

We could arrive in the same manner at the development in series of multiple arcs of the powers x^5, x^7, x^9, &c., and in general every function whose development contains only odd powers of the variable.

217. Equation (A), (Art. 216), can be put under a simpler form, which we may now indicate. We remark first, that part of the coefficient of $\sin x$ is the series

$$\phi'(0) + \frac{\pi^2}{\underline{|3}}\phi'''(0) + \frac{\pi^4}{\underline{|5}}\phi^{\text{v}}(0) + \frac{\pi^6}{\underline{|7}}\phi^{\text{vii}}(0) + \&c.,$$

which represents the quantity $\frac{1}{\pi}\phi(\pi)$. In fact, we have, in general,

$$\phi(x) = \phi(0) + x\phi'(0) + \frac{x^2}{\underline{|2}}\phi''(0) + \frac{x^3}{\underline{|3}}\phi'''(0) + \frac{x^4}{\underline{|4}}\phi^{\text{iv}}(0) + \&c.$$

Now, the function $\phi(x)$ containing by hypothesis only odd powers, we must have $\phi(0)=0$, $\phi''(0)=0$, $\phi^{\text{iv}}(0)=0$, and so on. Hence

$$\phi(x) = x\phi'(0) + \frac{x^3}{\underline{|3}}\phi'''(0) + \frac{x^5}{\underline{|5}}\phi^{\text{v}}(0) + \&c.;$$

a second part of the coefficient of $\sin x$ is found by multiplying by $-\frac{1}{2}$ the series

$$\phi'''(0) + \frac{\pi^3}{\underline{|3}}\phi^{\text{v}}(0) + \frac{\pi^4}{\underline{|5}}\phi^{\text{vii}}(0) + \frac{\pi^6}{\underline{|7}}\phi^{\text{ix}}(0) + \&c.,$$

whose value is $\frac{1}{\pi}\phi''(\pi)$. We can determine in this manner the different parts of the coefficient of $\sin x$, and the components of the coefficients of $\sin 2x$, $\sin 3x$, $\sin 4x$, &c. We may employ for this purpose the equations:

$$\phi'(0) + \frac{\pi^2}{\underline{|3}}\phi'''(0) + \frac{\pi^4}{\underline{|5}}\phi^{\text{v}}(0) + \&c. = \frac{1}{\pi}\phi(\pi);$$

$$\phi'''(0) + \frac{\pi^2}{\underline{|3}}\phi^{\text{v}}(0) + \frac{\pi^4}{\underline{|5}}\phi^{\text{vii}}(0) + \&c. = \frac{1}{\pi}\phi''(\pi);$$

$$\phi^{\text{v}}(0) + \frac{\pi^2}{\underline{|3}}\phi^{\text{vii}}(0) + \frac{\pi^4}{\underline{|5}}\phi^{\text{ix}}(0) + \&c. = \frac{1}{\pi}\phi^{\text{iv}}(\pi);$$

$$\phi^{\text{vii}}(0) + \frac{\pi^2}{\underline{|3}}\phi^{\text{ix}}(0) + \frac{\pi^4}{\underline{|5}}\phi^{\text{xi}}(0) + \&c. = \frac{1}{\pi}\phi^{\text{v}}(\pi).$$

By means of these reductions equation (A) takes the following form:

$$\begin{aligned}\frac{1}{2}\pi\phi(x) = {} & \sin x\left\{\phi(\pi) - \frac{1}{1^2}\phi''(\pi) + \frac{1}{1^4}\phi^{\text{iv}}(\pi) - \frac{1}{1^6}\phi^{\text{vi}}(\pi) + \&\text{c.}\right\}\\ & -\frac{1}{2}\sin 2x\left\{\phi(\pi) - \frac{1}{2^2}\phi''(\pi) + \frac{1}{2^4}\phi^{\text{iv}}(\pi) - \frac{1}{2^6}\phi^{\text{vi}}(\pi) + \&\text{c.}\right\}\\ & +\frac{1}{3}\sin 3x\left\{\phi(\pi) - \frac{1}{3^2}\phi''(\pi) + \frac{1}{3^4}\phi^{\text{iv}}(\pi) - \frac{1}{3^6}\phi^{\text{vi}}(\pi) + \&\text{c.}\right\}\\ & -\frac{1}{4}\sin 4x\left\{\phi(\pi) - \frac{1}{4^2}\phi''(\pi) + \frac{1}{4^4}\phi^{\text{iv}}(\pi) - \frac{1}{4^6}\phi^{\text{vi}}(\pi) + \&\text{c.}\right\}\\ & + \&\text{c.} \qquad\qquad \text{(B)};\end{aligned}$$

or this,

$$\begin{aligned}\frac{1}{2}\pi\phi(x) = {} & \phi(\pi)\left\{\sin x - \frac{1}{2}\sin 2x + \frac{1}{3}\sin 3x - \&\text{c.}\right\}\\ & -\phi''(\pi)\left\{\sin x - \frac{1}{2^3}\sin 2x + \frac{1}{3^3}\sin 3x - \&\text{c.}\right\}\\ & +\phi^{\text{iv}}(\pi)\left\{\sin x - \frac{1}{2^5}\sin 2x + \frac{1}{3^5}\sin 3x - \&\text{c.}\right\}\\ & -\phi^{\text{vi}}(\pi)\left\{\sin x - \frac{1}{2^7}\sin 2x + \frac{1}{3^7}\sin 3x - \&\text{c.}\right\}\\ & + \&\text{c.} \qquad\qquad \text{(C)}.\end{aligned}$$

218. We can apply one or other of these formulæ as often as we have to develope a proposed function in a series of sines of multiple arcs. If, for example, the proposed function is $e^x - e^{-x}$, whose development contains only odd powers of x, we shall have

$$\begin{aligned}\frac{1}{2}\pi\frac{e^x - e^{-x}}{e^\pi - e^{-\pi}} = {} & \left(\sin x - \frac{1}{2}\sin 2x + \frac{1}{3}\sin 3x - \&\text{c.}\right)\\ & -\left(\sin x - \frac{1}{2^3}\sin 2x + \frac{1}{3^3}\sin 3x - \&\text{c.}\right)\\ & +\left(\sin x - \frac{1}{2^5}\sin 2x + \frac{1}{3^5}\sin 3x - \&\text{c.}\right)\\ & -\left(\sin x - \frac{1}{2^7}\sin 2x + \frac{1}{3^7}\sin 3x - \&\text{c.}\right)\\ & + \&\text{c.}\end{aligned}$$

Collecting the coefficients of $\sin x$, $\sin 2x$, $\sin 3x$, $\sin 4x$, &c., and writing, instead of $\frac{1}{n} - \frac{1}{n^3} + \frac{1}{n^5} - \frac{1}{n^7} +$ etc., its value $\frac{n}{n^2+1}$, we have

$$\frac{1}{2}\pi \frac{(e^x - e^{-x})}{e^\pi - e^{-\pi}} = \frac{\sin x}{1+\frac{1}{1}} - \frac{\sin 2x}{2+\frac{1}{2}} + \frac{\sin 3x}{3+\frac{1}{3}} - \&c.$$

We might multiply these applications and derive from them several remarkable series. We have chosen the preceding example because it appears in several problems relative to the propagation of heat.

219. Up to this point we have supposed that the function whose development is required in a series of sines of multiple arcs can be developed in a series arranged according to powers of the variable x, and that only odd powers enter into that series. We can extend the same results to any functions, even to those which are discontinuous and entirely arbitrary. To establish clearly the truth of this proposition, we must follow the analysis which furnishes the foregoing equation (B), and examine what is the nature of the coefficients which multiply $\sin x$, $\sin 2x$, $\sin 3x$, &c. Denoting by $\frac{s}{n}$ the quantity which multiplies $\frac{1}{n}\sin nx$ in this equation when n is odd, and $-\frac{1}{n}\sin nx$ when n is even, we have

$$s = \phi(\pi) - \frac{1}{n^2}\phi''(\pi) + \frac{1}{n^4}\phi^{\text{iv}}(\pi) - \frac{1}{n^6}\phi^{\text{vi}}(\pi) + \&c.$$

Considering s as a function of π, differentiating twice, and comparing the results, we find $s + \frac{1}{n^2}\frac{d^2s}{d\pi^2} = \phi(\pi)$; an equation which the foregoing value of s must satisfy.

Now the integral of the equation $s + \frac{1}{n^2}\frac{d^2s}{dx^2} = \phi(x)$, in which s is considered to be a function of x, is

$$s = a\cos nx + b\sin nx$$
$$+ n\sin nx\int\cos nx\,\phi(x)\,dx - n\cos nx\int\sin nx\,\phi(x)\,dx.$$

If n is an integer, and the value of x is equal to π, we have $s = \pm n \int \phi(x) \sin nx dx$. The sign + must be chosen when n is odd, and the sign − when that number is even. We must make x equal to the semi-circumference π, after the integration indicated; the result may be verified by developing the term $\int \phi(x) \sin nx\, dx$, by means of integration by parts, remarking that the function $\phi(x)$ contains only odd powers of the variable x, and taking the integral from $x = 0$ to $x = \pi$.

We conclude at once that the term is equal to

$$\pm \left\{\phi(\pi) - \frac{1}{n^2}\phi''(\pi) + \frac{1}{n^4}\phi^{\text{iv}}(\pi) - \frac{1}{n^6}\phi^{\text{vi}}(\pi) + \frac{1}{n^8}\phi^{\text{viii}}(\pi) - \&\text{c.}\right\}.$$

If we substitute this value of $\frac{s}{n}$ in equation (B), taking the sign + when the term of this equation is of odd order, and the sign − when n is even, we shall have in general $\int \phi(x) \sin nx dx$ for the coefficient of $\sin nx$; in this manner we arrive at a very remarkable result expressed by the following equation:

$$\frac{1}{2}\pi\phi(x) = \sin x \int \sin x\phi(x)\, dx + \sin 2x \int \sin 2x\phi(x)\, dx + \&\text{c.}$$
$$+ \sin ix \int \sin ix\, \phi(x)\, dx + \&\text{c.} \quad \text{(D)},$$

the second member will always give the development required for the function $\phi(x)$, if we integrate from $x = 0$ to $x = \pi$.[1]

[1] Lagrange had already shewn (*Miscellanea Taurinensia*, Tom. III., 1766, pp. 260—1) that the function y given by the equation

$$y = 2\left(\sum_{r=1}^{r=n} Y_r \sin X_r\pi\,\Delta X\right)\sin x\pi + 2\left(\sum_{r=1}^{r=n} Y_r \sin 2X_r\pi\,\Delta X\right)\sin 2x\pi$$
$$+ 2\left(\sum_{r=1}^{r=n} Y_r \sin 3X_r\pi\,\Delta X\right)\sin 3x\pi + \ldots + 2\left(\sum_{r=1}^{r=n} Y_r \sin nX_r\pi\,\Delta X\right)\sin nx\pi$$

receives the values $Y_1, Y_2, Y_3 \ldots Y_n$ corresponding to the values $X_1, X_2, X_3 \ldots X_n$ of x, where $X_r = \frac{r}{n+1}$, and $\Delta X = \frac{1}{n+1}$.

Lagrange however abstained from the transition from this summation-formula to the integration-formula given by Fourier.

Cf. Riemann's *Gesammelte Mathematische Werke*, Leipzig, 1876, pp. 218—220 of his historical criticism, *Ueber die Darstellbarkeit einer Function durch eine Trigonometrische Reihe*. [A. F.]

220. We see by this that the coefficients a, b, c, d, e, f, &c., which enter into the equation

$$\frac{1}{2}\pi\phi(x) = a \sin x + b \sin 2x + c \sin 3x + d \sin 4x + \&c.,$$

and which we found formerly by way of successive eliminations, are the values of definite integrals expressed by the general term $\int \sin ix\, \phi(x)\, dx$, i being the number of the term whose coefficient is required. This remark is important, because it shews how even entirely arbitrary functions may be developed in series of sines of multiple arcs. In fact, if the function $\phi(x)$ be represented by the variable ordinate of any curve whatever whose abscissa extends from $x=0$ to $x=\pi$, and if on the same part of the axis the known trigonometric curve, whose ordinate is $y = \sin x$, be constructed, it is easy to represent the value of any integral term. We must suppose that for each abscissa x, to which corresponds one value of $\phi(x)$, and one value of $\sin x$, we multiply the latter value by the first, and at the same point of the axis raise an ordinate equal to the product $\phi(x) \sin x$. By this continuous operation a third curve is formed, whose ordinates are those of the trigonometric curve, reduced in proportion to the ordinates of the arbitary curve which represents $\phi(x)$. This done, the area of the reduced curve taken from $x=0$ to $x=\pi$ gives the exact value of the coefficient of $\sin x$; and whatever the given curve may be which corresponds to $\phi(x)$, whether we can assign to it an analytical equation, or whether it depends on no regular law, it is evident that it always serves to reduce in any manner whatever the trigonometric curve; so that the area of the reduced curve has, in all possible cases, a definite value, which is the value of the coefficient of $\sin x$ in the development of the function. The same is the case with the following coefficient b, or $\int \phi(x) \sin 2x dx$.

In general, to construct the values of the coefficients a, b, c, d, &c., we must imagine that the curves, whose equations are

$$y = \sin x, \quad y = \sin 2x, \quad y = \sin 3x, \quad y = \sin 4x, \text{ \&c.,}$$

have been traced for the same interval on the axis of x, from

$x=0$ to $x=\pi$; and then that we have changed these curves by multiplying all their ordinates by the corresponding ordinates of a curve whose equation is $y=\phi(x)$. The equations of the reduced curves are

$$y=\sin x\,\phi(x),\quad y=\sin 2x\,\phi(x),\quad y=\sin 3x\,\phi(x),\quad \&c.$$

The areas of the latter curves, taken from $x=0$ to $x=\pi$, are the values of the coefficients a, b, c, d, &c., in the equation

$$\frac{1}{2}\pi\phi(x)=a\sin x+b\sin 2x+c\sin 3x+d\sin 4x+\&c.$$

221. We can verify the foregoing equation (D), (Art. 220), by determining directly the quantities a_1, a_2, a_3, ... a_j, &c., in the equation

$$\phi(x)=a_1\sin x+a_2\sin 2x+a_3\sin 3x+\ldots a_j\sin jx+\&c.;$$

for this purpose, we multiply each member of the latter equation by $\sin ix\,dx$, i being an integer, and take the integral from $x=0$ to $x=\pi$, whence we have

$$\int\phi(x)\sin ix\,dx=a_1\int\sin x\sin ix\,dx+a_2\int\sin 2x\sin ix\,dx+\&c.$$

$$+a_j\int\sin jx\sin ix\,dx+\ldots\&c.$$

Now it can easily be proved, 1st, that all the integrals, which enter into the second member, have a nul value, except only the term $a_i\int\sin ix\sin ix\,dx$; 2nd, that the value of $\int\sin ix\sin ix\,dx$ is $\frac{1}{2}\pi$; whence we derive the value of a_i, namely

$$\frac{2}{\pi}\int\phi(x)\sin ix\,dx.$$

The whole problem is reduced to considering the value of the integrals which enter into the second member, and to demonstrating the two preceding propositions. The integral

$$2\int\sin jx\sin ix\,dx,$$

taken from $x=0$ to $x=\pi$, in which i and j are integers, is

$$\frac{1}{i-j}\sin(i-j)x - \frac{1}{i+j}\sin(i+j)x + C.$$

Since the integral must begin when $x=0$ the constant C is nothing, and the numbers i and j being integers, the value of the integral will become nothing when $x=\pi$; it follows that each of the terms, such as

$$a_1\int \sin x \sin ix\,dx, \quad a_2\int \sin 2x \sin ix\,dx, \quad a_3\int \sin 3x \sin ix\,dx, \quad \&c.,$$

vanishes, and that this will occur as often as the numbers i and j are different. The same is not the case when the numbers i and j are equal, for the term $\frac{1}{i-j}\sin(i-j)x$ to which the integral reduces, becomes $\frac{0}{0}$, and its value is π. Consequently we have

$$2\int \sin ix \sin ix\,dx = \pi;$$

thus we obtain, in a very brief manner, the values of $a_1, a_2, a_3, \ldots a_i$, &c., namely,

$$a_1 = \frac{2}{\pi}\int \phi(x)\sin x\,dx, \qquad a_2 = \frac{2}{\pi}\int \phi(x)\sin 2x\,dx,$$

$$a_3 = \frac{2}{\pi}\int \phi(x)\sin 3x\,dx, \qquad a_i = \frac{2}{\pi}\int \phi(x)\sin ix\,dx.$$

Substituting these we have

$$\tfrac{1}{2}\pi\phi(x) = \sin x\int \phi(x)\sin x\,dx + \sin 2x\int \phi(x)\sin 2x\,dx + \&c.$$
$$+ \sin ix\int \phi(x)\sin ix\,dx + \&c.$$

222. The simplest case is that in which the given function has a constant value for all values of the variable x included between 0 and π; in this case the integral $\int \sin ix\,dx$ is equal to $\frac{2}{i}$, if the number i is odd, and equal to 0 if the number i is even.

Hence we deduce the equation

$$\frac{1}{4}\pi = \sin x + \frac{1}{3}\sin 3x + \frac{1}{5}\sin 5x + \frac{1}{7}\sin 7x + \&c.,$$

which has been found before.

It must be remarked that when a function $\phi(x)$ has been developed in a series of sines of multiple arcs, the value of the series

$$a \sin x + b \sin 2x + c \sin 3x + d \sin 4x + \&c.$$

is the same as that of the function $\phi(x)$ so long as the variable x is included between 0 and π; but this equality ceases in general to hold good when the value of x exceeds the number π.

Suppose the function whose development is required to be x, we shall have, by the preceding theorem,

$$\frac{1}{2}\pi x = \sin x \int x \sin x \, dx + \sin 2x \int x \sin 2x \, dx$$
$$+ \sin 3x \int x \sin 3x \, dx + \&c.$$

The integral $\int_0^\pi x \sin ix dx$ is equal to $\pm\frac{\pi}{i}$; the indices 0 and π, which are connected with the sign $\int$, shew the limits of the integral; the sign $+$ must be chosen when i is odd, and the sign $-$ when i is even. We have then the following equation,

$$\frac{1}{2}x = \sin x - \frac{1}{2}\sin 2x + \frac{1}{3}\sin 3x - \frac{1}{4}\sin 4x + \frac{1}{5}\sin 5x - \&c.$$

223. We can develope also in a series of sines of multiple arcs functions different from those in which only odd powers of the variable enter. To instance by an example which leaves no doubt as to the possibility of this development, we select the function $\cos x$, which contains only even powers of x, and which may be developed under the following form:

$$a \sin x + b \sin 2x + c \sin 3x + d \sin 4x + e \sin 5x + \&c.,$$

although in this series only odd powers of the variable enter.

We have, in fact, by the preceding theorem,

$$\frac{1}{2}\pi\cos x = \sin x\int\cos x\sin x\,dx + \sin 2x\int\cos x\sin 2x\,dx$$
$$+ \sin 3x\int\cos x\sin 3x\,dx + \&c.$$

The integral $\int\cos x\sin ix\,dx$ is equal to zero when i is an odd number, and to $\frac{2i}{i^2-1}$ when i is an even number. Supposing successively $i = 2, 4, 6, 8$, etc., we have the always convergent series

$$\frac{1}{4}\pi\cos x = \frac{2}{1.3}\sin 2x + \frac{4}{3.5}\sin 4x + \frac{6}{5.7}\sin 6x + \&c.;$$

or,

$$\cos x = \frac{2}{\pi}\left\{\left(\frac{1}{1}+\frac{1}{3}\right)\sin 2x + \left(\frac{1}{3}+\frac{1}{5}\right)\sin 4x + \left(\frac{1}{5}+\frac{1}{7}\right)\sin 6x + \&c.\right\}.$$

This result is remarkable in this respect, that it exhibits the development of the cosine in a series of functions, each one of which contains only odd powers. If in the preceding equation x be made equal to $\frac{1}{4}\pi$, we find

$$\frac{1}{4}\frac{\pi}{\sqrt{2}} = \frac{1}{2}\left(\frac{1}{1}+\frac{1}{3}-\frac{1}{5}-\frac{1}{7}+\frac{1}{9}+\frac{1}{11}-\&c.\right).$$

This series is known (*Introd. ad analysin. infinit.* cap. x.).

224. A similar analysis may be employed for the development of any function whatever in a series of cosines of multiple arcs.

Let $\phi(x)$ be the function whose development is required, we may write

$$\phi(x) = a_0\cos 0x + a_1\cos x + a_2\cos 2x + a_3\cos 3x + \&c.$$
$$+ a_i\cos ix + \&c. \quad\ldots\ldots\ldots\ (m).$$

If the two members of this equation be multiplied by $\cos jx$, and each of the terms of the second member integrated from $x = 0$ to $x = \pi$; it is easily seen that the value of the integral will be nothing, save only for the term which already contains $\cos jx$. This remark gives immediately the coefficient a_j; it is sufficient in general to consider the value of the integral

$$\int\cos jx\cos ix\,dx,$$

taken from $x=0$ to $x=\pi$, supposing j and i to be integers. We have

$$\int \cos jx \cos ix\,dx = \frac{1}{2(j+i)}\sin(j+i)x + \frac{1}{2(j-i)}\sin(j-i)x + c.$$

This integral, taken from $x=0$ to $x=\pi$, evidently vanishes whenever j and i are two different numbers. The same is not the case when the two numbers are equal. The last term

$$\frac{1}{2(j-i)}\sin(j-i)x$$

becomes $\frac{0}{0}$, and its value is $\frac{1}{2}\pi$, when the arc x is equal to π. If then we multiply the two terms of the preceding equation (m) by $\cos ix$, and integrate it from 0 to π, we have

$$\int \phi(x)\cos ix\,dx = \tfrac{1}{2}\pi a_i,$$

an equation which exhibits the value of the coefficient a_i.

To find the first coefficient a_0, it may be remarked that in the integral

$$\frac{1}{2(j+i)}\sin(j+i)x + \frac{1}{2(j-i)}\sin(j-i)x,$$

if $j=0$ and $i=0$ each of the terms becomes $\frac{0}{0}$, and the value of each term is $\frac{1}{2}\pi$; thus the integral $\int \cos jx \cos ix\,dx$ taken from $x=0$ to $x=\pi$ is nothing when the two integers j and i are different: it is $\frac{1}{2}\pi$ when the two numbers j and i are equal but different from zero; it is equal to π when j and i are each equal to zero; thus we obtain the following equation,

$$\tfrac{1}{2}\pi\phi(x) = \frac{1}{2}\int_0^\pi \phi(x)\,dx + \cos x\int_0^\pi \phi(x)\cos x\,dx + \cos 2x\int_0^\pi \phi(x)\cos 2x\,dx$$
$$+\cos 3x\int_0^\pi \phi(x)\cos 3x\,dx + \&c. \qquad (n)^1.$$

[1] The process analogous to (A) in Art. 222 fails here; yet we see, Art. 177, that an analogous result exists. [R. L. E.]

This and the preceding theorem suit all possible functions, whether their character can be expressed by known methods of analysis, or whether they correspond to curves traced arbitrarily.

225. If the proposed function whose development is required in cosines of multiple arcs is the variable x itself; we may write down the equation

$$\frac{1}{2}\pi x = a_0 + a_1 \cos x + a_2 \cos 2x + a_3 \cos 3x + \ldots + a_i \cos ix + \&c.,$$

and we have, to determine any coefficient whatever a_i, the equation $a_i = \int_0^\pi x \cos ix\, dx$. This integral has a nul value when i is an even number, and is equal to $-\frac{2}{i^2}$ when i is odd. We have at the same time $a_0 = \frac{1}{4}\pi^2$. We thus form the following series,

$$x = \frac{1}{2}\pi - 4\frac{\cos x}{\pi} - 4\frac{\cos 3x}{3^2\pi} - 4\frac{\cos 5x}{5^2\pi} - 4\frac{\cos 7x}{7^2\pi} - \&c.$$

We may here remark that we have arrived at three different developments for x, namely,

$$\frac{1}{2}x = \sin x - \frac{1}{2}\sin 2x + \frac{1}{3}\sin 3x - \frac{1}{4}\sin 4x + \frac{1}{5}\sin 5x - \&c.,$$

$$\frac{1}{2}x = \frac{2}{\pi}\sin x - \frac{2}{3^2\pi}\sin 3x + \frac{2}{5^2\pi}\sin 5x - \&c. \text{ (Art. 181)},$$

$$\frac{1}{2}x = \frac{1}{4}\pi - \frac{2}{\pi}\cos x - \frac{2}{3^2\pi}\cos 3x - \frac{2}{5^2\pi}\cos 5x - \&c.$$

It must be remarked that these three values of $\frac{1}{2}x$ ought not to be considered as equal; with reference to all possible values of x, the three preceding developments have a common value only when the variable x is included between 0 and $\frac{1}{2}\pi$. The construction of the values of these three series, and the comparison of the lines whose ordinates are expressed by them, render sensible the alternate coincidence and divergence of values of these functions.

To give a second example of the development of a function in a series of cosines of multiple arcs, we choose the function $\sin x$, which contains only odd powers of the variable, and we may suppose it to be developed in the form

$$a + b\cos x + c\cos 2x + d\cos 3x + \&c.$$

Applying the general equation to this particular case, we find, as the equation required,

$$\frac{1}{4}\pi\sin x = \frac{1}{2} - \frac{\cos 2x}{1.3} - \frac{\cos 4x}{3.5} - \frac{\cos 6x}{5.7} - \&c.$$

Thus we arrive at the development of a function which contains only odd powers in a series of cosines in which only even powers of the variable enter. If we give to x the particular value $\frac{1}{2}\pi$, we find

$$\frac{1}{4}\pi = \frac{1}{2} + \frac{1}{1.3} - \frac{1}{3.5} + \frac{1}{5.7} - \frac{1}{7.9} + \&c.$$

Now, from the known equation,

$$\frac{1}{4}\pi = 1 - \frac{1}{3} + \frac{1}{5} - \frac{1}{7} + \frac{1}{9} - \frac{1}{11} + \&c.,$$

we derive

$$\frac{1}{8}\pi = \frac{1}{1.3} + \frac{1}{5.7} + \frac{1}{9.11} + \frac{1}{13.15} + \&c.,$$

and also

$$\frac{1}{8}\pi = \frac{1}{2} - \frac{1}{3.5} - \frac{1}{7.9} - \frac{1}{11.13} - \&c.$$

Adding these two results we have, as above,

$$\frac{1}{4}\pi = \frac{1}{2} + \frac{1}{1.3} - \frac{1}{3.5} + \frac{1}{5.7} - \frac{1}{7.9} + \frac{1}{9.11} - \&c.$$

226. The foregoing analysis giving the means of developing any function whatever in a series of sines or cosines of multiple arcs, we can easily apply it to the case in which the function to be developed has definite values when the variable is included between certain limits and has real values, or when the variable is included between other limits. We stop to examine this particular case, since it is presented in physical questions which depend on partial differential equations, and was proposed formerly as an example of functions which cannot be developed in sines or cosines

of multiple arcs. Suppose then that we have reduced to a series of this form a function whose value is constant, when x is included between 0 and α, and all of whose values are nul when x is included between α and π. We shall employ the general equation (D), in which the integrals must be taken from $x=0$ to $x=\pi$. The values of $\phi(x)$ which enter under the integral sign being nothing from $x=\alpha$ to $x=\pi$, it is sufficient to integrate from $x=0$ to $x=\alpha$. This done, we find, for the series required, denoting by h the constant value of the function,

$$\frac{1}{2}\pi\phi(x) = h\left\{(1-\cos\alpha)\sin x + \frac{1-\cos 2\alpha}{2}\sin 2x \right.$$
$$\left. + \frac{1-\cos 3\alpha}{3}\sin 3x + \&c.\right\}.$$

If we make $h=\frac{1}{2}\pi$, and represent the versed sine of the arc x by $\operatorname{versin} x$, we have

$$\phi(x) = \operatorname{versin}\alpha\sin x + \frac{1}{2}\operatorname{versin} 2\alpha\sin 2x + \frac{1}{3}\operatorname{versin} 3\alpha\sin 3x + \&c.$$[1]

This series, always convergent, is such that if we give to x any value whatever included between 0 and α, the sum of its terms will be $\frac{1}{2}\pi$; but if we give to x any value whatever greater than α and less than $\frac{1}{2}\pi$, the sum of the terms will be nothing.

In the following example, which is not less remarkable, the values of $\phi(x)$ are equal to $\sin\frac{\pi x}{\alpha}$ for all values of x included between 0 and α, and nul for values of x between α and π. To find what series satisfies this condition, we shall employ equation (D).

The integrals must be taken from $x=0$ to $x=\pi$; but it is sufficient, in the case in question, to take these integrals from $x=0$ to $x=\alpha$, since the values of $\phi(x)$ are supposed nul in the rest of the interval. Hence we find

$$\phi(x) = 2\alpha\left\{\frac{\sin\alpha\sin x}{\pi^2-\alpha^2} + \frac{\sin 2\alpha\sin 2x}{\pi^2-2^2\alpha^2} + \frac{\sin 3\alpha\sin 3x}{\pi^2-3^2\alpha^2} + \&c.\right\}.$$

[1] In what cases a function, arbitrary between certain limits, can be developed in a series of cosines, and in what cases in a series of sines, has been shewn by Sir W. Thomson, *Camb. Math. Journal*, Vol. II. pp. 258—262, in an article signed P. Q. R., *On Fourier's Expansions of Functions in Trigonometrical Series.*

If α be supposed equal to π, all the terms of the series vanish, except the first, which becomes $\frac{0}{0}$, and whose value is $\sin x$; we have then $\phi(x) = \sin x$.

227. The same analysis could be extended to the case in which the ordinate represented by $\phi(x)$ was that of a line composed of different parts, some of which might be arcs of curves and others straight lines. For example, let the value of the function, whose development is required in a series of cosines of multiple arcs, be $\left(\frac{\pi}{2}\right)^2 - x^2$, from $x = 0$ to $x = \frac{1}{2}\pi$, and be nothing from $x = \frac{1}{2}\pi$ to $x = \pi$. We shall employ the general equation (n), and effecting the integrations within the given limits, we find that the general term[1] $\int\left[\left(\frac{\pi}{2}\right)^2 - x^2\right]\cos ix\,dx$ is equal to $\frac{2}{i^3}$ when i is even, to $+\frac{\pi}{i^2}$ when i is the double of an odd number, and to $-\frac{\pi}{i^2}$ when i is four times an odd number. On the other hand, we find $\frac{1}{3}\frac{\pi^3}{2^3}$ for the value of the first term $\frac{1}{2}\int\phi(x)\,dx$. We have then the following development:

$$\frac{1}{2}\phi(x) = \frac{1}{2.3}\left(\frac{\pi}{2}\right)^2 + \frac{2}{\pi}\left\{\frac{\cos x}{1^3} + \frac{\cos 3x}{3^3} + \frac{\cos 5x}{5^3} + \frac{\cos 7x}{7^3} + \&\text{c.}\right\}$$

$$+\frac{\cos 2x}{2^2} - \frac{\cos 4x}{4^2} + \frac{\cos 6x}{6^2} - \&\text{c.}$$

The second member is represented by a line composed of parabolic arcs and straight lines.

228. In the same manner we can find the development of a function of x which expresses the ordinate of the contour of a trapezium. Suppose $\phi(x)$ to be equal to x from $x = 0$ to $x = \alpha$, that the function is equal to α from $x = \alpha$ to $x = \pi - \alpha$, and lastly equal to $\pi - x$, from $x = \pi - \alpha$ to $x = \pi$. To reduce it to a series

[1] $\int\left[\left(\frac{\pi}{2}\right)^2 - x\right]\cos ix\,dx = \left(\frac{\pi}{2}\right)^2\frac{\sin ix}{i} - \frac{x^2}{i}\sin ix - \frac{2}{i^2}x\cos ix + 2\frac{\sin ix}{i^3}$.

[R. L. E.]

of sines of multiple arcs, we employ the general equation (D). The general term $\int \phi(x) \sin ix \, dx$ is composed of three different parts, and we have, after the reductions, $\frac{2}{i^2} \sin i\alpha$ for the coefficient of $\sin ix$, when i is an odd number; but the coefficient vanishes when i is an even number. Thus we arrive at the equation

$$\frac{1}{2}\pi\phi(x) = 2\left\{\sin\alpha \sin x + \frac{1}{3^2}\sin 3\alpha \sin 3x + \frac{1}{5^2}\sin 5\alpha \sin 5x + \frac{1}{7^2}\sin 7\alpha \sin 7x + \&c.\right\} \quad (\lambda).$$[1]

If we supposed $\alpha = \frac{1}{2}\pi$, the trapezium would coincide with an isosceles triangle, and we should have, as above, for the equation of the contour of this triangle,

$$\frac{1}{2}\pi\phi(x) = 2\left(\sin x - \frac{1}{3^2}\sin 3x + \frac{1}{5^2}\sin 5x - \frac{1}{7^2}\sin 7x + \&c.\right),$$[2]

a series which is always convergent whatever be the value of x. In general, the trigonometric series at which we have arrived, in developing different functions are always convergent, but it has not appeared to us necessary to demonstrate this here; for the terms which compose these series are only the coefficients of terms of series which give the values of the temperature; and these coefficients are affected by certain exponential quantities which decrease very rapidly, so that the final series are very convergent. With regard to those in which only the sines and cosines of multiple arcs enter, it is equally easy to prove that they are convergent, although they represent the ordinates of discontinuous lines. This does not result solely from the fact that the values of the terms diminish continually; for this condition is not sufficient to establish the convergence of a series. It is necessary that the values at which we arrive on increasing continually the number of terms, should approach more and more a fixed limit,

[1] The accuracy of this and other series given by Fourier is maintained by Sir W. Thomson in the article quoted in the note, p. 194.

[2] Expressed in cosines between the limits 0 and π,

$$\tfrac{1}{2}\pi\phi(x) = \frac{\pi^2}{8} - \left(\cos 2x + \frac{1}{3^2}\cos 6x + \frac{1}{5^2}\cos 10x + \&c.\right).$$

Cf. De Morgan's *Diff. and Int. Calc.*, p. 622. [A. F.]

and should differ from it only by a quantity which becomes less than any given magnitude: this limit is the value of the series. Now we may prove rigorously that the series in question satisfy the last condition.

229. Take the preceding equation (λ) in which we can give to x any value whatever; we shall consider this quantity as a new ordinate, which gives rise to the following construction.

Fig. 8.

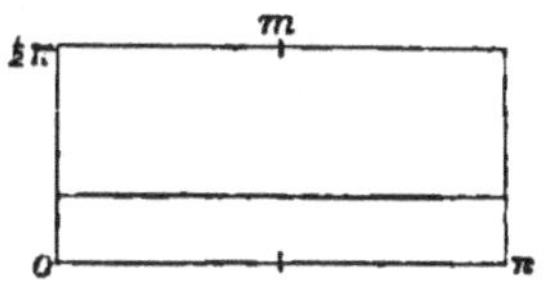

Having traced on the plane of x and y (see fig. 8) a rectangle whose base $O\pi$ is equal to the semi-circumference, and whose height is $\frac{1}{2}\pi$; on the middle point m of the side parallel to the base, let us raise perpendicularly to the plane of the rectangle a line equal to $\frac{1}{2}\pi$, and from the upper end of this line draw straight lines to the four corners of the rectangle. Thus will be formed a quadrangular pyramid. If we now measure from the point O on the shorter side of the rectangle, any line equal to α, and through the end of this line draw a plane parallel to the base $O\pi$, and perpendicular to the plane of the rectangle, the section common to this plane and to the solid will be the trapezium whose height is equal to α. The variable ordinate of the contour of this trapezium is equal, as we have just seen, to

$$\frac{4}{\pi}\left(\sin\alpha\sin x + \frac{1}{3^2}\sin 3\alpha\sin 3x + \frac{1}{5^2}\sin 5\alpha\sin 5x + \&c.\right).$$

It follows from this that calling x, y, z the co-ordinates of any point whatever of the upper surface of the quadrangular pyramid which we have formed, we have for the equation of the surface of the polyhedron, between the limits

$$x = 0,\quad x = \pi,\quad y = 0,\quad y = \tfrac{1}{2}\pi,$$

$$\frac{1}{2}\pi z = \frac{\sin x\sin y}{1^2} + \frac{\sin 3x\sin 3y}{3^2} + \frac{\sin 5x\sin 5y}{5^2} + \&c.$$

This convergent series gives always the value of the ordinate z or the distance of any point whatever of the surface from the plane of x and y.

The series formed of sines or cosines of multiple arcs are therefore adapted to represent, between definite limits, all possible functions, and the ordinates of lines or surfaces whose form is discontinuous. Not only has the possibility of these developments been demonstrated, but it is easy to calculate the terms of the series; the value of any coefficient whatever in the equation

$$\phi(x) = a_1 \sin x + a_2 \sin 2x + a_3 \sin 3x + \ldots + a_i \sin ix + \text{etc.},$$

is that of a definite integral, namely,

$$\frac{2}{\pi}\int \phi(x) \sin ix\, dx.$$

Whatever be the function $\phi(x)$, or the form of the curve which it represents, the integral has a definite value which may be introduced into the formula. The values of these definite integrals are analogous to that of the whole area $\int \phi(x)\, dx$ included between the curve and the axis in a given interval, or to the values of mechanical quantities, such as the ordinates of the centre of gravity of this area or of any solid whatever. It is evident that all these quantities have assignable values, whether the figure of the bodies be regular, or whether we give to them an entirely arbitrary form.

230. If we apply these principles to the problem of the motion of vibrating strings, we can solve difficulties which first appeared in the researches of Daniel Bernoulli. The solution given by this geometrician assumes that any function whatever may always be developed in a series of sines or cosines of multiple arcs. Now the most complete of all the proofs of this proposition is that which consists in actually resolving a given function into such a series with determined coefficients.

In researches to which partial differential equations are applied, it is often easy to find solutions whose sum composes a more general integral; but the employment of these integrals requires us to determine their extent, and to be able to distinguish clearly the cases in which they represent the general integral from those in which they include only a part. It is necessary above all to assign the values of the constants, and the difficulty of the application consists in the discovery of the coefficients. It is remarkable that we can express by convergent series, and, as we shall see in the sequel, by definite integrals, the ordinates of lines and surfaces which are not subject to a continuous law[1]. We see by this that we must admit into analysis functions which have equal values, whenever the variable receives any values whatever included between two given limits, even though on substituting in these two functions, instead of the variable, a number included in another interval, the results of the two substitutions are not the same. The functions which enjoy this property are represented by different lines, which coincide in a definite portion only of their course, and offer a singular species of finite osculation. These considerations arise in the calculus of partial differential equations; they throw a new light on this calculus, and serve to facilitate its employment in physical theories.

231. The two general equations which express the development of any function whatever, in cosines or sines of multiple arcs, give rise to several remarks which explain the true meaning of these theorems, and direct the application of them.

If in the series

$$a + b \cos x + c \cos 2x + d \cos 3x + e \cos 4x + \&c.,$$

we make the value of x negative, the series remains the same; it also preserves its value if we augment the variable by any multiple whatever of the circumference 2π. Thus in the equation

$$\frac{1}{2}\pi\phi(x) = \frac{1}{2}\int\phi(x)\,dx + \cos x\int\phi(x)\cos x\,dx$$

$$+ \cos 2x\int\phi(x)\cos 2x\,dx + \cos 3x\int\phi(x)\cos 3x\,dx + \&c.\ldots(\nu),$$

the function ϕ is periodic, and is represented by a curve composed of a multitude of equal arcs, each of which corresponds to an

[1] Demonstrations have been supplied by Poisson, Deflers, Dirichlet, Dirksen, Bessel, Hamilton, Boole, De Morgan, Stokes. See note, pp. 208, 209. [A. F.]

interval equal to 2π on the axis of the abscissæ. Further, each of these arcs is composed of two symmetrical branches, which correspond to the halves of the interval equal to 2π.

Suppose then that we trace a line of any form whatever $\phi\phi\alpha$ (see fig. 9.), which corresponds to an interval equal to π.

Fig. 9.

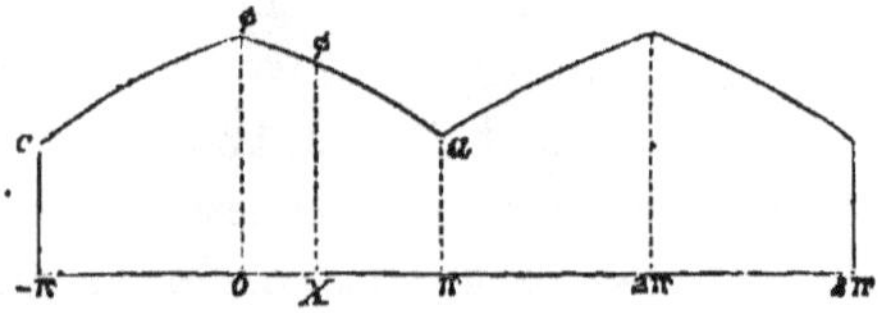

If a series be required of the form

$$a + b\cos x + c\cos 2x + d\cos 3x + \&c.,$$

such that, substituting for x any value whatever X included between 0 and π, we find for the value of the series that of the ordinate $X\phi$, it is easy to solve the problem: for the coefficients given by the equation (ν) are

$$\frac{1}{\pi}\int \phi(x)\,dx, \quad \frac{2}{\pi}\int \phi(x)\cos 2x\,dx, \quad \frac{2}{\pi}\int \phi(x)\cos 3x\,dx, \quad \&c.$$

These integrals, which are taken from $x = 0$ to $x = \pi$, having always measurable values like that of the area $0\phi\alpha\pi$, and the series formed by these coefficients being always convergent, there is no form of the line $\phi\phi\alpha$, for which the ordinate $X\phi$ is not exactly represented by the development

$$a + b\cos x + c\cos 2x + d\cos 3x + e\cos 4x + \&c.$$

The arc $\phi\phi\alpha$ is entirely arbitrary; but the same is not the case with other parts of the line, they are, on the contrary, determinate; thus the arc $\phi\alpha$ which corresponds to the interval from 0 to $-\pi$ is the same as the arc ϕa; and the whole arc $a\phi a$ is repeated on consecutive parts of the axis, whose length is 2π.

We may vary the limits of the integrals in equation (ν). If they are taken from $x = -\pi$ to $x = \pi$ the result will be doubled: it would also be doubled if the limits of the integrals were 0 and 2π, instead of being 0 and π. We denote in general by the sign $\int_a^b$ an integral which begins when the variable is equal to a, and is completed when the variable is equal to b; and we write equation (n) under the following form:

$$\frac{1}{2}\pi\phi(x)=\frac{1}{2}\int_0^\pi \phi(x)\,dx+\cos x\int_0^\pi \phi(x)\cos x\,dx+\cos 2x\int_0^\pi \phi(x)\cos 2x\,dx$$

$$+\cos 3x\int_0^\pi \phi(x)\cos 3x\,dx+\text{etc.}\ \ldots\ldots\ldots\ldots(\nu).$$

Instead of taking the integrals from $x=0$ to $x=\pi$, we might take them from $x=0$ to $x=2\pi$, or from $x=-\pi$ to $x=\pi$; but in each of these two cases, $\pi\phi(x)$ must be written instead of $\frac{1}{2}\pi\phi(x)$ in the first member of the equation.

232. In the equation which gives the development of any function whatever in sines of multiple arcs, the series changes sign and retains the same absolute value when the variable x becomes negative; it retains its value and its sign when the variable is increased or diminished by any multiple whatever of

Fig. 10.

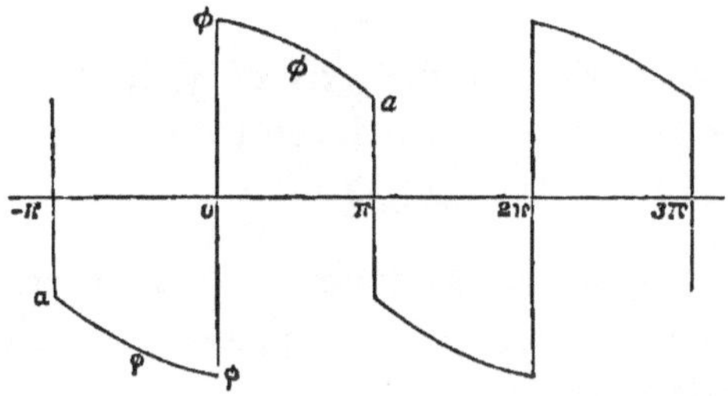

the circumference 2π. The arc $\phi\phi a$ (see fig. 10), which corresponds to the interval from 0 to π is arbitrary; all the other parts of the line are determinate. The arc $\phi\phi\alpha$, which corresponds to the interval from 0 to $-\pi$, has the same form as the given arc $\phi\phi a$; but it is in the opposite position. The whole arc $a\phi\phi\phi\phi a$ is repeated in the interval from π to 3π, and in all similar intervals. We write this equation as follows:

$$\frac{1}{2}\pi\phi(x)=\sin x\int_0^\pi \phi(x)\sin x\,dx+\sin 2x\int_0^\pi \phi(x)\sin 2x\,dx$$

$$+\sin 3x\int_0^\pi \phi(x)\sin 3x\,dx+\text{\&c.}\ \ldots\ldots\ldots(\mu).$$

We might change the limits of the integrals and write

$$\int_0^{2\pi} \text{ or } \int_{-\pi}^{+\pi} \text{ instead of } \int_0^{\pi};$$

but in each of these two cases it would be necessary to substitute in the first member $\pi\phi(x)$ for $\frac{1}{2}\pi\phi(x)$.

233. The function $\phi(x)$ developed in cosines of multiple arcs, is represented by a line formed of two equal arcs placed sym-

Fig. 11.

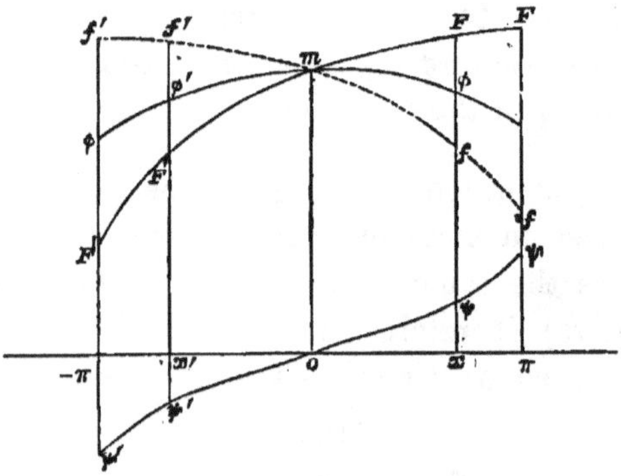

metrically on each side of the axis of y, in the interval from $-\pi$ to $+\pi$ (see fig. 11); this condition is expressed thus,

$$\phi(x) = \phi(-x).$$

The line which represents the function $\psi(x)$ is, on the contrary, formed in the same interval of two opposed arcs, which is what is expressed by the equation

$$\psi(x) = -\psi(-x).$$

Any function whatever $F(x)$, represented by a line traced arbitrarily in the interval from $-\pi$ to $+\pi$, may always be divided into two functions such as $\phi(x)$ and $\psi(x)$. In fact, if the line $F'F'mFF$ represents the function $F(x)$, and we raise at the point o the ordinate om, we can draw through the point m to the right of the axis om the arc mff similar to the arc $mF'F$ of the given curve, and to the left of the same axis we may trace the arc $mf'f'$ similar to the arc mFF; we must then draw through the point m a line $\phi'\phi'm\phi\phi$ which shall divide into two equal parts the difference between each ordinate xF or $x'f'$ and the corresponding ordinate xf or $x'F'$. We must draw also the line $\psi'\psi'0\psi\psi$, whose ordinate measures the half-difference between the ordinate of $F'F'mFF$ and that of $f'f'mff$. This done the ordinate of the lines $F'F'mFF$, and $f'f'mff$ being denoted by $F(x)$ and $f(x)$ respectively, we evidently have $f(x) = F(-x)$; denoting also the ordinate of $\phi'\phi'm\phi\phi$ by $\phi(x)$, and that of $\psi'\psi'0\psi\psi$ by $\psi(x)$, we have

$$F(x) = \phi(x) + \psi(x) \text{ and } f(x) = \phi(x) - \psi(x) = F(-x),$$

hence

$$\phi(x) = \frac{1}{2}F(x) + \frac{1}{2}F(-x) \text{ and } \psi(x) = \frac{1}{2}F(x) - \frac{1}{2}F(-x),$$

whence we conclude that

$$\phi(x) = \phi(-x) \text{ and } \psi(x) = -\psi(-x),$$

which the construction makes otherwise evident.

Thus the two functions $\phi(x)$ and $\psi(x)$, whose sum is equal to $F(x)$ may be developed, one in cosines of multiple arcs, and the other in sines.

If to the first function we apply equation (ν), and to the second the equation (μ), taking the integrals in each case from $x = -\pi$ to $x = \pi$, and adding the two results, we have

$$\pi[\phi(x) + \psi(x)] = \pi F(x)$$

$$= \frac{1}{2}\int \phi(x)\,dx + \cos x \int \phi(x) \cos x\,dx + \cos 2x \int \phi(x) \cos 2x\,dx + \&c.$$

$$+ \sin x \int \psi(x) \sin x\,dx + \sin 2x \int \psi(x) \sin 2x\,dx + \&c.$$

The integrals must be taken from $x = -\pi$ to $x = \pi$. It may now be remarked, that in the integral $\int_{-\pi}^{+\pi} \phi(x) \cos x\,dx$ we could, without changing its value, write $\phi(x) + \psi(x)$ instead of $\phi(x)$: for the function $\cos x$ being composed, to right and left of the axis of x, of two similar parts, and the function $\psi(x)$ being, on the contrary, formed of two opposite parts, the integral $\int_{-\pi}^{+\pi} \psi(x) \cos x\,dx$ vanishes. The same would be the case if we wrote $\cos 2x$ or $\cos 3x$, and in general $\cos ix$ instead of $\cos x$, i being any integer from 0 to infinity. Thus the integral $\int_{-\pi}^{+\pi}\phi(x)\cos ix\,dx$ is the same as the integral

$$\int_{-\pi}^{+\pi}[\psi(x)+\psi(x)]\cos ix\,dx,\quad\text{or}\quad\int_{-\pi}^{+\pi}F(x)\cos ix\,dx.$$

It is evident also that the integral $\int_{-\pi}^{+\pi}\psi(x)\sin ix\,dx$ is equal to the integral $\int_{-\pi}^{+\pi}F(x)\sin ix\,dx$, since the integral $\int_{-\pi}^{+\pi}\phi(x)\sin ix\,dx$ vanishes. Thus we obtain the following equation (p), which serves to develope any function whatever in a series formed of sines and cosines of multiple arcs:

$$\begin{aligned}\pi F(x)=\frac{1}{2}\int F(x)\,dx&\qquad\qquad(\text{p})\\ +\cos x\int F(x)\cos x\,dx+\cos 2x\int F(x)\cos 2x\,dx+\&\text{c.}\\ +\sin x\int F(x)\sin x\,dx+\sin 2x\int F(x)\sin 2x\,dx+\&\text{c.}\end{aligned}$$

234. The function $F(x)$, which enters into this equation, is represented by a line $F'F'FF$, of any form whatever. The arc $F'F'FF$, which corresponds to the interval from $-\pi$ to $+\pi$, is arbitrary; all the other parts of the line are determinate, and the arc $F'F'FF$ is repeated in each consecutive interval whose length is 2π. We shall make frequent applications of this theorem, and of the preceding equations (μ) and (ν).

If it be supposed that the function $F(x)$ in equation (p) is represented, in the interval from $-\pi$ to $+\pi$, by a line composed of two equal arcs symmetrically placed, all the terms which contain sines vanish, and we find equation (ν). If, on the contrary, the line which represents the given function $F(x)$ is formed of two equal arcs opposed in position, all the terms which do not contain sines disappear, and we find equation (μ). Submitting the function $F(x)$ to other conditions, we find other results.

If in the general equation (p) we write, instead of the variable x, the quantity $\frac{\pi x}{r}$, x denoting another variable, and $2r$ the length of the interval which includes the arc which represents $F(x)$; the function becomes $F\left(\frac{\pi x}{r}\right)$, which we may denote by $f(x)$. The limits $x = -\pi$ and $x = \pi$ become $\frac{\pi x}{r} = -\pi$, $\frac{\pi x}{r} = \pi$; we have therefore, after the substitution,

$$rf(x) = \frac{1}{2}\int_{-r}^{+r} f(x)\,dx \qquad \text{(P)}.$$

$$+ \cos \pi \frac{x}{r} \int f(x) \cos \frac{\pi x}{r}\,dx + \cos \frac{2\pi x}{r} \int f(x) \cos \frac{2\pi x}{r}\,dx + \text{etc.}$$

$$+ \sin \pi \frac{x}{r} \int f(x) \sin \frac{\pi x}{r}\,dx + \sin \frac{2\pi x}{r} \int f(x) \sin \frac{2\pi x}{r}\,dx + \text{etc.}$$

All the integrals must be taken like the first from $x = -r$ to $x = +r$. If the same substitution be made in the equations (ν) and (μ), we have

$$\frac{1}{2} rf(x) = \frac{1}{2}\int_0^r f(x)\,dx + \cos \frac{\pi x}{r} \int f(x) \cos \frac{\pi x}{r}\,dx$$

$$+ \cos \frac{2\pi x}{r} \int f(x) \cos \frac{2\pi x}{r}\,dx + \&c. \ldots\ldots\ldots\ldots \text{(N)},$$

and

$$\frac{1}{2} rf(x) = \sin \frac{\pi x}{r} \int_0^r f(x) \sin \frac{\pi x}{r}\,dx$$

$$+ \sin \frac{2\pi x}{r} \int f(x) \sin \frac{2\pi x}{r}\,dx + \&c. \ldots\ldots\ldots\ldots \text{(M)}.$$

In the first equation (P) the integrals might be taken from from $x = 0$ to $x = 2r$, and representing by x the whole interval $2r$, we should have[1]

[1] It has been shewn by Mr J. O'Kinealy that if the values of the arbitrary function $f(x)$ be imagined to recur for every range of x over successive intervals λ, we have the symbolical equation

$$\left(e^{\lambda \frac{d}{dx}} - 1\right) f(x) = 0;$$

and the roots of the auxiliary equation being

$$\pm n \frac{2\pi\sqrt{-1}}{\lambda}, \quad n = 0, 1, 2, 3 \ldots \infty,$$

[*Turn over.*

$$\frac{1}{2}Xf(x) = \frac{1}{2}\int f(x)\,dx \qquad (\Pi)$$

$$+ \cos\frac{2\pi x}{X}\int f(x)\cos\frac{2\pi x}{X}\,dx + \cos\frac{4\pi x}{X}\int f(x)\cos\frac{4\pi x}{X}\,dx + \&c.$$

$$+ \sin\frac{2\pi x}{X}\int f(x)\sin\frac{2\pi x}{X}\,dx + \sin\frac{4\pi x}{X}\int f(x)\sin\frac{4\pi x}{X}\,dx + \&c.$$

235. It follows from that which has been proved in this section, concerning the development of functions in trigonometrical series, that if a function $f(x)$ be proposed, whose value in a definite interval from $x=0$ to $x=X$ is represented by the ordinate of a curved line arbitrarily drawn; we can always develope this function in a series which contains only sines or only cosines, or the sines and cosines of multiple arcs, or the cosines only of odd multiples. To ascertain the terms of these series we must employ equations (M), (N), (P).

The fundamental problems of the theory of heat cannot be completely solved, without reducing to this form the functions which represent the initial state of the temperatures.

These trigonometric series, arranged according to cosines or sines of multiples of arcs, belong to elementary analysis, like the series whose terms contain the successive powers of the variable. The coefficients of the trigonometric series are definite areas, and those of the series of powers are functions given by differentiation, in which, moreover, we assign to the variable a definite value. We could have added several remarks concerning the use and properties of trigonometrical series; but we shall limit ourselves to enunciating briefly those which have the most direct relation to the theory with which we are concerned.

it follows that

$$f(x) = A_0 + A_1\cos\frac{2\pi x}{\lambda} + A_2\cos 2\frac{2\pi x}{\lambda} + A_3\cos 3\frac{2\pi x}{\lambda} + \&c.$$

$$+ B_1\sin\frac{2\pi x}{\lambda} + B_2\sin 2\frac{2\pi x}{\lambda} + B_3\sin 3\frac{2\pi x}{\lambda} + \&c.$$

The coefficients being determined in Fourier's manner by multiplying both sides by $\genfrac{}{}{0pt}{}{\cos}{\sin} n\frac{2\pi x}{\lambda}$ and integrating from 0 to λ. (*Philosophical Magazine*, August 1874, pp. 95, 96). [A. F.]

1st. The series arranged according to sines or cosines of multiple arcs are always convergent; that is to say, on giving to the variable any value whatever that is not imaginary, the sum of the terms converges more and more to a single fixed limit, which is the value of the developed function.

2nd. If we have the expression of a function $f(x)$ which corresponds to a given series

$$a + b\cos x + c\cos 2x + d\cos 3x + e\cos 4x + \&c.,$$

and that of another function $\phi(x)$, whose given development is

$$\alpha + \beta\cos x + \gamma\cos 2x + \delta\cos 3x + \epsilon\cos 4x + \&c.,$$

it is easy to find in real terms the sum of the compound series

$$a\alpha + b\beta + c\gamma + d\delta + e\epsilon + \&c.,$$ [1]

and more generally that of the series

$$a\alpha + b\beta\cos x + c\gamma\cos 2x + d\delta\cos 3x + e\epsilon\cos 4x + \&c.,$$

which is formed by comparing term by term the two given series. This remark applies to any number of series.

3rd. The series (P) (Art. 234) which gives the development of a function $F(x)$ in a series of sines and cosines of multiple arcs, may be arranged under the form

$$\pi F(x) = \frac{1}{2}\int F(\alpha)\,d\alpha$$
$$+ \cos x\int F(\alpha)\cos\alpha\,d\alpha + \cos 2x\int F(\alpha)\cos 2\alpha\,d\alpha + \&c.$$
$$+ \sin x\int F(\alpha)\sin\alpha\,d\alpha + \sin 2x\int F(\alpha)\sin 2\alpha\,d\alpha + \&c.$$

α being a new variable which disappears after the integrations.

We have then

$$\pi F(x) = \int_{-\pi}^{+\pi} F(\alpha)\,d\alpha \left\{\frac{1}{2}\right.$$
$$+ \cos x\cos\alpha + \cos 2x\cos 2\alpha + \cos 3x\cos 3\alpha + \&c.$$
$$\left. + \sin x\sin\alpha + \sin 2x\sin 2\alpha + \sin 3x\sin 3\alpha + \&c.\right\},$$

[1] We shall have

$$\int_0^\pi \psi(x)\,\phi(x)\,dx = a\alpha\pi + \tfrac{1}{2}\pi\,\{b\beta + c\gamma + \ldots\}.$$ [R. L. E.]

or

$$F(x)=\frac{1}{\pi}\int_{-\pi}^{+\pi}F(\alpha)\,d\alpha\left\{\frac{1}{2}+\cos(x-\alpha)+\cos 2(x-\alpha)+\&c.\right\}.$$

Hence, denoting the sum of the preceding series by

$$\Sigma \cos i(x-\alpha)$$

taken from $i=1$ to $i=\infty$, we have

$$F(x)=\frac{1}{\pi}\int F(\alpha)\,d\alpha\left\{\frac{1}{2}+\Sigma\cos i(x-\alpha)\right\}.$$

The expression $\frac{1}{2}+\Sigma\cos i(x-\alpha)$ represents a function of x and α, such that if it be multiplied by any function whatever $F(\alpha)$, and integrated with respect to α between the limits $\alpha=-\pi$ and $\alpha=\pi$, the proposed function $F(\alpha)$ becomes changed into a like function of x multiplied by the semi-circumference π. It will be seen in the sequel what is the nature of the quantities, such as $\frac{1}{2}+\Sigma\cos i(x-\alpha)$, which enjoy the property we have just enunciated.

4th. If in the equations (M), (N), and (P) (Art 234), which on being divided by r give the development of a function $f(x)$, we suppose the interval r to become infinitely large, each term of the series is an infinitely small element of an integral; the sum of the series is then represented by a definite integral. When the bodies have determinate dimensions, the arbitrary functions which represent the initial temperatures, and which enter into the integrals of the partial differential equations, ought to be developed in series analogous to those of the equations (M), (N), (P); but these functions take the form of definite integrals, when the dimensions of the bodies are not determinate, as will be explained in the course of this work, in treating of the free diffusion of heat (Chapter IX.).

Note on Section VI. On the subject of the development of a function whose values are arbitrarily assigned between certain limits, in series of sines and cosines of multiple arcs, and on questions connected with the values of such series at the limits, on the convergency of the series, and on the discontinuity of their values, the principal authorities are

Poisson, *Théorie mathématique de la Chaleur*, Paris, 1835, Chap. VII. Arts. 92—102, *Sur la manière d'exprimer les fonctions arbitraires par des séries de quantités périodiques*. Or, more briefly, in his *Traité de Mécanique*, Arts. 325—328. Poisson's original memoirs on the subject were published in the *Journal de l'École Polytechnique*, Cahier 18, pp. 417—489, year 1820, and Cahier 19, pp. 404—509, year 1823.

De Morgan, *Differential and Integral Calculus*. London, 1842, pp. 609—617. The proofs of the developments appear to be original. In the verification of the developments the author follows Poisson's methods.

Stokes, *Cambridge Philosophical Transactions*, 1847, Vol. VIII. pp. 533—556. *On the Critical values of the sums of Periodic Series*. Section I. *Mode of ascertaining the nature of the discontinuity of a function which is expanded in a series of sines or cosines, and of obtaining the developments of the derived functions*. Graphically illustrated.

Thomson and Tait, *Natural Philosophy*, Oxford, 1867, Vol. I. Arts. 75—77.

Donkin, *Acoustics*, Oxford, 1870, Arts. 72—79, and Appendix to Chap. IV.

Matthieu, *Cours de Physique Mathématique*, Paris, 1873, pp. 33—36.

Entirely different methods of discussion, not involving the introduction of arbitrary multipliers to the successive terms of the series were originated by

Dirichlet, *Crelle's Journal*, Berlin, 1829, Band IV. pp. 157—169. *Sur la convergence des séries trigonométriques qui servent à représenter une fonction arbitraire entre les limites données*. The methods of this memoir thoroughly deserve attentive study, but are not yet to be found in English text-books. Another memoir, of greater length, by the same author appeared in Dove's *Repertorium der Physik*, Berlin, 1837, Band I. pp. 152—174. *Ueber die Darstellung ganz willkührlicher Functionen durch Sinus- und Cosinusreihen*. Von G. Lejeune Dirichlet.

Other methods are given by

Dirksen, *Crelle's Journal*, 1829, Band IV. pp. 170—178. *Ueber die Convergenz einer nach den Sinussen und Cosinussen der Vielfachen eines Winkels fortschreitenden Reihe*.

Bessel, *Astronomische Nachrichten*, Altona, 1839, pp. 230—238. *Ueber den Ausdruck einer Function $\phi(x)$ durch Cosinusse und Sinusse der Vielfachen von x.*

The writings of the last three authors are criticised by Riemann, *Gesammelte Mathematische Werke*, Leipzig, 1876, pp. 221—225. *Ueber die Darstellbarkeit einer Function durch eine Trigonometrische Reihe*.

On *Fluctuating Functions* and their properties, a memoir was published by Sir W. R. Hamilton, *Transactions of the Royal Irish Academy*, 1843, Vol. XIX. pp. 264—321. The introductory and concluding remarks may at this stage be studied.

The writings of Deflers, Boole, and others, on the subject of the expansion of an arbitrary function by means of a double integral (*Fourier's Theorem*) will be alluded to in the notes on Chap. IX. Arts. 361, 362. [A. F.]

SECTION VII.

Application to the actual problem.

236. We can now solve in a general manner the problem of the propagation of heat in a rectangular plate BAC, whose end A is constantly heated, whilst its two infinite edges B and C are maintained at the temperature 0.

Suppose the initial temperature at all points of the slab BAC to be nothing, but that the temperature at each point m of the edge A is preserved by some external cause, and that its fixed value is a function $f(x)$ of the distance of the point m from the end O of the edge A whose whole length is $2r$; let v be the constant temperature of the point m whose co-ordinates are x and y, it is required to determine v as a function of x and y.

The value $v = ae^{-my} \sin mx$ satisfies the equation

$$\frac{d^2v}{dx^2} + \frac{d^2v}{dy^2} = 0;$$

a and m being any quantities whatever. If we take $m = i\frac{\pi}{r}$, i being an integer, the value $ae^{-i\pi\frac{y}{r}} \sin \frac{i\pi x}{r}$ vanishes, when $x = r$, whatever the value of y may be. We shall therefore assume, as a more general value of v,

$$v = a_1 e^{-\pi\frac{y}{r}} \sin \frac{\pi x}{r} + a_2 e^{-2\pi\frac{y}{r}} \sin \frac{2\pi x}{r} + a_3 e^{-3\pi\frac{y}{r}} \sin \frac{3\pi x}{r} + \&c.$$

If y be supposed nothing, the value of v will by hypothesis be equal to the known function $f(x)$. We then have

$$f(x) = a_1 \sin \frac{\pi x}{r} + a_2 \sin \frac{2\pi x}{r} + a_3 \sin \frac{3\pi x}{r} + \&c.$$

The coefficients a_1, a_2, a_3, &c. can be determined by means of equation (M), and on substituting them in the value of v we have

$$\frac{1}{2} rv = e^{-\pi\frac{y}{r}} \sin \frac{\pi x}{r} \int f(x) \sin \frac{\pi x}{r} dx + e^{-2\pi\frac{y}{r}} \sin \frac{2\pi x}{r} \int f(x) \sin \frac{2\pi x}{r} dx$$

$$+ e^{-3\pi\frac{y}{r}} \sin \frac{3\pi x}{r} \int f(x) \sin \frac{3\pi x}{r} dx + \&c.$$

237. Assuming $r = \pi$ in the preceding equation, we have the solution under a more simple form, namely

$$\frac{1}{2} \pi v = e^{-y} \sin x \int f(x) \sin x dx + e^{-2y} \sin 2x \int f(x) \sin 2x dx$$

$$+ e^{-3y} \sin 3x \int f(x) \sin 3x dx + \&c. \ldots\ldots\ldots (a),$$

or

$$\frac{1}{2}\pi v = \int_0^\pi f(\alpha)\,d\alpha\,(e^{-y}\sin x\sin\alpha + e^{-2y}\sin 2x\sin 2\alpha + e^{-3y}\sin 3x\sin 3\alpha + \&c.)$$

α is a new variable, which disappears after integration.

If the sum of the series be determined, and if it be substituted in the last equation, we have the value of v in a finite form. The double of the series is equal to

$$e^{-y}[\cos(x-\alpha)-\cos(x+\alpha)] + e^{-2y}[\cos 2(x-\alpha)-\cos 2(x+\alpha)] + e^{-3y}[\cos 3(x-\alpha)-\cos 3(x+\alpha)] + \&c.;$$

denoting by $F(y, p)$ the sum of the infinite series

$$e^{-y}\cos p + e^{-2y}\cos 2p + e^{-3y}\cos 3p + \&c.,$$

we find

$$\pi v = \int_0^\pi f(\alpha)\,d\alpha\,\{F(y, x-\alpha) - F(y, x+\alpha)\}.$$

We have also

$$2F(y,p) = \begin{cases} e^{-(y+p\sqrt{-1})} + e^{-2(y+p\sqrt{-1})} + e^{-3(y+p\sqrt{-1})} + \&c. \\ e^{-(y-p\sqrt{-1})} + e^{-2(y-p\sqrt{-1})} + e^{-3(y-p\sqrt{-1})} + \&c. \end{cases}$$

$$= \frac{e^{-(y+p\sqrt{-1})}}{1-e^{-(y+p\sqrt{-1})}} + \frac{e^{-(y-p\sqrt{-1})}}{1-e^{-(y-p\sqrt{-1})}},$$

or $$F(y,p) = \frac{\cos p - e^{-y}}{e^{y} - 2\cos p + e^{-y}},$$

whence

$$\pi v = \int_0^\pi f(\alpha)\,d\alpha\left\{\frac{\cos(x-\alpha)-e^{-y}}{e^{y}-2\cos(x-\alpha)+e^{-y}} - \frac{\cos(x+\alpha)-e^{-y}}{e^{y}-2\cos(x+\alpha)+e^{-y}}\right\},$$

or

$$\pi v = \int_0^\pi f(\alpha)\,d\alpha\left\{\frac{2(e^{y}-e^{-y})\sin x\sin\alpha}{[e^{y}-2\cos(x-\alpha)+e^{-y}]\,[e^{y}-2\cos(x+\alpha)+e^{-y}]}\right\},$$

or, decomposing the coefficient into two fractions,

$$\pi v = \frac{e^{y}-e^{-y}}{2}\int_0^\pi f(\alpha)\,d\alpha\left\{\frac{1}{e^{y}-2\cos(x-\alpha)+e^{-y}} - \frac{1}{e^{y}-2\cos(x+\alpha)+e^{-y}}\right\}.$$

This equation contains, in real terms under a finite form, the integral of the equation $\frac{d^2v}{dx^2}+\frac{d^2v}{dy^2}=0$, applied to the problem of the uniform movement of heat in a rectangular solid, exposed at its extremity to the constant action of a single source of heat.

It is easy to ascertain the relations of this integral to the general integral, which has two arbitrary functions; these functions are by the very nature of the problem determinate, and nothing arbitrary remains but the function $f(\alpha)$, considered between the limits $\alpha=0$ and $\alpha=\pi$. Equation (a) represents, under a simple form, suitable for numerical applications, the same value of v reduced to a convergent series.

If we wished to determine the quantity of heat which the solid contains when it has arrived at its permanent state, we should take the integral $\int dx \int dy\, v$ from $x=0$ to $x=\pi$, and from $y=0$ to $y=\infty$; the result would be proportianal to the quantity required. In general there is no property of the uniform movement of heat in a rectangular plate, which is not exactly represented by this solution.

We shall next regard problems of this kind from another point of view, and determine the varied movement of heat in different bodies.

CHAPTER IV.

OF THE LINEAR AND VARIED MOVEMENT OF HEAT IN A RING.

SECTION I.

General solution of the problem.

238. THE equation which expresses the movement of heat in a ring has been stated in Article 105; it is

$$\frac{dv}{dt}=\frac{K}{CD}\frac{d^2v}{dx^2}-\frac{hl}{CDS}v \text{ .... } (b).$$

The problem is now to integrate this equation: we may write it simply

$$\frac{dv}{dt}=k\frac{d^2v}{dx^2}-hv,$$

wherein k represents $\frac{K}{CD}$, and h represents $\frac{hl}{CDS}$, x denotes the length of the arc included between a point m of the ring and the origin O, and v is the temperature which would be observed at the point m after a given time t. We first assume $v=e^{-ht}u$, u being a new unknown, whence we deduce $\frac{du}{dt}=k\frac{d^2u}{dx^2}$; now this equation belongs to the case in which the radiation is nul at the surface, since it may be derived from the preceding equation by making $h=0$: we conclude from it that the different points of the ring are cooled successively, by the action of the medium, without this circumstance disturbing in any manner the law of the distribution of the heat.

In fact on integrating the equation $\frac{du}{dt}=k\frac{d^2u}{dx^2}$, we should find the values of u which correspond to different points of the ring at the same instant, and we should ascertain what the state of the solid would be if heat were propagated in it without any loss at the surface; to determine then what would be the state of the solid at the same instant if this loss had occurred, it will be sufficient to multiply all the values of u taken at different points, at the same instant, by the same fraction e^{-ht}. Thus the cooling which is effected at the surface does not change the law of the distribution of heat; the only result is that the temperature of each point is less than it would have been without this circumstance, and the temperature diminishes from this cause according to the successive powers of the fraction e^{-ht}.

239. The problem being reduced to the integration of the equation $\frac{du}{dt}=k\frac{d^2u}{dx^2}$, we shall, in the first place, select the simplest particular values which can be attributed to the variable u; from them we shall then compose a general value, and we shall prove that this value is as extensive as the integral, which contains an arbitrary function of x, or rather that it is this integral itself, arranged under the form which the problem requires, so that there cannot be any different solution.

It may be remarked first, that the equation is satisfied if we give to u the particular value $ae^{mt}\sin nx$, m and n being subject to the condition $m=-kn^2$. Take then as a particular value of u the function $e^{-kn^2t}\sin nx$.

In order that this value may belong to the problem, it must not change when the distance x is increased by the quantity $2\pi r$, r denoting the mean radius of the ring. Hence $2\pi nr$ must be a multiple i of the circumference 2π; which gives $n=\frac{i}{r}$.

We may take i to be any integer; we suppose it to be always positive, since, if it were negative, it would suffice to change the sign of the coefficient a in the value $ae^{-kn^2t}\sin nx$. The particular value $ae^{-k\frac{i^2t}{r^2}}\sin\frac{ix}{r}$ could not satisfy the problem proposed unless it represented the initial state of the solid. Now on making $t=0$, we find $u=a\sin\frac{ix}{r}$: suppose then that the initial values of u are actually expressed by $a\sin\frac{x}{r}$; that is to say, that the primitive temperatures at the different points are proportional to the sines of angles included between the radii which pass through those points and that which passes through the origin, the movement of heat in the interior of the ring will be exactly represented by the equation $u = ae^{-\frac{kt}{r^2}}\sin\frac{x}{r}$, and if we take account of the loss of heat through the surface, we find

$$v = ae^{-\left(h+\frac{k}{r^2}\right)t}\sin\frac{x}{r}.$$

In the case in question, which is the simplest of all those which we can imagine, the variable temperatures preserve their primitive ratios, and the temperature at any point diminishes according to the successive powers of a fraction which is the same for every point.

The same properties would be noticed if we supposed the initial temperatures to be proportional to the sines of the double of the arc $\frac{x}{r}$; and in general the same happens when the given temperatures are represented by $a\sin\frac{ix}{r}$, i being any integer whatever.

We should arrive at the same results on taking for the particular value of u the quantity $ae^{-kn^2t}\cos nx$: here also we have $2n\pi r = 2i\pi$, and $n = \frac{i}{r}$; hence the equation

$$u = ae^{-k\frac{i^2t}{r^2}}\cos\frac{ix}{r}$$

expresses the movement of heat in the interior of the ring if the initial temperatures are represented by $\cos\frac{ix}{r}$.

In all these cases, where the given temperatures are proportional to the sines or to the cosines of a multiple of the arc $\frac{x}{r}$, the ratios established between these temperatures exist continually during the infinite time of the cooling. The same would be the case if the initial temperatures were represented by the function $a\sin\frac{ix}{r}+b\cos\frac{ix}{r}$, i being any integer, a and b any coefficients whatever.

240. Let us pass now to the general case in which the initial temperatures have not the relations which we have just supposed, but are represented by any function whatever $F(x)$. Let us give to this function the form $\phi\left(\frac{x}{r}\right)$, so that we have $F(x)=\phi\left(\frac{x}{r}\right)$, and imagine the function $\phi\left(\frac{x}{r}\right)$ to be decomposed into a series of sines or cosines of multiple arcs affected by suitable coefficients. We write down the equation

$$\phi\left(\frac{x}{r}\right)=\left\{\begin{array}{l} a_0\sin\left(0\frac{x}{r}\right)+a_1\sin\left(1\frac{x}{r}\right)+a_2\sin\left(2\frac{x}{r}\right)+\&c. \\ +b_0\cos\left(0\frac{x}{r}\right)+b_1\cos\left(1\frac{x}{r}\right)+b_2\cos\left(2\frac{x}{r}\right)+\&c.\end{array}\right\} \quad (\epsilon).$$

The numbers $a_0, a_1, a_2 \ldots, b_0, b_1, b_2 \ldots$ are regarded as known and calculated beforehand. It is evident that the value of u will then be represented by the equation

$$u=b_0+\left.\begin{array}{r} a_1\sin\frac{x}{r} \\ \\ b_1\cos\frac{x}{r}\end{array}\right| e^{-k\frac{t}{r^2}} + \left.\begin{array}{r} a_2\sin 2\frac{x}{r} \\ \\ b_2\cos 2\frac{x}{r}\end{array}\right| e^{-k\frac{2^2t}{r^2}} + \&c.$$

In fact, 1st, this value of u satisfies the equation $\frac{du}{dt}=k\frac{d^2u}{dx^2}$, since it is the sum of several particular values; 2nd, it does not change when we increase the distance x by any multiple whatever of the circumference of the ring; 3rd, it satisfies the initial state, since on making $t=0$, we find the equation (ϵ). Hence all the conditions of the problem are fulfilled, and it remains only to multiply the value of u by e^{-ht}.

241. As the time increases, each of the terms which compose the value of u becomes smaller and smaller; the system of temperatures tends therefore continually towards the regular and con-

stant state in which the difference of the temperature u from the constant b_0 is represented by

$$\left(a \sin \frac{x}{r} + b \cos \frac{x}{r}\right) e^{-\frac{kt}{r^2}}.$$

Thus the particular values which we have previously considered, and from which we have composed the general value, derive their origin from the problem itself. Each of them represents an elementary state which could exist of itself as soon as it is supposed to be formed; these values have a natural and necessary relation with the physical properties of heat.

To determine the coefficients a_0, a_1, a_2, &c., b_0, b_1, b_2, &c., we must employ equation (II), Art. 234, which was proved in the last section of the previous Chapter.

Let the whole abscissa denoted by X in this equation be $2\pi r$, let x be the variable abscissa, and let $f(x)$ represent the initial state of the ring, the integrals must be taken from $x=0$ to $x=2\pi r$; we have then

$$\pi r f(x) = \frac{1}{2}\int f(x)\,dx$$

$$+ \cos\left(\frac{x}{r}\right)\int \cos\left(\frac{x}{r}\right) f(x)\,dx + \cos\left(2\frac{x}{r}\right)\int \cos\left(2\frac{x}{r}\right) f(x)\,dx + \text{\&c.}$$

$$+ \sin\left(\frac{x}{r}\right)\int \sin\left(\frac{x}{r}\right) f(x)\,dx + \sin\left(2\frac{x}{r}\right)\int \sin\left(2\frac{x}{r}\right) f(x)\,dx + \text{\&c.}$$

Knowing in this manner the values of a_0, a_1, a_2, &c., b_0, b_1, b_2, &c., if they be substituted in the equation we have the following equation, which contains the complete solution of the problem:

$$\pi r v = e^{-ht}\left\{\frac{1}{2}\int f(x)\,dx \right.$$

$$+ \left|\begin{matrix} \sin\frac{x}{r}\int\left(\sin\frac{x}{r} f(x)\,dx\right) \\ \cos\frac{x}{r}\int\left(\cos\frac{x}{r} f(x)\,dx\right)\end{matrix}\right| e^{-\frac{kt}{r^2}}$$

$$\left. + \left|\begin{matrix} \sin 2\frac{x}{r}\int\left(\sin\frac{2x}{r} f(x)\,dx\right) \\ \cos 2\frac{x}{r}\int\left(\cos\frac{2x}{r} f(x)\,dx\right)\end{matrix}\right| e^{-\frac{2^2 kt}{r^2}} + \text{\&c.}\right\} \ldots\ldots \text{(E)}.$$

All the integrals must be taken from $x=0$ to $x=2\pi r$.

The first term $\frac{1}{2\pi r}\int f(x)\,dx$, which serves to form the value of v, is evidently the mean initial temperature, that is to say, that which each point would have if all the initial heat were distributed equally throughout.

242. The preceding equation (E) may be applied, whatever the form of the given function $f(x)$ may be. We shall consider two particular cases, namely: 1st, that which occurs when the ring having been raised by the action of a source of heat to its permanent temperatures, the source is suddenly suppressed; 2nd, the case in which half the ring, having been equally heated throughout, is suddenly joined to the other half, throughout which the initial temperature is 0.

We have seen previously that the permanent temperatures of the ring are expressed by the equation $v=a\alpha^x+b\alpha^{-x}$; the value of quantity α being $e^{-\sqrt{\frac{hl}{KS}}}$, where l is the perimeter of the generating section, and S the area of that section.

If it be supposed that there is but a single source of heat, the equation $\frac{dv}{dx}=0$ must necessarily hold at the point opposite to that which is occupied by the source. The condition $a\alpha^x-b\alpha^{-x}=0$ will therefore be satisfied at this point. For convenience of calculation let us consider the fraction $\frac{hl}{KS}$ to be equal to unity, and let us take the radius r of the ring to be the radius of the trigonometrical tables, we shall then have $v=ae^x+be^{-x}$; hence the initial state of the ring is represented by the equation

$$v=be^{-\pi}\left(e^{-\pi+x}+e^{\pi-x}\right).$$

It remains only to apply the general equation (E), and denoting by M the mean initial heat (Art. 241), we shall have

$$v=2e^{-ht}M\left(\frac{1}{2}-\frac{\cos x}{1^2+1}e^{-kt}+\frac{\cos 2x}{2^2+1}e^{-2^2kt}-\frac{\cos 3x}{3^2+1}e^{-3^2kt}+\&c.\right).$$

This equation expresses the variable state of a solid ring, which having been heated at one of its points and raised to stationary temperatures, cools in air after the suppression of the source of heat.

243. In order to make a second application of the general equation (E), we shall suppose the initial heat to be so distributed that half the ring included between $x=0$ and $x=\pi$ has throughout the temperature 1, the other half having the temperature 0. It is required to determine the state of the ring after the lapse of a time t.

The function $f(x)$, which represents the initial state, is in this case such that its value is 1 so long as the variable is included between 0 and π. It follows from this that we must suppose $f(x)=1$, and take the integrals only from $x=0$ to $x=\pi$, the other parts of the integrals being nothing by hypothesis. We obtain first the following equation, which gives the development of the function proposed, whose value is 1 from $x=0$ to $x=\pi$ and nothing from $x=\pi$ to $x=2\pi$,

$$f(x)=\frac{1}{2}+\frac{2}{\pi}\left(\sin x+\frac{1}{3}\sin 3x+\frac{1}{5}\sin 5x+\frac{1}{7}\sin 7x+\&c.\right).$$

If now we substitute in the general equation the values which we have just found for the constant coefficients, we shall have the equation

$$\frac{1}{2}\pi v=e^{-ht}\left(\frac{1}{4}\pi+\sin x\,e^{-kt}+\frac{1}{3}\sin 3x\,e^{-3^2kt}+\frac{1}{5}\sin 5x\,e^{-5^2kt}+\&c.\right),$$

which expresses the law according to which the temperature at each point of the ring varies, and indicates its state after any given time: we shall limit ourselves to the two preceding applications, and add only some observations on the general solution expressed by the equation (E).

244. 1st. If k is supposed infinite, the state of the ring is expressed thus, $\pi r v=e^{-ht}\frac{1}{2}\int f(x)\,dx$, or, denoting by M the mean initial temperature (Art. 241), $v=e^{-ht}M$. The temperature at every point becomes suddenly equal to the mean temperature, and all the different points retain always equal temperatures, which is a necessary consequence of the hypothesis in which we admit infinite conducibility.

2nd. We should have the same result if the radius of the ring were infinitely small.

3rd. To find the mean temperature of the ring after a time t we must take the integral $\int f(x)\,dx$ from $x=0$ to $x=2\pi r$, and divide by $2\pi r$. Integrating between these limits the different parts of the value of u, and then supposing $x=2\pi r$, we find the total values of the integrals to be nothing except for the first term; the value of the mean temperature is therefore, after the time t, the quantity $e^{-ht}M$. Thus the mean temperature of the ring decreases in the same manner as if its conducibility were infinite; the variations occasioned by the propagation of heat in the solid have no influence on the value of this temperature.

In the three cases which we have just considered, the temperature decreases in proportion to the powers of the fraction e^{-h}, or, which is the same thing, to the ordinate of a logarithmic curve, the abscissa being equal to the time which has elapsed. This law has been known for a long time, but it must be remarked that it does not generally hold unless the bodies are of small dimensions. The previous analysis tells us that if the diameter of a ring is not very small, the cooling at a definite point would not be at first subject to that law; the same would not be the case with the mean temperature, which decreases always in proportion to the ordinates of a logarithmic curve. For the rest, it must not be forgotten that the generating section of the ring is supposed to have dimensions so small that different points of the same section do not differ sensibly in temperature.

4th. If we wished to ascertain the quantity of heat which escapes in a given time through the surface of a given portion of the ring, the integral $hl\int dt \int v\,dx$ must be employed, and must be taken between limits relative to the time. For example, if we took 0 and 2π to be the limits of x, and 0, ∞, to be the limits of t; that is to say, if we wished to determine the whole quantity of heat which escapes from the entire surface, during the complete course of the cooling, we ought to find after the integrations a result equal to the whole quantity of the initial heat, or $2\pi rM$, M being the mean initial temperature.

5th. If we wish to ascertain how much heat flows in a given time, across a definite section of the ring, we must employ the integral $-KS\int dt\frac{dv}{dx}$, writing for $\frac{dv}{dx}$ the value of that function, taken at the point in question.

245. Heat tends to be distributed in the ring according to a law which ought to be noticed. The more the time which has elapsed increases the smaller do the terms which compose the value of v in equation (E) become with respect to those which precede them. There is therefore a certain value of t for which the movement of heat begins to be represented sensibly by the equation

$$u = a_0 + \left(a_1 \sin\frac{x}{r} + b_1 \cos\frac{x}{r}\right) e^{-\frac{kt}{r^2}}.$$

The same relation continues to exist during the infinite time of the cooling. In this state, if we choose two points of the ring situated at the ends of the same diameter, and represent their respective distances from the origin by x_1 and x_2, and their corresponding temperatures at the time t by v_1 and v_2; we shall have

$$v_1 = \left\{a_0 + \left(a_1 \sin\frac{x_1}{r} + b_1 \cos\frac{x_1}{r}\right) e^{-\frac{kt}{r^2}}\right\} e^{-ht},$$

$$v_2 = \left\{a_0 + \left(a_1 \sin\frac{x_2}{r} + b_1 \cos\frac{x_2}{r}\right) e^{-\frac{kt}{r^2}}\right\} e^{-ht}.$$

The sines of the two arcs $\frac{x_1}{r}$ and $\frac{x_2}{r}$ differ only in sign; the same is the case with the quantities $\cos\frac{x_1}{r}$ and $\cos\frac{x_2}{r}$; hence

$$\frac{v_1 + v_2}{2} = a_0 e^{-ht},$$

thus the half-sum of the temperatures at opposite points gives a quantity $a_0 e^{-ht}$, which would remain the same if we chose two points situated at the ends of another diameter. The quantity $a_0 e^{-ht}$, as we have seen above, is the value of the mean temperature after the time t. Hence the half-sum of the temperature at any two opposite points decreases continually with the mean temperature of the ring, and represents its value without sensible error, after the cooling has lasted for a certain time. Let us examine more particularly in what the final state consists which is expressed by the equation

$$v = \left\{a_0 + \left(a_1 \sin\frac{x}{r} + b_1 \cos\frac{x}{r}\right) e^{-\frac{kt}{r^2}}\right\} e^{-ht}.$$

If first we seek the point of the ring at which we have the condition

$$a_1 \sin\frac{x}{r} + b_1 \cos\frac{x}{r} = 0, \text{ or } \frac{x}{r} = -\text{arc tan}\left(\frac{b_1}{a_1}\right),$$

we see that the temperature at this point is at every instant the mean temperature of the ring: the same is the case with the point diametrically opposite; for the abscissa x of the latter point will also satisfy the above equation

$$\frac{x}{r} = \text{arc tan}\left(-\frac{b_1}{a_1}\right).$$

Let us denote by X the distance at which the first of these points is situated, and we shall have

$$b_1 = -a_1 \frac{\sin\frac{X}{r}}{\cos\frac{X}{r}};$$

and substituting this value of b_1, we have

$$v = \left\{a_0 + \frac{a_1}{\cos\frac{X}{r}} \sin\left(\frac{x}{r} - \frac{X}{r}\right) e^{-\frac{kt}{r^2}}\right\} e^{-ht}.$$

If we now take as origin of abscissæ the point which corresponds to the abscissa X, and if we denote by u the new abscissa $x - X$, we shall have

$$v = e^{-ht}\left(a_0 + b \sin\frac{u}{r} e^{-\frac{kt}{r^2}}\right).$$

At the origin, where the abscissa u is 0, and at the opposite point, the temperature v is always equal to the mean temperature; these two points divide the circumference of the ring into two parts whose state is similar, but of opposite sign; each point of one of these parts has a temperature which exceeds the mean temperature, and the amount of that excess is proportional to the sine of the distance from the origin. Each point of the other part has a temperature less than the mean temperature, and the defect is the same as the excess at the opposite point. This symmetrical distribution of heat exists throughout the whole duration of the cooling. At the two ends of the heated half, two flows of heat are established in direction towards the cooled half, and their effect is continually to bring each half of the ring towards the mean temperature.

246. We may now remark that in the general equation which gives the value of v, each of the terms is of the form

$$\left(a_i \sin i\frac{x}{r} + b_i \cos i\frac{x}{r}\right) e^{-i^2\frac{kt}{r^2}}.$$

We can therefore derive, with respect to each term, consequences analogous to the foregoing. In fact denoting by X the distance for which the coefficient

$$a_i \sin i\frac{x}{r} + b_i \cos i\frac{x}{r}$$

is nothing, we have the equation $b_i = -a_i \tan i\frac{X}{r}$, and this substitution gives, as the value of the coefficient,

$$a \sin i\left(\frac{x-X}{r}\right),$$

a being a constant. It follows from this that taking the point whose abscissa is X as the origin of co-ordinates, and denoting by u the new abscissa $x-X$, we have, as the expression of the changes of this part of the value of v, the function

$$ae^{-ht} \sin i\frac{u}{r} e^{-i^2\frac{kt}{r^2}}.$$

If this particular part of the value of v existed alone, so as to make the coefficients of all the other parts nul, the state of the ring would be represented by the function

$$ae^{-ht}e^{-i^2\frac{kt}{r^2}} \sin\left(i\frac{u}{r}\right),$$

and the temperature at each point would be proportional to the sine of the multiple i of the distance of this point from the origin. This state is analogous to that which we have already described:

it differs from it in that the number of points which have always the same temperature equal to the mean temperature of the ring is not 2 only, but in general equal to $2i$. Each of these points or nodes separates two adjacent portions of the ring which are in a similar state, but opposite in sign. The circumference is thus found to be divided into several equal parts whose state is alternately positive and negative. The flow of heat is the greatest possible in the nodes, and is directed towards that portion which is in the negative state, and it is nothing at the points which are equidistant from two consecutive nodes. The ratios which exist then between the temperatures are preserved during the whole of the cooling, and the temperatures vary together very rapidly in proportion to the successive powers of the fraction

$$e^{-h}e^{-i^2\frac{k}{r^2}}.$$

If we give successively to i the values 0, 1, 2, 3, &c., we shall ascertain all the regular and elementary states which heat can assume whilst it is propagated in a solid ring. When one of these simple modes is once established, it is maintained of itself, and the ratios which exist between the temperatures do not change; but whatever the primitive ratios may be, and in whatever manner the ring may have been heated, the movement of heat can be decomposed into several simple movements, similar to those which we have just described, and which are accomplished all together without disturbing each other. In each of these states the temperature is proportional to the sine of a certain multiple of the distance from a fixed point. The sum of all these partial temperatures, taken for a single point at the same instant, is the actual temperature of that point. Now some of the parts which compose this sum decrease very much more rapidly than the others. It follows from this that the elementary states of the ring which correspond to different values of i, and whose superposition determines the total movement of heat, disappear in a manner one after the other. They cease soon to have any sensible influence on the value of the temperature, and leave only the first among them to exist, in which i is the least of all. In this manner we form an exact idea of the law according to which heat is distributed in a ring, and is dissipated at its surface. The state of the ring becomes more and more symmetrical; it soon becomes confounded with that towards which it has a natural tendency, and which consists in this, that the temperatures of the different points become proportional to the sine of the same multiple of the arc which measures the distance from the origin. The initial distribution makes no change in these results.

SECTION II.

Of the communication of heat between separate masses.

247. We have now to direct attention to the conformity of the foregoing analysis with that which must be employed to determine the laws of propagation of heat between separate masses; we shall thus arrive at a second solution of the problem of the movement of heat in a ring. Comparison of the two results will indicate the true foundations of the method which we have followed, in integrating the equations of the propagation of heat in continuous bodies. We shall examine, in the first place, an extremely simple case, which is that of the communication of heat between two equal masses.

Suppose two cubical masses m and n of equal dimensions and of the same material to be unequally heated; let their respective temperatures be a and b, and let them be of infinite conducibility. If we placed these two bodies in contact, the temperature in each would suddenly become equal to the mean temperature $\frac{1}{2}(a+b)$. Suppose the two masses to be separated by a very small interval, that an infinitely thin layer of the first is detached so as to be joined to the second, and that it returns to the first immediately after the contact. Continuing thus to be transferred alternately, and at equal infinitely small intervals, the interchanged layer causes the heat of the hotter body to pass gradually into that which is less heated; the problem is to determine what would be, after a given time, the heat of each body, if they lost at their surface no part of the heat which they contained. We do not suppose the transfer of heat in solid continuous bodies to be effected in a manner similar to that which we have just described: we wish only to determine by analysis the result of such an hypothesis.

Each of the two masses possessing infinite conducibility, the quantity of heat contained in an infinitely thin layer, is suddenly added to that of the body with which it is in contact; and a common temperature results which is equal to the quotient of the sum of the quantities of heat divided by the sum of the masses. Let ω be the mass of the infinitely small layer which is separated from the hotter body, whose temperature is a; let α and β be the variable temperatures which correspond to the time t, and whose initial values are a and b. When the layer ω is separated from the mass m which becomes $m-\omega$, it has like this mass the temperature α, and as soon as it touches the second body affected with the temperature β, it assumes at the same time with that body a temperature equal to $\dfrac{m\beta+\alpha\omega}{m+\omega}$. The layer ω, retaining the last temperature, returns to the first body whose mass is $m-\omega$ and temperature α. We find then for the temperature after the second contact

$$\frac{\alpha(m-\omega)+\left(\dfrac{m\beta+\alpha\omega}{m+\omega}\right)\omega}{m} \quad \text{or} \quad \frac{\alpha m+\beta\omega}{m+\omega}.$$

The variable temperatures α and β become, after the interval dt, $\alpha+(\alpha-\beta)\dfrac{\omega}{m}$, and $\beta+(\alpha-\beta)\dfrac{\omega}{m}$; these values are found by suppressing the higher powers of ω. We thus have

$$d\alpha=-(\alpha-\beta)\frac{\omega}{m} \text{ and } d\beta=(\alpha-\beta)\frac{\omega}{m};$$

the mass which had the initial temperature β has received in one instant a quantity of heat equal to $md\beta$ or $(\alpha-\beta)\,\omega$, which has been lost in the same time by the first mass. We see by this that the quantity of heat which passes in one instant from the most heated body into that which is less heated, is, all other things being equal, proportional to the actual difference of temperature of the two bodies. The time being divided into equal intervals, the infinitely small quantity ω may be replaced by kdt, k being the number of units of mass whose sum contains ω as many times as the unit of time contains dt, so that we have $\dfrac{k}{\omega}=\dfrac{1}{dt}$. We thus obtain the equations

$$d\alpha=-(\alpha-\beta)\frac{k}{m}dt \text{ and } d\beta=(\alpha-\beta)\frac{k}{m}dt.$$

248. If we attributed a greater value to the volume ω, which serves, it may be said, to draw heat from one of the bodies for the purpose of carrying it to the other, the transfer would be quicker; in order to express this condition it would be necessary to increase in the same ratio the quantity k which enters into the equations. We might also retain the value of ω and suppose the layer to accomplish in a given time a greater number of oscillations, which again would be indicated by a greater value of k. Hence this coefficient represents in some respects the velocity of transmission, or the facility with which heat passes from one of the bodies into the other, that is to say, their reciprocal conducibility.

249. Adding the two preceding equations, we have

$$d\alpha + d\beta = 0,$$

and if we subtract one of the equations from the other, we have

$$d\alpha - d\beta + 2(\alpha - \beta)\frac{k}{m}dt = 0, \text{ and, making } \alpha - \beta = y,$$

$$dy + 2\frac{k}{m}ydt = 0.$$

Integrating and determining the constant by the condition that the initial value is $a - b$, we have $y = (a-b)e^{-\frac{2kt}{m}}$. The difference y of the temperatures diminishes as the ordinate of a logarithmic curve, or as the successive powers of the fraction $e^{-\frac{2k}{m}}$. As the values of α and β, we have

$$\alpha = \frac{1}{2}(a+b) - \frac{1}{2}(a-b)e^{-\frac{2kt}{m}}, \quad \beta = \frac{1}{2}(a+b) + \frac{1}{2}(a-b)e^{-\frac{2kt}{m}}.$$

250. In the preceding case, we suppose the infinitely small mass ω, by means of which the transfer is effected, to be always the same part of the unit of mass, or, which is the same thing, we suppose the coefficient k which measures the reciprocal conducibility to be a constant quantity. To render the investigation in question more general, the constant k must be considered as a function of the two actual temperatures α and β. We should then have the two equations $d\alpha = -(\alpha - \beta)\frac{k}{m}dt$, and

$$d\beta = (\alpha - \beta)\frac{k}{m}\,dt,$$

in which k would be equal to a function of α and β, which we denote by $\phi(\alpha, \beta)$. It is easy to ascertain the law which the variable temperatures α and β follow, when they approach extremely near to their final state. Let y be a new unknown equal to the difference between α and the final value which is $\frac{1}{2}(a+b)$ or c. Let z be a second unknown equal to the difference $c-\beta$. We substitute in place of α and β their values $c-y$ and $c-z$; and, as the problem is to find the values of y and z, when we suppose them very small, we need retain in the results of the substitutions only the first power of y and z. We therefore find the two equations,

$$-dy = -(z-y)\frac{k}{m}\phi(c-y,\ c-z)\,dt$$

and $$-dz = \frac{k}{m}(z-y)\,\phi(c-y,\ c-z)\,dt,$$

developing the quantities which are under the sign ϕ and omitting the higher powers of y and z. We find $dy = (z-y)\frac{k}{m}\phi dt$, and $dz = -(z-y)\frac{k}{m}\phi dt$. The quantity ϕ being constant, it follows that the preceding equations give for the value of the difference $z-y$, a result similar to that which we found above for the value of $\alpha-\beta$.

From this we conclude that if the coefficient k, which was at first supposed constant, were represented by any function whatever of the variable temperatures, the final changes which these temperatures would experience, during an infinite time, would still be subject to the same law as if the reciprocal conducibility were constant. The problem is actually to determine the laws of the propagation of heat in an indefinite number of equal masses whose actual temperatures are different.

251. Prismatic masses n in number, each of which is equal to m, are supposed to be arranged in the same straight line, and affected with different temperatures a, b, c, d, &c.; infinitely

thin layers, each of which has a mass ω, are supposed to be separated from the different bodies except the last, and are conveyed in the same time from the first to the second, from the second to the third, from the third to the fourth, and so on; immediately after contact, these layers return to the masses from which they were separated; the double movement taking place as many times as there are infinitely small instants dt; it is required to find the law to which the changes of temperature are subject.

Let $\alpha, \beta, \gamma, \delta, \ldots \omega$, be the variable values which correspond to the same time t, and which have succeeded to the initial values a, b, c, d, &c. When the layers ω have been separated from the $\overline{n-1}$ first masses, and put in contact with the neighbouring masses, it is easy to see that the temperatures become

$$\frac{\alpha(m-\omega)}{m-\omega}, \quad \frac{\beta(m-\omega)+\alpha\omega}{m}, \quad \frac{\gamma(m-\omega)+\beta\omega}{m},$$

$$\frac{\delta(m-\omega)+\gamma\omega}{m}, \ldots \quad \frac{m\omega+\psi\omega}{m+\omega};$$

or,

$$\alpha, \quad \beta+(\alpha-\beta)\frac{\omega}{m}, \quad \gamma+(\beta-\gamma)\frac{\omega}{m}, \quad \delta+(\gamma-\delta)\frac{\omega}{m}, \ldots \omega+(\psi-\omega)\frac{\omega}{m}.$$

When the layers ω have returned to their former places, we find new temperatures according to the same rule, which consists in dividing the sum of the quantities of heat by the sum of the masses, and we have as the values of $\alpha, \beta, \gamma, \delta$, &c., after the instant dt,

$$\alpha-(\alpha-\beta)\frac{\omega}{m}, \quad \beta+(\alpha-\beta-\overline{\beta-\gamma})\frac{\omega}{m},$$

$$\gamma+(\beta-\gamma-\overline{\gamma-\delta})\frac{\omega}{m}, \ldots \omega+(\psi-\omega)\frac{\omega}{m}.$$

The coefficient of $\frac{\omega}{m}$ is the difference of two consecutive differences taken in the succession $\alpha, \beta, \gamma, \ldots \psi, \omega$. As to the first and last coefficients of $\frac{\omega}{m}$, they may be considered also as differences of the second order. It is sufficient to suppose the term α to be preceded by a term equal to α, and the term ω to be followed by a term equal to ω. We have then, as formerly, on substituting kdt for ω, the following equations:

$$d\alpha = \frac{k}{m} dt \{(\beta - \alpha) - (\alpha - \alpha)\},$$

$$d\beta = \frac{k}{m} dt \{(\gamma - \beta) - (\beta - \alpha)\},$$

$$d\gamma = \frac{k}{m} dt \{(\delta - \gamma) - (\gamma - \beta)\},$$

....

$$d\omega = \frac{k}{m} dt \{(\omega - \omega) - (\omega - \psi)\}.$$

252. To integrate these equations, we assume, according to the known method,

$$\alpha = a_1 e^{ht}, \quad \beta = a_2 e^{ht}, \quad \gamma = a_3 e^{ht}, \ldots \quad \omega = a_n e^{ht};$$

$h_1, a_1, a_2, a_3, \ldots a_n$, being constant quantities which must be determined. The substitutions being made, we have the following equations:

$$a_1 h = \frac{k}{m} (a_2 - a_1),$$

$$a_2 h = \frac{k}{m} \{(a_3 - a_2) - (a_2 - a_1)\},$$

$$a_3 h = \frac{k}{m} \{(a_4 - a_3) - (a_3 - a_2)\},$$

....

$$a_n h = \frac{k}{m} \{(a_{n+1} - a_n) - (a_n - a_{n-1})\}.$$

If we regard a_1 as a known quantity, we find the expression for a_2 in terms of a_1 and h, then that of a_3 in a_2 and h; the same is the case with all the other unknowns, a_4, a_5, &c. The first and last equations may be written under the form

$$a_1 h = \frac{k}{m} \{(a_2 - a_1) - (a_1 - a_0)\},$$

and

$$a_n h = \frac{k}{m} \{(a_{n+1} - a_n) - (a_n - a_{n-1})\}.$$

Retaining the two conditions $a = a_1$ and $a_n = a_{n+1}$, the value of a_2 contains the first power of h, the value of a_3 contains the second power of h, and so on up to a_{n+1}, which contains the nth power of h. This arranged, a_{n+1} becoming equal to a_n, we have, to determine h, an equation of the nth degree, and a_1 remains undetermined.

It follows from this that we shall find n values for h, and in accordance with the nature of linear equations, the general value of α is composed of n terms, so that the quantities $\alpha, \beta, \gamma, \ldots$ &c. are determined by means of equations such as

$$\alpha = a_1 e^{ht} + a_1' e^{h't} + a_1'' e^{h''t} + \text{\&c.},$$

$$\beta = a_2 e^{ht} + a_2' e^{h't} + a_2'' e^{h''t} + \text{\&c.},$$

$$\gamma = a_3 e^{ht} + a_3' e^{h't} + a_3'' e^{h''t} + \text{\&c.},$$

$$\cdots\cdots\cdots\cdots\cdots\cdots$$

$$\omega = a_n e^{ht} + a_n' e^{h't} + a_n'' e^{h''t} + \text{\&c.}$$

The values of h, h', h'', &c. are n in number, and are equal to the n roots of the algebraical equation of the nth degree in h, which has, as we shall see further on, all its roots real.

The coefficients of the first equation a_1, a_1', a_1'', a_1''', &c., are arbitrary; as for the coefficients of the lower lines, they are determined by a number n of systems of equations similar to the preceding equations. The problem is now to form these equations.

253. Writing the letter q instead of $\dfrac{hm}{k}$, we have the following equations

$$a_0 = a_0,$$

$$a_1 = a_1,$$

$$a_2 = a_1(q+2) - a_0,$$

$$a_3 = a_2(q+2) - a_1,$$

$$\cdots\cdots\cdots\cdots$$

$$a_{n+1} = a_n(q+2) - a_{n-1}.$$

We see that these quantities belong to a recurrent series whose scale of relation consists of two terms $(q+2)$ and -1. We can therefore express the general term a_m by the equation

$$a_m = A \sin mu + B \sin (m-1) u,$$

determining suitably the quantities A, B, and u. First we find A and B by supposing m equal to 0 and then equal to 1, which gives $a_0 = B \sin u$, and $a_1 = A \sin u$, and consequently

$$a_m = a_1 \sin mu - \frac{a_1}{\sin u} \sin (m-1) u.$$

Substituting then the values of

$$a_m, a_{m-1}, a_{m-2}, \&c.$$

in the general equation

$$a_m = a_{m-1}(q+2) - a_{m-2},$$

we find

$$\sin mu = (q+2) \sin (m-1) u - \sin (m-2) u,$$

comparing which equation with the next,

$$\sin mu = 2 \cos u \sin (m-1) u - \sin (m-2) u,$$

which expresses a known property of the sines of arcs increasing in arithmetic progression, we conclude that $q + 2 = \cos u$, or $q = -2 \operatorname{versin} u$; it remains only to determine the value of the arc u.

The general value of a_m being

$$\frac{a_1}{\sin u} [\sin mu - \sin (m-1) u],$$

we must have, in order to satisfy the condition $a_{n+1} = a_n$, the equation

$$\sin (n+1) u - \sin u = \sin nu - \sin (n-1) u,$$

whence we deduce $\sin nu = 0$, or $u = i\frac{\pi}{n}$, π being the semi-circumference and i any integer, such as 0, 1, 2, 3, 4, ... $(n-1)$; thence we deduce the n values of q or $\frac{hm}{k}$. Thus all the roots of the equation in h, which give the values of h, h', h'', h''', &c. are real and negative, and are furnished by the equations

$$\dot{h} = -2\frac{k}{m}\operatorname{versin}\left(0\,\frac{\pi}{n}\right),$$

$$h' = -2\frac{k}{m}\operatorname{versin}\left(1\,\frac{\pi}{n}\right),$$

$$h'' = -2\frac{k}{m}\operatorname{versin}\left(2\,\frac{\pi}{n}\right),$$

$$\ldots = \ldots\ldots\ldots\ldots\ldots$$

$$h^{(n-1)} = -2\frac{k}{m}\operatorname{versin}\left\{(n-1)\,\frac{\pi}{n}\right\}.$$

Suppose then that we have divided the semi-circumference π into n equal parts, and that in order to form u, we take i of those parts, i being less than n, we shall satisfy the differential equations by taking a_1 to be any quantity whatever, and making

$$\alpha = a_1\frac{\sin u - \sin 0u}{\sin u}e^{-\frac{2kt}{m}\operatorname{versin} u},$$

$$\beta = a_1\frac{\sin 2u - \sin 1u}{\sin u}e^{-\frac{2kt}{m}\operatorname{versin} u},$$

$$\gamma = a_1\frac{\sin 3u - \sin 2u}{\sin u}e^{-\frac{2kt}{m}\operatorname{versin} u},$$

$$\ldots\ldots\ldots\ldots\ldots\ldots\ldots\ldots$$

$$\omega = a_1\frac{\sin nu - \sin(n-1)u}{\sin u}e^{-\frac{2kt}{m}\operatorname{versin} u}.$$

As there are n different arcs which we may take for u, namely,

$$0\frac{\pi}{n},\quad 1\frac{\pi}{n},\quad 2\frac{\pi}{n},\quad \ldots\ldots,\quad (n-1)\frac{\pi}{n},$$

there are also n systems of particular values for α, β, γ, &c., and the general values of these variables are the sums of the particular values.

254. We see first that if the arc u is nothing, the quantities which multiply a, in the values of α, β, γ, &c., become all equal to unity, since $\dfrac{\sin u - \sin 0u}{\sin u}$ takes the value 1 when the arc u vanishes; and the same is the case with the quantities which are

found in the following equations. From this we conclude that constant terms must enter into the general values of $\alpha, \beta, \gamma, \ldots \omega$.

Further, adding all the particular values corresponding to $\alpha, \beta, \gamma, \ldots$ &c., we have

$$\alpha+\beta+\gamma+\&c. = a_1 \frac{\sin nu}{\sin u} e^{-\frac{2kt}{m}\operatorname{versin} u};$$

an equation whose second member is reduced to 0 provided the arc u does not vanish; but in that case we should find n to be the value of $\frac{\sin nu}{\sin u}$. We have then in general

$$\alpha+\beta+\gamma+\&c. = na_1;$$

now the initial values of the variables being a, b, c, &c., we must necessarily have

$$na_1 = a+b+c+\&c.;$$

it follows that the constant term which must enter into each of the general values of

$$\alpha, \beta, \gamma, \ldots \omega \text{ is } \frac{1}{n}(a+b+c+\&c.),$$

that is to say, the mean of all the initial temperatures.

As to the general values of $\alpha, \beta, \gamma, \ldots \omega$, they are expressed by the following equations:

$$\begin{aligned}
\alpha = \frac{1}{n}(a+b+c+\&c.) &+ a_1 \frac{\sin u - \sin 0u}{\sin u} e^{-\frac{2kt}{m}\operatorname{versin} u} \\
&+ b_1 \frac{\sin u' - \sin 0u'}{\sin u'} e^{-\frac{2kt}{m}\operatorname{versin} u'} \\
&+ c_1 \frac{\sin u'' - \sin 0u''}{\sin u''} e^{-\frac{2kt}{m}\operatorname{versin} u''} \\
&+ \&c.,
\end{aligned}$$

$$\begin{aligned}
\beta = \frac{1}{n}(a+b+c+\&c.) &+ a_1 \frac{\sin 2u - \sin u}{\sin u} e^{-\frac{2kt}{m}\operatorname{versin} u} \\
&+ b_1 \frac{\sin 2u' - \sin u'}{\sin u'} e^{-\frac{2kt}{m}\operatorname{versin} u'} \\
&+ c_1 \frac{\sin 2u'' - \sin u''}{\sin u''} e^{-\frac{2kt}{m}\operatorname{versin} u''} \\
&+ \&c.,
\end{aligned}$$

$$\gamma = \frac{1}{n}(a+b+c+\&c.) + a_1 \frac{\sin 3u - \sin 2u}{\sin u} e^{-\frac{2kt}{m}\operatorname{versin} u}$$

$$+ b_1 \frac{\sin 3u' - \sin 2u'}{\sin u'} e^{-\frac{2kt}{m}\operatorname{versin} u'}$$

$$+ c_1 \frac{\sin 3u'' - \sin 2u''}{\sin u''} e^{-\frac{2kt}{m}\operatorname{versin} u''}$$

$$+ \&c.,$$

$$\omega = \frac{1}{n}(a+b+c+\&c.) + a_1 \left(\frac{\sin nu - \sin (n-1) u}{\sin u}\right) e^{-\frac{2kt}{m}\operatorname{versin} u}$$

$$+ b_1 \left(\frac{\sin nu' - \sin (n-1) u'}{\sin u'}\right) e^{-\frac{2kt}{m}\operatorname{versin} u'}$$

$$+ c_1 \left(\frac{\sin nu'' - \sin (n-1) u''}{\sin u''}\right) e^{-\frac{2kt}{m}\operatorname{versin} u''}$$

$$+ \&c.$$

255. To determine the constants a, b, c, d ... &c., we must consider the initial state of the system. In fact, when the time is nothing, the values of α, β, γ, &c. must be equal to a, b, c, &c.; we have then n similar equations to determine the n constants. The quantities

$$\sin u - \sin 0u,\ \sin 2u - \sin u,\ \sin 3u - \sin 2u, \ldots,\ \sin nu - \sin (n-1) u,$$

may be indicated in this manner,

$$\Delta \sin 0u,\ \Delta \sin u,\ \Delta \sin 2u,\ \Delta \sin 3u, \ldots \Delta \sin (n-1) u;$$

the equations proper for the determination of the constants are, if the initial mean temperature be represented by C,

$$a = C + a_1 + b_1 + c_1 + \&c.$$

$$b = C + a_1 \frac{\Delta \sin u}{\sin u} + b_1 \frac{\Delta \sin u'}{\sin u'} + c_1 \frac{\Delta \sin u''}{\sin u''} + \&c.,$$

$$c = C + a_1 \frac{\Delta \sin 2u}{\sin u} + b_1 \frac{\Delta \sin 2u'}{\sin u'} + c \frac{\Delta \sin 2u''}{\sin u''} + \&c.,$$

$$d = C + a_1 \frac{\Delta \sin 3u}{\sin u} + b \frac{\Delta \sin 3u'}{\sin u'} + c \frac{\Delta \sin 3u''}{\sin u''} + \&c.,$$

&c.

The quantities a_1, b_1, c_1, d_1, and C being determined by these equations, we know completely the values of the variables

$$\alpha,\ \beta,\ \gamma,\ \delta,\ \ldots\ \omega.$$

We can in general effect the elimination of the unknowns in these equations, and determine the values of the quantities a, b, c, d, &c., even when the number of equations is infinite; we shall employ this process of elimination in the following articles.

256. On examining the equations which give the general values of the variables $\alpha, \beta, \gamma \ldots\ldots \omega$, we see that as the time increases the successive terms in the value of each variable decrease very unequally: for the values of u, u', u'', u''', &c. being

$$1\frac{\pi}{n},\ 2\frac{\pi}{n},\ 3\frac{\pi}{n},\ 4\frac{\pi}{n},\ \text{\&c.},$$

the exponents versin u, versin u', versin u'', versin u''', &c. become greater and greater. If we suppose the time t to be infinite, the first term of each value alone exists, and the temperature of each of the masses becomes equal to the mean temperature $\frac{1}{n}(a+b+c+\ldots\text{\&c.})$. Since the time t continually increases, each of the terms of the value of one of the variables diminishes proportionally to the successive powers of a fraction which, for the second term, is $e^{-\frac{2k}{m}\text{versin}\,u}$, for the third term $e^{-\frac{2k}{m}\text{versin}\,u'}$, and so on. The greatest of these fractions being that which corresponds to the least of the values of u, it follows that to ascertain the law which the ultimate changes of temperature follow, we need consider only the two first terms; all the others becoming incomparably smaller according as the time t increases. The ultimate variations of the temperatures α, β, γ, &c. are therefore expressed by the following equations:

$$\alpha = \frac{1}{n}(a+b+c+\text{\&c.}) + a_1\frac{\sin u - \sin 0u}{\sin u}e^{-\frac{2kt}{m}\text{versin}\,u},$$

$$\beta = \frac{1}{n}(a+b+c+\text{\&c.}) + a_1\frac{\sin 2u - \sin u}{\sin u}e^{-\frac{2kt}{m}\text{versin}\,u},$$

$$\gamma = \frac{1}{n}(a+b+c+\text{\&c.}) + a_1\frac{\sin 3u - \sin 2u}{\sin u}e^{-\frac{2kt}{m}\text{versin}\,u}.$$

257. If we divide the semi-circumference into n equal parts, and, having drawn the sines, take the difference between two consecutive sines, the n differences are proportional to the coefficients of $e^{-\frac{2kt}{m}\operatorname{versin} u}$, or to the second terms of the values of $\alpha, \beta, \gamma, \ldots \omega$. For this reason the later values of $\alpha, \beta, \gamma \ldots \omega$ are such that the differences between the final temperatures and the mean initial temperature $\frac{1}{n}(a+b+c+\&c.)$ are always proportional to the differences of consecutive sines. In whatever manner the masses have first been heated, the distribution of heat is effected finally according to a constant law. If we measured the temperatures in the last stage, when they differ little from the mean temperature, we should observe that the difference between the temperature of any mass whatever and the mean temperature decreases continually according to the successive powers of the same fraction; and comparing amongst themselves the temperatures of the different masses taken at the same instant, we should see that the differences between the actual temperatures and the mean temperature are proportional to the differences of consecutive sines, the semi-circumference having been divided into n equal parts.

258. If we suppose the masses which communicate heat to each other to be infinite in number, we find for the arc u an infinitely small value; hence the differences of consecutive sines, taken on the circle, are proportional to the cosines of the corresponding arcs; for $\frac{\sin mu - \sin(m-1)u}{\sin u}$ is equal to $\cos mu$, when the arc u is infinitely small. In this case, the quantities whose temperatures taken at the same instant differ from the mean temperature to which they all must tend, are proportional to the cosines which correspond to different points of the circumference divided into an infinite number of equal parts. If the masses which transmit heat are situated at equal distances from each other on the perimeter of the semi-circumference π, the cosine of the arc at the end of which any one mass is placed is the measure of the quantity by which the temperature of that mass differs yet from the mean temperature. Thus the body placed in the middle of all the others is that which arrives most quickly at that mean

temperature; those which are situated on one side of the middle, all have an excessive temperature, which surpasses the mean temperature the more, according as they are more distant from the middle; the bodies which are placed on the other side, all have a temperature lower than the mean temperature, and they differ from it as much as those on the opposite side, but in contrary sense. Lastly, these differences, whether positive or negative, all decrease at the same time, proportionally to the successive powers of the same fraction; so that they do not cease to be represented at the same instant by the values of the cosines of the same semi-circumference. Such in general, singular cases excepted, is the law to which the ultimate temperatures are subject. The initial state of the system does not change these results. We proceed now to deal with a third problem of the same kind as the preceding, the solution of which will furnish us with many useful remarks.

259. Suppose n equal prismatic masses to be placed at equal distances on the circumference of a circle. All these bodies, enjoying perfect conducibility, have known actual temperatures, different for each of them; they do not permit any part of the heat which they contain to escape at their surface; an infinitely thin layer is separated from the first mass to be united to the second, which is situated towards the right; at the same time a parallel layer is separated from the second mass, carried from left to right, and joined to the third; the same is the case with all the other masses, from each of which an infinitely thin layer is separated at the same instant, and joined to the following mass. Lastly, the same layers return immediately afterwards, and are united to the bodies from which they had been detached.

Heat is supposed to be propagated between the masses by means of these alternate movements, which are accomplished twice during each instant of equal duration; the problem is to find according to what law the temperatures vary: that is to say, the initial values of the temperatures being given, it is required to ascertain after any given time the new temperature of each of the masses.

We shall denote by $a_1, a_2, a_3, \ldots a_i \ldots a_n$ the initial temperatures whose values are arbitrary, and by $\alpha_1, \alpha_2, \alpha_3 \ldots \alpha_i \ldots \alpha_n$ the values of the same temperatures after the time t has elapsed. Each of the quantities a is evidently a function of the time t and of all the initial values $a_1, a_2, a_3 \ldots a_n$: it is required to determine the functions α.

260. We shall represent the infinitely small mass of the layer which is carried from one body to the other by ω. We may remark, in the first place, that when the layers have been separated from the masses of which they have formed part, and placed respectively in contact with the masses situated towards the right, the quantities of heat contained in the different bodies become $(m-\omega)\,\alpha_1 + \omega\alpha_n$, $(m-\omega)\,\alpha_2 + \omega\alpha_1$, $(m-\omega)\,\alpha_3 + \omega\alpha_2, \ldots$, $(m-\omega)\,\alpha_n + \omega\alpha_{n-1}$; dividing each of these quantities of heat by the mass m, we have for the new values of the temperatures

$$\alpha_1 + \frac{\omega}{m}(\alpha_n - \alpha_1),\ \alpha_2 + \frac{\omega}{m}(\alpha_1 - \alpha_2),\ \alpha_3 + \frac{\omega}{m}(\alpha_2 - \alpha)_3,\ \ldots$$

$$\alpha_i + \frac{\omega}{m}(\alpha_{i-1} - \alpha_i), \ldots \text{ and } \alpha_n + \frac{\omega}{m}(\alpha_{n-1} - \alpha_n);$$

that is to say, to find the new state of the temperature after the first contact, we must add to the value which it had formerly the product of $\frac{\omega}{m}$ by the excess of the temperature of the body from which the layer has been separated over that of the body to which it has been joined. By the same rule it is found that the temperatures, after the second contact, are

$$\alpha_1 + \frac{\omega}{m}(\alpha_n - \alpha_1) + \frac{\omega}{m}(\alpha_2 - \alpha_1),$$

$$\alpha_2 + \frac{\omega}{m}(\alpha_1 - \alpha_2) + \frac{\omega}{m}(\alpha_3 - \alpha_2),$$

$$\alpha_i + \frac{\omega}{m}(\alpha_{i-1} - \alpha_i) + \frac{\omega}{m}(\alpha_{i+1} - \alpha_i),$$

$$\alpha_n + \frac{\omega}{m}(\alpha_{n-1} - \alpha_n) + \frac{\omega}{m}(\alpha_1 - \alpha_n).$$

The time being divided into equal instants, denote by dt the duration of the instant, and suppose ω to be contained in k units of mass as many times as dt is contained in the units of time, we thus have $\omega = k\,dt$. Calling $d\alpha_1, d\alpha_2, d\alpha_3 \ldots d\alpha_i \ldots d\alpha_n$ the infinitely small increments which the temperatures $\alpha_1, \alpha_2, \ldots \alpha_i \ldots \alpha_n$ receive during the instant dt, we have the following differential equations:

$$d\alpha_1 = \frac{k}{m} dt\,(\alpha_n - 2\alpha_1 + \alpha_2),$$

$$d\alpha_2 = \frac{k}{m} dt\,(\alpha_1 - 2\alpha_2 + \alpha_3),$$

$$d\alpha_i = \frac{k}{m} dt\,(\alpha_{i-1} - 2\alpha_i + \alpha_{i+1}),$$

$$d\alpha_{n-1} = \frac{k}{m} dt\,(\alpha_{n-2} - 2\alpha_{n-1} + \alpha_n),$$

$$d\alpha_n = \frac{k}{m} dt\,(\alpha_{n-1} - 2\alpha_n + \alpha_1).$$

261. To solve these equations, we suppose in the first place, according to the known method,

$$\alpha_1 = b_1 e^{ht}, \quad \alpha_2 = b_2 e^{ht}, \quad \alpha_i = b_i e^{ht}, \quad \alpha_n = b_n e^{ht}.$$

The quantities $b_1, b_2, b_3, \ldots b_n$ are undetermined constants, as also is the exponent h. It is easy to see that the values of $\alpha_1, \alpha_2, \ldots \alpha_n$ satisfy the differential equations if they are subject to the following conditions:

$$b_1 h = \frac{k}{m}(b_n - 2b_1 + b_2),$$

$$b_2 h = \frac{k}{m}(b_1 - 2b_2 + b_3),$$

$$b_i h = \frac{k}{m}(b_{i-1} - 2b_i + b_{i+1}),$$

$$b_{n-1} h = \frac{k}{m}(b_{n-2} - 2b_{n-1} + b_n),$$

$$b_n h = \frac{k}{m}(b_{n-1} - 2b_n + b_1).$$

Let $q = \frac{hm}{k}$, we have, beginning at the last equation,

$$b_1 = b_n(q+2) - b_{n-1},$$
$$b_2 = b_1(q+2) - b_n,$$
$$b_3 = b_2(q+2) - b_1,$$
$$b_i = b_{i-1}(q+2) - b_{i-2},$$
$$b_n = b_{n-1}(q+2) - b_{n-2}.$$

It follows from this that we may take, instead of $b_1, b_2, b_3, \ldots b_i, \ldots b_n$, the n consecutive sines which are obtained by dividing the whole circumference 2π into n equal parts. In fact, denoting the arc $2\frac{\pi}{n}$ by u, the quantities

$$\sin 0u, \ \sin 1u, \ \sin 2u, \ \sin 3u, \ldots, \ \sin (n-1)u,$$

whose number is n, belong, as it is said, to a recurring series whose scale of relation has two terms, $2 \cos u$ and -1: so that we always have the condition

$$\sin iu = 2 \cos u \sin (i-1) u - \sin (i-2) u.$$

Take then, instead of $b_1, b_2, b_3, \ldots b_n$, the quantities

$$\sin 0u, \ \sin 1u, \ \sin 2u, \ldots \sin (n-1) u,$$

and we have

$$q + 2 = 2 \cos u, \quad q = -2 \operatorname{versin} u, \quad \text{or } q = -2 \operatorname{versin} \frac{2\pi}{n}.$$

We have previously written q instead of $\frac{hm}{k}$, so that the value of h is $-\frac{2k}{m} \operatorname{versin} \frac{2\pi}{n}$; substituting in the equations these values of b_i and h we have

$$\alpha_1 = \sin 0u e^{-\frac{2kt}{m} \operatorname{versin} \frac{2\pi}{n}},$$

$$\alpha_2 = \sin 1u e^{-\frac{2kt}{m} \operatorname{versin} \frac{2\pi}{n}},$$

$$\alpha_3 = \sin 2u e^{-\frac{2kt}{m} \operatorname{versin} \frac{2\pi}{n}},$$

$$\ldots\ldots\ldots\ldots\ldots$$

$$\alpha_n = \sin (n-1) u e^{-\frac{2kt}{m} \operatorname{versin} \frac{2\pi}{n}}.$$

262. The last equations furnish only a very particular solution of the problem proposed; for if we suppose $t = 0$ we have, as the initial values of $\alpha_1, \alpha_2, \alpha_3, \ldots \alpha_n$, the quantities

$$\sin 0u, \ \sin 1u, \ \sin 2u, \ldots \sin (n-1) u,$$

which in general differ from the given values $a_1, a_2, a_3, \ldots a_n$: but the foregoing solution deserves to be noticed because it expresses, as we shall see presently, a circumstance which belongs to all possible cases, and represents the ultimate variations of the temperatures. We see by this solution that, if the initial temperatures $a_1, a_2, a_3, \ldots a_n$, were proportional to the sines

$$\sin 0\frac{2\pi}{n}, \quad \sin 1\frac{2\pi}{n}, \quad \sin 2\frac{2\pi}{n}, \ \ldots \ \sin(n-1)\frac{2\pi}{n},$$

they would remain continually proportional to the same sines, and we should have the equations

$$\left.\begin{array}{l} a_1 = a_1 e^{-ht}, \\ a_2 = a_2 e^{-ht}, \\ a_3 = a_3 e^{-ht}, \\ \ldots\ldots\ldots \\ a_n = a_n e^{-ht}, \end{array}\right\} \text{ where } h = \frac{2k}{m}\operatorname{versin}\frac{2\pi}{n}.$$

For this reason, if the masses which are situated at equal distances on the circumference of a circle had initial temperatures proportional to the perpendiculars let fall on the diameter which passes through the first point, the temperatures would vary with the time, but remain always proportional to those perpendiculars, and the temperatures would diminish simultaneously as the terms of a geometrical progression whose ratio is the fraction $e^{-\frac{2k}{m}\operatorname{versin}\frac{2\pi}{n}}$.

263. To form the general solution, we may remark in the first place that we could take, instead of $b_1, b_2, b_3, \ldots b_n$, the n cosines corresponding to the points of division of the circumference divided into n equal parts. The quantities $\cos 0u$, $\cos 1u$, $\cos 2u, \ldots \cos(n-1)u$, in which u denotes the arc $\frac{2\pi}{n}$, form also a recurring series whose scale of relation consists of two terms, $2\cos u$ and -1, for which reason we could satisfy the differential equations by means of the following equations,

$$\begin{aligned} a_1 &= \cos 0u\, e^{-\frac{2kt}{m}\operatorname{versin} u}, \\ a_2 &= \cos 1u\, e^{-\frac{2kt}{m}\operatorname{versin} u}, \\ a_3 &= \cos 2u\, e^{-\frac{2kt}{m}\operatorname{versin} u}, \\ &\ldots\ldots\ldots\ldots\ldots \\ a_n &= \cos(n-1)u\, e^{-\frac{2kt}{m}\operatorname{versin} u}. \end{aligned}$$

Independently of the two preceding solutions we could select for the values of $b_1, b_2, b_3, \ldots b_n$, the quantities

$$\sin 0.2u,\ \sin 1.2u,\ \sin 2.2u,\ \sin 3.2u, \ldots,\ \sin (n-1)\,2u;$$

or else

$$\cos 0.2u,\ \cos 1.2u,\ \cos 2.2u,\ \cos 3.2u, \ldots,\ \cos (n-1)\,2u.$$

In fact, each of these series is recurrent and composed of n terms; in the scale of relation are two terms, $2\cos 2u$ and -1; and if we continued the series beyond n terms, we should find n others respectively equal to the n preceding.

In general, if we denote the arcs

$$0\frac{2\pi}{n},\quad 1\frac{2\pi}{n},\quad 2\frac{2\pi}{n}, \ldots,\quad (n-1)\frac{2\pi}{n},\quad \text{\&c.},$$

by $u_1, u_2, u_3, \ldots, u_n$, we can take for the values of $b_1, b_2, b_3, \ldots b_n$ the n quantities,

$$\sin 0u_i,\ \sin 1u_i,\ \sin 2u_i,\ \sin 3u_i, \ldots,\ \sin (n-1)\,u_i;$$

or else

$$\cos 0u_i,\ \cos 1u_i,\ \cos 2u_i,\ \cos 3u_i, \ldots,\ \cos (n-1)\,u_i.$$

The value of h corresponding to each of these series is given by the equation

$$h = -\frac{2k}{m} \operatorname{versin} u_i.$$

We can give n different values to i, from $i=1$ to $i=n$.

Substituting these values of $b_1, b_2, b_3 \ldots b_n$, in the equations of Art. 261, we have the differential equations of Art. 260 satisfied by the following results:

$$\alpha_1 = \sin 0u_i e^{-\frac{2kt}{m}\operatorname{versin} u_i}, \qquad \text{or } \alpha_1 = \cos 0u_i e^{-\frac{2kt}{m}\operatorname{versin} u_i},$$

$$\alpha_2 = \sin 1u_i e^{-\frac{2kt}{m}\operatorname{versin} u_i}, \qquad \alpha_2 = \cos 1u_i e^{-\frac{2kt}{m}\operatorname{versin} u_i},$$

$$\alpha_3 = \sin 2u_i e^{-\frac{2kt}{m}\operatorname{versin} u_i}, \qquad \alpha_3 = \cos 2u_i e^{-\frac{2kt}{m}\operatorname{versin} u_i},$$

$$\cdots\cdots\cdots\cdots \qquad \cdots\cdots\cdots\cdots$$

$$\alpha_n = \sin (n-1)\,u_i e^{-\frac{2kt}{m}\operatorname{versin} u_i}, \qquad \alpha_n = \cos (n-1)\,u_i e^{-\frac{2kt}{m}\operatorname{versin} u_i}.$$

264. The equations of Art. 260 could equally be satisfied by constructing the values of each one of the variables $\alpha_1, \alpha_2, \alpha_3, \ldots \alpha_n$ out of the sum of the several particular values which have been found for that variable; and each one of the terms which enter into the general value of one of the variables may also be multiplied by any constant coefficient. It follows from this that, denoting by $A_1, B_1, A_2, B_2, A_3, B_3, \ldots A_n, B_n$, any coefficients whatever, we may take to express the general value of one of the variables, α_{m+1} for example, the equation

$$\alpha_{m+1} = (A_1 \sin mu_1 + B_1 \cos mu_2)\, e^{-\frac{2kt}{m}\operatorname{versin} u_1}$$
$$+ (A_1 \sin mu_2 + B_2 \cos mu_2)\, e^{-\frac{2kt}{m}\operatorname{versin} u_2}$$
$$\ldots\ldots\ldots\ldots$$
$$+ (A_n \sin mu_n + B_n \cos mu_n)\, e^{-\frac{2kt}{m}\operatorname{versin} u_n}.$$

The quantities $A_1, A_2, A_3, \ldots A_n$, $B_1, B_2, B_3, \ldots B_n$, which enter into this equation, are arbitrary, and the arcs $u_1, u_2, u_3, \ldots u_n$ are given by the equations:

$$u_1 = 0\frac{2\pi}{n}, \quad u_2 = 1\frac{2\pi}{n}, \quad u_3 = 2\frac{2\pi}{n}, \quad \ldots, \quad u_n = (n-1)\frac{2\pi}{n}.$$

The general values of the variables $\alpha_1, \alpha_2, \alpha_3, \ldots \alpha_n$ are then expressed by the following equations:

$$\alpha_1 = (A_1 \sin 0u_1 + B_1 \cos 0u_1)\, e^{-\frac{2kt}{m}\operatorname{versin} u_1}$$
$$+ (A_2 \sin 0u_2 + B_2 \cos 0u_2)\, e^{-\frac{2k}{m}t\operatorname{versin} u_2}$$
$$+ (A_3 \sin 0u_3 + B_3 \cos 0u_3)\, e^{-\frac{2kt}{m}\operatorname{versin} u_3}$$
$$+ \&c.;$$

$$\alpha_2 = (A_1 \sin 1u_1 + B_1 \cos 1u_1)\, e^{-\frac{2kt}{m}\operatorname{versin} u_1}$$
$$+ (A_2 \sin 1u_2 + B_2 \cos 1u_2)\, e^{-\frac{2kt}{m}\operatorname{versin} u_2}$$
$$+ (A_3 \sin 1u_3 + B_3 \cos 1u_3)\, e^{-\frac{2kt}{m}\operatorname{versin} u_3}$$
$$+ \&c.;$$

$$\alpha_3 = (A_1 \sin 2u_1 + B_1 \cos 2u_1)\, e^{-\frac{2kt}{m}\operatorname{versin} u_1}$$
$$+ (A_2 \sin 2u_2 + B_2 \cos 2u_2)\, e^{-\frac{2kt}{m}\operatorname{versin} u_2}$$
$$+ (A_3 \sin 2u_3 + B_3 \cos 2u_3)\, e^{-\frac{2kt}{m}\operatorname{versin} u_3}$$
$$+ \&c.;$$

$$\alpha_n = \{A_1 \sin (n-1)\, u_1 + B_1 \cos (n-1)\, u_1\}\, e^{-\frac{2kt}{m}\operatorname{versin} u_1}$$
$$+ \{A_2 \sin (n-1)\, u_2 + B_2 \cos (n-1)\, u_2\}\, e^{-\frac{2kt}{m}\operatorname{versin} u_2}$$
$$+ \{A_3 \sin (n-1)\, u_3 + B_3 \cos (n-1)\, u_3\}\, e^{-\frac{2kt}{m}\operatorname{versin} u_3}$$
$$+ \&c.$$

265. If we suppose the time nothing, the values $\alpha_1, \alpha_2, \alpha_3, \ldots \alpha_n$ must become the same as the initial values $a_1, a_2, a_3, \ldots a_n$. We derive from this n equations, which serve to determine the coefficients $A_1, B_1, A_2, B_2, A_3, B_3, \ldots$. It will readily be perceived that the number of unknowns is always equal to the number of equations. In fact, the number of terms which enter into the value of one of these variables depends on the number of different quantities versin u_1, versin u_2, versin u_3, &c., which we find on dividing the circumference 2π into n equal parts. Now the number of quantities versin $0\frac{2\pi}{n}$, versin $1\frac{2\pi}{n}$, versin $2\frac{2\pi}{n}$, &c., is very much less than n, if we count only those that are different. Denoting the number n by $2i+1$ if it is odd, and by $2i$ if it is even, $i+1$ always denotes the number of different versed sines. On the other hand, when in the series of quantities versin $0\frac{2\pi}{n}$, versin $1\frac{2\pi}{n}$, versin $2\frac{2\pi}{n}$, &c., we come to a versed sine, versin $\lambda\frac{2\pi}{n}$, equal to one of the former versin $\lambda'\frac{2\pi}{n}$, the two terms of the equations which contain this versed sine form only one term; the two different arcs u_λ and $u_{\lambda'}$, which have the same versed sine, have also the same cosine, and the sines differ only in sign. It is easy to see that the arcs u_λ and $u_{\lambda'}$, which have the same versed sine, are such that

the cosine of any multiple whatever of u_λ is equal to the cosine of the same multiple of $u_{\lambda'}$, and that the sine of any multiple of u_λ differs only in sign from the sine of the same multiple of $u_{\lambda'}$. It follows from this that when we unite into one the two corresponding terms of each of the equations, the two unknowns A_λ and $A_{\lambda'}$, which enter into these equations, are replaced by a single unknown, namely $A_\lambda - A_{\lambda'}$. As to the two unknown B_λ and $B_{\lambda'}$ they also are replaced by a single one, namely $B_\lambda + B_{\lambda'}$: it follows from this that the number of unknowns is equal in all cases to the number of equations; for the number of terms is always $i+1$. We must add that the unknown A disappears of itself from the first terms, since it is multiplied by the sine of a nul arc. Further, when the number n is even, there is found at the end of each equation a term in which one of the unknowns disappears of itself, since it multiplies a nul sine; thus the number of unknowns which enter into the equations is equal to $2(i+1)-2$, when the number n is even; consequently the number of unknowns is the same in all these cases as the number of equations.

266. To express the general values of the temperatures $\alpha_1, \alpha_2, \alpha_3 \ldots \alpha_n$, the foregoing analysis furnishes us with the equations

$$\alpha_1 = \left(A_1 \sin 0.0\frac{2\pi}{n} + B_1 \cos 0.0\frac{2\pi}{n}\right) e^{-\frac{2kt}{m}\text{ versin } 0\frac{2\pi}{n}}$$

$$+\left(A_2 \sin 0.1\frac{2\pi}{n} + B_2 \cos 0.1\frac{2\pi}{n}\right) e^{-\frac{2kt}{m}\text{ versin } 1\frac{2\pi}{n}}$$

$$+\left(A_3 \sin 0.2\frac{2\pi}{n} + B_3 \cos 0.2\frac{2\pi}{n}\right) e^{-\frac{2kt}{m}\text{ versin } 2\frac{2\pi}{n}}$$

$$+ \&c.,$$

$$\alpha_2 = \left(A_1 \sin 1.0\frac{2\pi}{n} + B_1 \cos 1.0\frac{2\pi}{n}\right) e^{-\frac{2kt}{m}\text{ versin } 0\frac{2\pi}{n}}$$

$$+\left(A_2 \sin 1.1\frac{2\pi}{n} + B_2 \cos 1.1\frac{2\pi}{n}\right) e^{-\frac{2kt}{m}\text{ versin } 1\frac{2\pi}{n}}$$

$$+\left(A_3 \sin 2.2\frac{2\pi}{n} + B_3 \cos 2.2\frac{2\pi}{n}\right) e^{-\frac{2kt}{m}\text{ versin } 2\frac{2\pi}{n}}$$

$$+ \&c.,$$

$$\alpha_3 = \left(A_1 \sin 2 \,.\, 0 \frac{2\pi}{n} + B_1 \cos 2 \,.\, 0 \frac{2\pi}{n}\right) e^{-\frac{2kt}{m} \text{versin } 0 \frac{2\pi}{n}}$$

$$+ \left(A_2 \sin 2 \,.\, 1 \frac{2\pi}{n} + B_2 \cos 2 \,.\, 1 \frac{2\pi}{n}\right) e^{-\frac{2kt}{m} \text{versin } 1 \frac{2\pi}{n}}$$

$$+ \left(A_3 \sin 2 \,.\, 2 \frac{2\pi}{n} + B_3 \cos 2 \,.\, 2 \frac{2\pi}{n}\right) e^{-\frac{2kt}{m} \text{versin } 2 \frac{2\pi}{n}}$$

$$+ \text{\&c.},$$

$$\alpha_n = \left\{A_1 \sin (n-1)\, 0 \frac{2\pi}{n} + B_1 \cos (n-1) 0 \frac{2\pi}{n}\right\} e^{-\frac{2kt}{m} \text{versin } 0 \frac{2\pi}{n}}$$

$$+ \left\{A_2 \sin (n-1)\, 1 \frac{2\pi}{n} + B_2 \cos (n-1) 1 \frac{2\pi}{n}\right\} e^{-\frac{2kt}{m} \text{versin } 1 \frac{2\pi}{n}}$$

$$+ \left\{A_3 \sin (n-1)\, 2 \frac{2\pi}{n} + B_3 \cos (n-1) 2 \frac{2\pi}{n}\right\} e^{-\frac{2kt}{m} \text{versin } 2 \frac{2\pi}{n}}$$

$$+ \text{\&c.} \ldots\ldots\ldots\ldots (\mu).$$

To form these equations, we must continue in each equation the succession of terms which contain versin $0\frac{2\pi}{n}$, versin $1\frac{2\pi}{n}$, versin $2\frac{2\pi}{n}$, &c. until we have included every different versed sine; and we must omit all the subsequent terms, commencing with that in which a versed sine appears equal to one of the preceding.

The number of these equations is n. If n is an even number equal to $2i$, the number of terms of each equation is $i+1$; if n the number of equations is an odd number represented by $2i+1$, the number of terms is still equal to $i+1$. Lastly, among the quantities A_1, B_1, A_2, B_2, &c., which enter into these equations, there are some which must be omitted because they disappear of themselves, being multiplied by nul sines.

267. To determine the quantities A_1, B_1, A_2, B_2, A_3, B_3, &c., which enter into the preceding equations, we must consider the initial state which is known: suppose $t = 0$; and instead of α_1, α_2, α_3, &c., write the given quantities a_1, a_2, a_3, &c., which are the initial values of the temperatures. We have then to determine A_1, B_1, A_2, B_2, A_3, B_3, &c., the following equations:

$$a_1 = A_1 \sin 0 . 0 \frac{2\pi}{n} + A_2 \sin 0 . 1 \frac{2\pi}{n} + A_3 \sin 0 . 2 \frac{2\pi}{n} + \&c.$$

$$+ B_1 \cos 0 . 0 \frac{2\pi}{n} + B_2 \cos 0 . 1 \frac{2\pi}{n} + B_3 \cos 0 . 2 \frac{2\pi}{n} + \&c.$$

$$a_2 = A_1 \sin 1 . 0 \frac{2\pi}{n} + A_2 \sin 1 . 1 \frac{2\pi}{n} + A_3 \sin 1 . 2 \frac{2\pi}{n} + \&c.$$

$$+ B_1 \cos 1 . 0 \frac{2\pi}{n} + B_2 \cos 1 . 1 \frac{2\pi}{n} + B_3 \cos 1 . 2 \frac{2\pi}{n} + \&c.$$

$$a_3 = A_1 \sin 2 . 0 \frac{2\pi}{n} + A_2 \sin 2 . 1 \frac{2\pi}{n} + A_3 \sin 2 . 2 \frac{2\pi}{n} + \&c.$$

$$+ B_1 \cos 2 . 0 \frac{2\pi}{n} + B_2 \cos 2 . 1 \frac{2\pi}{n} + B_3 \cos 2 . 2 \frac{2\pi}{n} + \&c.$$

$$a_n = A_1 \sin (n-1) 0 \frac{2\pi}{n} + A_2 \sin (n-1) 1 \frac{2\pi}{n} + A_3 \sin (n-1) 2 \frac{2\pi}{n} + \&c.$$

$$+ B_1 \cos (n-1) 0 \frac{2\pi}{n} + B_2 \cos (n-1) 1 \frac{2\pi}{n} + B_3 \cos (n-1) 2 \frac{2\pi}{n} + \&c.$$

$$\ldots\ldots (m).$$

268. In these equations, whose number is n, the unknown quantities are A_1, B_1, A_2, B_2, A_3, B_3, &c., and it is required to effect the eliminations and to find the values of these unknowns. We may remark, first, that the same unknown has a different multiplier in each equation, and that the succession of multipliers composes a recurring series. In fact this succession is that of the sines of arcs increasing in arithmetic progression, or of the cosines of the same arcs; it may be represented by

$$\sin 0u, \quad \sin 1u, \quad \sin 2u, \quad \sin 3u, \ldots \sin (n-1)\, u,$$

or by $$\cos 0u, \quad \cos 1u, \quad \cos 2u, \quad \cos 3u, \ldots \cos (n-1)\, u.$$

The arc u is equal to $i \left(\frac{2\pi}{n}\right)$ if the unknown in question is A_{i+1} or B_{i+1}. This arranged, to determine the unknown A_{i+1} by means of the preceding equations, we must combine the succession of equations with the series of multipliers, $\sin 0u$, $\sin 1u$, $\sin 2u$, $\sin 3u, \ldots \sin (n-1)\, u$, and multiply each equation by the corresponding term of the series. If we take the sum of the equations thus multiplied, we eliminate all the unknowns, except that which is required to be determined. The same is the case if we wish to find the value of B_{i+1}; we must multiply each equation by the multiplier of B_{i+1} in that equation, and then take the sum of all the equations. It is requisite to prove that by operating in this manner we do in fact make all the unknowns disappear except one only. For this purpose it is sufficient to shew, firstly, that if we multiply term by term the two following series

$$\sin 0u,\ \sin 1u,\ \sin 2u,\ \sin 3u, \ldots \sin (n-1)\, u,$$

$$\sin 0v,\ \sin 1v,\ \sin 2v,\ \sin 3v, \ldots \sin (n-1)\, v,$$

the sum of the products

$$\sin 0u \sin 0v + \sin 1u \sin 1v + \sin 2u \sin 2v + \&c.$$

is nothing, except when the arcs u and v are the same, each of these arcs being otherwise supposed to be a multiple of a part of the circumference equal to $\frac{2\pi}{n}$; secondly, that if we multiply term by term the two series

$$\cos 0u,\ \cos 1u,\ \cos 2u, \ldots \cos (n-1)\, u,$$

$$\cos 0v,\ \cos 1v,\ \cos 2v, \ldots \cos (n-1)\, v,$$

the sum of the products is nothing, except in the case when u is equal to v; thirdly, that if we multiply term by term the two series

$$\sin 0u,\ \sin 1u,\ \sin 2u,\ \sin 3u, \ldots \sin (n-1)\, u,$$

$$\cos 0v,\ \cos 1v,\ \cos 2v,\ \cos 3v, \ldots \cos (n-1)\, v,$$

the sum of the products is always nothing.

269. Let us denote by q the arc $\frac{2\pi}{n}$, by μq the arc u, and by νq the arc v; μ and ν being positive integers less than n. The product of two terms corresponding to the two first series will be represented by

$$\sin j\mu q \sin j\nu q, \text{ or } \frac{1}{2} \cos j(\mu-\nu)\, q - \frac{1}{2} \cos j(\mu+\nu)\, q,$$

the letter j denoting any term whatever of the series 0, 1, 2, 3...

$(n-1)$; now it is easy to prove that if we give to j its n successive values, from 0 to $(n-1)$, the sum

$$\frac{1}{2}\cos 0\,(\mu-\nu)\,q+\frac{1}{2}\cos 1\,(\mu-\nu)\,q+\frac{1}{2}\cos 2\,(\mu-\nu)\,q$$
$$+\frac{1}{2}\cos 3\,(\mu-\nu)\,q+\ldots+\frac{1}{2}\cos(n-1)\,(\mu-\nu)\,q$$

has a nul value, and that the same is the case with the series

$$\frac{1}{2}\cos 0\,(\mu+\nu)\,q+\frac{1}{2}\cos 1\,(\mu+\nu)\,q+\frac{1}{2}\cos 2\,(\mu+\nu)\,q$$
$$+\frac{1}{2}\cos 3\,(\mu+\nu)\,q+\ldots+\frac{1}{2}\cos(n-1)\,(\mu+\nu)\,q.$$

In fact, representing the arc $(\mu-\nu)\,q$ by α, which is consequently a multiple of $\frac{2\pi}{n}$, we have the recurring series

$$\cos 0\alpha,\quad \cos 1\alpha,\quad \cos 2\alpha,\ldots\quad \cos(n-1)\,\alpha,$$

whose sum is nothing.

To shew this, we represent the sum by s, and the two terms of the scale of relation being $2\cos\alpha$ and -1, we multiply successively the two members of the equation

$$s=\cos 0\alpha+\cos 2\alpha+\cos 3\alpha+\ldots+\cos(n-1)\,\alpha$$

by $-2\cos\alpha$ and by $+1$; then on adding the three equations we find that the intermediate terms cancel after the manner of recurring series.

If we now remark that $n\alpha$ being a multiple of the whole circumference, the quantities $\cos(n-1)\,\alpha$, $\cos(n-2)\,\alpha$, $\cos(n-3)\,\alpha$, &c. are respectively the same as those which have been denoted by $\cos(-\alpha)$, $\cos(-2\alpha)$, $\cos(-3\alpha)$, ... &c. we conclude that

$$2s-2s\cos\alpha=0;$$

thus the sum sought must in general be nothing. In the same way we find that the sum of the terms due to the development of $\frac{1}{2}\cos j\,(\mu+\nu)\,q$ is nothing. The case in which the arc represented by α is 0 must be excepted; we then have $1-\cos\alpha=0$; that is to say, the arcs u and v are the same. In this case the term $\frac{1}{2}\cos j\,(\mu+\nu)\,q$ still gives a development whose sum is nothing;

but the quantity $\frac{1}{2}\cos j(\mu-\nu)q$ furnishes equal terms, each of which has the value $\frac{1}{2}$; hence the sum of the products term by term of the two first series is $\frac{1}{2}n$.

In the same manner we can find the value of the sum of the products term by term of the two second series, or

$$\Sigma(\cos j\mu q \cos j\nu q);$$

in fact, we can substitute for $\cos j\mu q \cos j\nu q$ the quantity

$$\tfrac{1}{2}\cos j(\mu-\nu)q + \tfrac{1}{2}\cos j(\mu+\nu)q,$$

and we then conclude, as in the preceding case, that $\Sigma\frac{1}{2}\cos j(\mu+\nu)q$ is nothing, and that $\Sigma\frac{1}{2}\cos j(\mu-\nu)q$ is nothing, except in the case where $\mu=\nu$. It follows from this that the sum of the products term by term of the two second series, or $\Sigma(\cos j\mu q \cos j\nu q)$, is always 0 when the arcs u and v are different, and equal to $\frac{1}{2}n$ when $u=v$. It only remains to notice the case in which the arcs μq and νq are both nothing, when we have 0 as the value of

$$\Sigma(\sin j\mu q \sin j\nu q),$$

which denotes the sum of the products term by term of the two first series.

The same is not the case with the sum $\Sigma(\cos j\mu q \cos j\nu q)$ taken when μq and νq are both nothing; the sum of the products term by term of the two second series is evidently equal to n.

As to the sum of the products term by term of the two series

$$\sin 0u,\ \sin 1u,\ \sin 2u,\ \sin 3u, \ldots \sin(n-1)u,$$

$$\cos 0u,\ \cos 1u,\ \cos 2u,\ \cos 3u, \ldots \cos(n-1)u,$$

it is nothing in all cases, as may easily be ascertained by the foregoing analysis.

270. The comparison then of these series furnishes the following results. If we divide the circumference 2π into n equal parts, and take an arc u composed of an integral number μ of these parts, and mark the ends of the arcs $u, 2u, 3u, \ldots (n-1)u$, it follows from the known properties of trigonometrical quantities that the quantities

$$\sin 0u,\ \sin 1u,\ \sin 2u,\ \sin 3u, \ldots \sin(n-1)u,$$

or indeed

$$\cos 0u,\ \cos 1u,\ \cos 2u,\ \cos 3u, \ldots \cos (n-1)\,u,$$

form a recurring periodic series composed of n terms: if we compare one of the two series corresponding to an arc u or $\mu \frac{2\pi}{n}$ with a series corresponding to another arc v or $\nu \frac{2\pi}{n}$, and multiply term by term the two compared series, the sum of the products will be nothing when the arcs u and v are different. If the arcs u and v are equal, the sum of the products is equal to $\frac{1}{2}n$, when we combine two series of sines, or when we combine two series of cosines; but the sum is nothing if we combine a series of sines with a series of cosines. If we suppose the arcs u and v to be nul, it is evident that the sum of the products term by term is nothing whenever one of the two series is formed of sines, or when both are so formed, but the sum of the products is n if the combined series both consist of cosines. In general, the sum of the products term by term is equal to 0, or $\frac{1}{2}n$ or n; known formulæ would, moreover, lead directly to the same results. They are produced here as evident consequences of elementary theorems in trigonometry.

271. By means of these remarks it is easy to effect the elimination of the unknowns in the preceding equations. The unknown A_1 disappears of itself through having nul coefficients; to find B_1 we must multiply the two members of each equation by the coefficient of B_1 in that equation, and on adding all the equations thus multiplied, we find

$$a_1 + a_2 + a_3 + \ldots + a_n = B_1.$$

To determine A_2 we must multiply the two members of each equation by the coefficient of A_2 in that equation, and denoting the arc $\frac{2\pi}{n}$ by q, we have, after adding the equations together,

$$a_1 \sin 0q + a_2 \sin 1q + a_3 \sin 2q + \ldots + a_n \sin (n-1)\, q = \frac{1}{2} n A_2.$$

Similarly to determine B_2 we have

$$a_1 \cos 0q + a_2 \cos 1q + a_3 \cos 2q + \ldots + a_n \cos (n-1)\, q = \frac{1}{2} n B_2.$$

In general we could find each unknown by multiplying the two members of each equation by the coefficient of the unknown in that equation, and adding the products. Thus we arrive at the following results:

$$nB_1 = a_1 \quad + a_2 \quad + a_3 \quad +\&c. = \Sigma a_i,$$

$$\frac{n}{2}A_2 = a_1 \sin 0\frac{2\pi}{n} + a_2 \sin 1\frac{2\pi}{n} + a_3 \sin 2\frac{2\pi}{n} + \&c. = \Sigma a_i \sin(i-1)1\frac{2\pi}{n},$$

$$\frac{n}{2}B_2 = a_1 \cos 0\frac{2\pi}{n} + a_2 \cos 1\frac{2\pi}{n} + a_3 \cos 2\frac{2\pi}{n} + \&c. = \Sigma a_i \cos(i-1)1\frac{2\pi}{n},$$

$$\frac{n}{2}A_3 = a_1 \sin 0.2\frac{2\pi}{n} + a_2 \sin 1.2\frac{2\pi}{n} + a_3 \sin 2.2\frac{2\pi}{n} + \&c. = \Sigma a_i \sin(i-1)2\frac{2\pi}{n}$$

$$\frac{n}{2}B_3 = a_1 \cos 0.2\frac{2\pi}{n} + a_2 \cos 1.2\frac{2\pi}{n} + a_3 \cos 2.2\frac{2\pi}{n} + \&c. = \Sigma a_i \cos(i-1)2\frac{2\pi}{n},$$

$$\frac{n}{2}A_4 = a_1 \sin 0.3\frac{2\pi}{n} + a_2 \sin 1.3\frac{2\pi}{n} + a_3 \sin 2.3\frac{2\pi}{n} + \&c. = \Sigma a_i \sin(i-1)3\frac{2\pi}{n},$$

$$\frac{n}{2}B_4 = a_1 \cos 0.3\frac{2\pi}{n} + a_2 \cos 1.3\frac{2\pi}{n} + a_3 \cos 2.3\frac{2\pi}{n} + \&c. = \Sigma a_i \cos(i-1)3\frac{2\pi}{n},$$

&c. .... (M).

To find the development indicated by the symbol Σ, we must give to i its n successive values 1, 2, 3, 4, &c., and take the sum, in which case we have in general

$$\frac{n}{2}A_j = \Sigma a_i \sin(i-1)(j-1)\frac{2\pi}{n} \text{ and } \frac{n}{2}B = \Sigma a_i \cos(i-1)(j-1)\frac{2\pi}{n}.$$

If we give to the integer j all the successive values 1, 2, 3, 4, &c. which it can take, the two formulæ give our equations, and if we develope the term under the sign Σ, by giving to i its n values 1, 2, 3, ... n, we have the values of the unknowns A_1, B_1, A_2, B_2, A_3, B_3, &c., and the equations (m), Art. 267, are completely solved.

272. We now substitute the known values of the coefficients A_1, B_1, A_2, B_2, A_3, B_3, &c., in equations (μ), Art. 266, and obtain the following values:

$$\alpha_1 = N_0 \qquad + N_1 \epsilon^{t \operatorname{versin} q_1} \qquad + N_2 \epsilon^{t \operatorname{versin} q_2} + \&c.$$

$$\alpha_2 = N_0 + (M_1 \sin q_1 + N_1 \cos q_1)\, \epsilon^{t \operatorname{versin} q_1} + (M_2 \sin q_2 + N_2 \cos q_2)\, \epsilon^{t \operatorname{versin} q_2} + \&c.$$

$$\alpha_3 = N_0 + (M_1 \sin 2q_1 + N_1 \cos 2q_1)\, \epsilon^{t \operatorname{versin} q_1} + (M_2 \sin 2q_2 + N_2 \cos 2q_2)\, \epsilon^{t \operatorname{versin} q_2} + \&c.$$

....

$$\alpha_j = N_0 + \{M_1 \sin (j-1)\, q_1 + N_1 \cos (j-1)\, q_1\}\, \epsilon^{t \operatorname{versin} q_1} + \{M_2 \sin (j-1)\, q_2 + N_2 \cos (j-1)\, q_2\}\, \epsilon^{t \operatorname{versin} q_2} + \&c.$$

....

$$\alpha_n = N_0 + \{M_1 \sin (n-1)\, q_1 + N_1 \cos (n-1)\, q_1\}\, \epsilon^{t \operatorname{versin} q_1} + \{M_2 \sin (n-1)\, q_2 + N_2 \cos (n-1)\, q_2\}\, \epsilon^{t \operatorname{versin} q_2} + \&c.$$

In these equations

$$\epsilon = e^{-\frac{2k}{m}}, \quad q_1 = 1\frac{2\pi}{n}, \quad q_2 = 2\frac{2\pi}{n}, \quad q_3 = 3\frac{2\pi}{n}, \ \&c.,$$

$$N_0 = \frac{1}{n}\Sigma\, a_i,$$

$$N_1 = \frac{2}{n}\Sigma a_i \cos (i-1)\, q_1, \qquad M_1 = \frac{2}{n}\Sigma a_i \sin (i-1)\, q_1,$$

$$N_2 = \frac{2}{n}\Sigma a_i \cos (i-1)\, q_2, \qquad M_2 = \frac{2}{n}\Sigma a_i \sin (i-1)\, q_2,$$

$$N_3 = \frac{2}{n}\Sigma a_i \cos (i-1)\, q_3, \qquad M_3 = \frac{2}{n}\Sigma a_i \sin (i-1)\, q_3,$$

&c. &c.

273. The equations which we have just set down contain the complete solution of the proposed problem; it is represented by the general equation

$$\alpha_j = \frac{1}{n}\Sigma a_i + \left[\frac{2}{n}\sin (j-1)\frac{2\pi}{n}\,\Sigma a_i \sin (i-1)\frac{2\pi}{n} + \frac{2}{n}\cos (j-1)\frac{2\pi}{n}\,\Sigma a_i \cos (i-1)\frac{2\pi}{n}\right] e^{-\frac{2kt}{m}\operatorname{versin} 1\frac{2\pi}{n}}$$

$$+ \left[\frac{2}{n}\sin (j-1)\, 2\frac{2\pi}{n}\,\Sigma a_i \sin (i-1)\, 2\frac{2\pi}{n} + \frac{2}{n}\cos (j-1)\, 2\frac{2\pi}{n}\,\Sigma a_i \cos (i-1)\, 2\frac{2\pi}{n}\right] e^{-\frac{2kt}{m}\operatorname{versin} 2\frac{2\pi}{n}}$$

$$+ \&c. \ldots\ldots (\epsilon),$$

in which only known quantities enter, namely, $a_1, a_2, a_3 \ldots a_n$, which are the initial temperatures, k the measure of the conducibility, m the value of the mass, n the number of masses heated, and t the time elapsed.

From the foregoing analysis it follows, that if several equal bodies n in number are arranged in a circle, and, having received any initial temperatures, begin to communicate heat to each other in the manner we have supposed; the mass, of each body being denoted by m, the time by t, and a certain constant coefficient by k, the variable temperature of each mass, which must be a function of the quantities t, m, and k, and of all the initial temperatures, is given by the general equation (ϵ). We first substitute instead of j the number which indicates the place of the body whose temperature we wish to ascertain, that is to say, 1 for the first body, 2 for the second, &c.; then with respect to the letter i which enters under the sign Σ, we give to it the n successive values $1, 2, 3, \ldots n$, and take the sum of all the terms. As to the number of terms which enter into this equation, there must be as many of them as there are different versed sines belonging to the successive arcs

$$0\frac{2\pi}{n},\ 1\frac{2\pi}{n},\ 2\frac{2\pi}{n},\ 3\frac{2\pi}{n},\ \&c.,$$

that is to say, whether the number n be equal to $(2\lambda+1)$ or 2λ, according as it is odd or even, the number of terms which enter into the general equation is always $\lambda+1$.

274. To give an example of the application of this formula, let us suppose that the first mass is the only one which at first was heated, so that the initial temperatures $a_1, a_2, a_3 \ldots a_n$ are all nul, except the first. It is evident that the quantity of heat contained in the first mass is distributed gradually among all the others. Hence the law of the communication of heat is expressed by the equation

$$\begin{aligned} a_j = \frac{1}{n}a_1 &+ \frac{2}{n}a_1 \cos(j-1)\frac{2\pi}{n}e^{-\frac{2kt}{m}\operatorname{versin} 1\frac{2\pi}{n}} \\ &+ \frac{2}{n}a_1 \cos(j-1)2\frac{2\pi}{n}e^{-\frac{2kt}{m}\operatorname{versin} 2\frac{2\pi}{n}} \\ &+ \frac{2}{n}a_1 \cos(j-1)3\frac{2\pi}{n}e^{-\frac{2kt}{m}\operatorname{versin} 3\frac{2\pi}{n}} + \&c. \end{aligned}$$

If the second mass alone had been heated and the temperatures $a_1, a_3, a_4, \ldots a_n$ were nul, we should have

$$a_j = \frac{1}{n} a_2 + \frac{2}{n} a_2 \left\{ \sin (j-1) \frac{2\pi}{n} \sin \frac{2\pi}{n} + \cos (j-1) \frac{2\pi}{n} \cos \frac{2\pi}{n} \right\} e^{-\frac{2kt}{m} \operatorname{versin} 1 \frac{2\pi}{n}}$$

$$+ \frac{2}{n} a_2 \left\{ \sin (j-1) 2 \frac{2\pi}{n} \sin 2 \frac{2\pi}{n} + \cos (j-1) 2 \frac{2\pi}{n} \cos 2 \frac{2\pi}{n} \right\} e^{-\frac{2kt}{m} \operatorname{versin} 2 \frac{2\pi}{n}}$$

$$+ \&c.,$$

and if all the initial temperatures were supposed nul, except a_1 and a_2, we should find for the value of a_j the sum of the values found in each of the two preceding hypotheses. In general it is easy to conclude from the general equation (ϵ), Art. 273, that in order to find the law according to which the initial quantities of heat are distributed between the masses, we may consider separately the cases in which the initial temperatures are nul, one only excepted. The quantity of heat contained in one of the masses may be supposed to communicate itself to all the others, regarding the latter as affected with nul temperatures; and having made this hypothesis for each particular mass with respect to the initial heat which it has received, we can ascertain the temperature of any one of the bodies, after a given time, by adding all the temperatures which the same body ought to have received on each of the foregoing hypotheses.

275. If in the general equation (ϵ) which gives the value of a_j, we suppose the time to be infinite, we find $a_j = \frac{1}{n} \Sigma a_i$, so that each of the masses has acquired the mean temperature; a result which is self-evident.

As the value of the time increases, the first term $\frac{1}{n} \Sigma a_i$ becomes greater and greater relatively to the following terms, or to their sum. The same is the case with the second with respect to the terms which follow it; and, when the time has become considerable, the value of a_j is represented without sensible error by the equation,

$$a_j = \frac{1}{n}\Sigma a_i + \frac{2}{n}\left\{\sin(j-1)\frac{2\pi}{n}\Sigma a_i \sin(i-1)\frac{2\pi}{n}\right.$$
$$\left. + \cos(j-1)\frac{2\pi}{n}\Sigma a_i \cos(i-1)\frac{2\pi}{n}\right\} e^{-\frac{2kt}{m}\operatorname{versin}\frac{2\pi}{n}}.$$

Denoting by a and b the coefficients of $\sin(j-1)\frac{2\pi}{n}$ and of $\cos(j-1)\frac{2\pi}{n}$, and the fraction $e^{-\frac{2k}{m}\operatorname{versin}\frac{2\pi}{n}}$ by ω, we have

$$a_j = \frac{1}{n}\Sigma a_i + \left\{a\sin(j-1)\frac{2\pi}{n} + b\cos(j-1)\frac{2\pi}{n}\right\}\omega^t.$$

The quantities a and b are constant, that is to say, independent of the time and of the letter j which indicates the order of the mass whose variable temperature is a_j. These quantities are the same for all the masses. The difference of the variable temperature a_j from the final temperature $\frac{1}{n}\Sigma a_i$ decreases therefore for each of the masses, in proportion to the successive powers of the fraction ω. Each of the bodies tends more and more to acquire the final temperature $\frac{1}{n}\Sigma a_i$, and the difference between that final limit and the variable temperature of the same body ends always by decreasing according to the successive powers of a fraction. This fraction is the same, whatever be the body whose changes of temperature are considered; the coefficient of ω^t or $(a\sin u_j + b\cos u_j)$, denoting by u_j the arc $(j-1)\frac{2\pi}{n}$, may be put under the form $A\sin(u_j + B)$, taking A and B so as to have $a = A\cos B$, and $b = A\sin B$. If we wish to determine the coefficient of ω^t with regard to the successive bodies whose temperature is a_{j+1}, a_{j+2}, a_{j+3}, &c., we must add to u_j the arc $\frac{2\pi}{n}$ or $2\frac{2\pi}{n}$, and so on; so that we have the equations

$$a_j - \frac{1}{n}\Sigma a_i = A\sin(B + u_j)\,\omega^t + \&c.$$

$$a_{j+1} - \frac{1}{n}\Sigma a_i = A\sin\left(B + u_j + 1\frac{2\pi}{n}\right)\omega^t + \&c.$$

$$a_{j+2} - \frac{1}{n} \Sigma a_i = A \sin \left(B + u_j + 2 \frac{2\pi}{n} \right) \omega^t + \&c.$$

$$a_{j+3} - \frac{1}{n} \Sigma a_i = A \sin \left(B + u_j + 3 \frac{2\pi}{n} \right) \omega^t + \&c.$$

&c.

276. We see, by these equations, that the later differences between the actual temperatures and the final temperatures are represented by the preceding equations, preserving only the first term of the second member of each equation. These later differences vary then according to the following law: if we consider only one body, the variable difference in question, that is to say, the excess of the actual temperature of the body over the final and common temperature, diminishes according to the successive powers of a fraction, as the time increases by equal parts; and, if we compare at the same instant the temperatures of all the bodies, the difference in question varies proportionally to the successive sines of the circumference divided into equal parts. The temperature of the same body, taken at different successive equal instants, is represented by the ordinates of a logarithmic curve, whose axis is divided into equal parts, and the temperature of each of these bodies, taken at the same instant for all, is represented by the ordinates of a circle whose circumference is divided into equal parts. It is easy to see, as we have remarked before, that if the initial temperatures are such, that the differences of these temperatures from the mean or final temperature are proportional to the successive sines of multiple arcs, these differences will all diminish at the same time without ceasing to be proportional to the same sines. This law, which governs also the initial temperatures, will not be disturbed by the reciprocal action of the bodies, and will be maintained until they have all acquired a common temperature. The difference will diminish for each body according to the successive powers of the same fraction. Such is the simplest law to which the communication of heat between a succession of equal masses can be submitted. When this law has once been established between the initial temperatures, it is maintained of itself; and when it does not govern the initial temperatures, that is to say, when the differences of these temperatures from the mean temperature are not proportional to successive sines of multiple arcs, the law in question tends always to be set

up, and the system of variable temperatures ends soon by coinciding sensibly with that which depends on the ordinates of a circle and those of a logarithmic curve.

Since the later differences between the excess of the temperature of a body over the mean temperature are proportional to the sine of the arc at the end of which the body is placed, it follows that if we regard two bodies situated at the ends of the same diameter, the temperature of the first will surpass the mean and constant temperature as much as that constant temperature surpasses the temperature of the second body. For this reason, if we take at each instant the sum of the temperatures of two masses whose situation is opposite, we find a constant sum, and this sum has the same value for any two masses situated at the ends of the same diameter.

277. The formulæ which represent the variable temperatures of separate masses are easily applied to the propagation of heat in continuous bodies. To give a remarkable example, we will determine the movement of heat in a ring, by means of the general equation which has been already set down.

Let it be supposed that n the number of masses increases successively, and that at the same time the length of each mass decreases in the same ratio, so that the length of the system has a constant value equal to 2π. Thus if n the number of masses be successively 2, 4, 8, 16, to infinity, each of the masses will be π, $\frac{\pi}{2}$, $\frac{\pi}{4}$, $\frac{\pi}{8}$, &c. It must also be assumed that the facility with which heat is transmitted increases in the same ratio as the number of masses m; thus the quantity which k represents when there are only two masses becomes double when there are four, quadruple when there are eight, and so on. Denoting this quantity by g, we see that the number k must be successively replaced by g, $2g$, $4g$, &c. If we pass now to the hypothesis of a continuous body, we must write instead of m, the value of each infinitely small mass, the element dx; instead of n, the number of masses, we must write $\frac{2\pi}{dx}$; instead of k write $g\frac{n}{2}$ or $\frac{\pi g}{dx}$.

As to the initial temperatures $a_1, a_2, a_3 \ldots a_n$, they depend on the value of the arc x, and regarding these temperatures as the successive states of the same variable, the general value a_i represents an arbitrary function of x. The index i must then be replaced by $\frac{x}{dx}$. With respect to the quantities $\alpha_1, \alpha_2, \alpha_3, \ldots$, these are variable temperatures depending on two quantities x and t. Denoting the variable by v, we have $v = \phi(x, t)$. The index j, which marks the place occupied by one of the bodies, should be replaced by $\frac{x}{dx}$. Thus, to apply the previous analysis to the case of an infinite number of layers, forming a continuous body in the form of a ring, we must substitute for the quantities n, m, k, a_i, i, a_j, j, their corresponding quantities, namely, $\frac{2\pi}{dx}$, dx, $\frac{\pi g}{dx}$, $f(x)$, $\frac{x}{dx}$, $\phi(x, t)$, $\frac{x}{dx}$. Let these substitutions be made in equation (ϵ) Art. 273, and let $\frac{1}{2}\,dx^2$ be written instead of versin dx, and i and j instead of $i-1$ and $j-1$. The first term $\frac{1}{n}\Sigma a_i$ becomes the value of the integral $\frac{1}{2\pi}\int f(x)\,dx$ taken from $x = 0$ to $x = 2\pi$; the quantity $\sin(j-1)\frac{2\pi}{n}$ becomes $\sin jdx$ or $\sin x$; the value of $\cos(j-1)\frac{2\pi}{dx}$ is $\cos x$; that of $\frac{2}{n}\Sigma a_i \sin(i-1)\frac{2\pi}{n}$ is $\frac{1}{\pi}\int f(x)\sin x dx$, the integral being taken from $x = 0$ to $x = 2\pi$; and the value of $\frac{2}{n}\Sigma a_i \cos\,(i-1)\,\frac{2\pi}{n}$ is $\frac{1}{\pi}\int f(x)\cos x\,dx$, the integral being taken between the same limits. Thus we obtain the equation

$$\phi(x,t) = v = \frac{1}{2\pi}\int f(x)\,dx$$

$$+\frac{1}{\pi}\left(\sin x\int f(x)\sin x dx + \cos x\int f(x)\cos x dx\right)e^{-g\pi t}$$

$$+\frac{1}{\pi}\left(\sin 2x\int f(x)\sin 2x dx + \cos 2x\int f(x)\cos 2x\,dx\right)e^{-2^2 g\pi t}$$

$$+\,\&c. \quad\ldots\ldots\ldots\ldots\ldots\ldots\ldots\ldots \quad (E)$$

and representing the quantity $g\pi$ by k, we have

$$\pi v = \frac{1}{2}\int f(x)\,dx + \left(\sin x\int f(x)\sin x\,dx + \cos x\int f(x)\cos x\,dx\right)e^{-kt}$$
$$+\left(\sin 2x\int f(x)\sin 2x\,dx + \cos 2x\int f(x)\cos 2x\,dx\right)e^{-2^2kt}$$
$$+ \&c.$$

278. This solution is the same as that which was given in the preceding section, Art. 241; it gives rise to several remarks. 1st. It is not necessary to resort to the analysis of partial differential equations in order to obtain the general equation which expresses the movement of heat in a ring. The problem may be solved for a definite number of bodies, and that number may then be supposed infinite. This method has a clearness peculiar to itself, and guides our first researches. It is easy afterwards to pass to a more concise method by a process indicated naturally. We see that the discrimination of the particular values, which, satisfying the partial differential equation, compose the general value, is derived from the known rule for the integration of linear differential equations whose coefficients are constant. The discrimination is moreover founded, as we have seen above, on the physical conditions of the problem. 2nd. To pass from the case of separate masses to that of a continuous body, we supposed the coefficient k to be increased in proportion to n, the number of masses. This continual change of the number k follows from what we have formerly proved, namely, that the quantity of heat which flows between two layers of the same prism is proportional to the value of $\frac{dv}{dx}$, x denoting the abscissa which corresponds to the section, and v the temperature. If, indeed, we did not suppose the coefficient k to increase in proportion to the number of masses, but were to retain a constant value for that coefficient, we should find, on making n infinite, a result contrary to that which is observed in continuous bodies. The diffusion of heat would be infinitely slow, and in whatever manner the mass was heated, the temperature at a point would suffer no sensible change during a finite time, which is contrary to fact. Whenever we resort to the consideration of an infinite number of separate masses which

transmit heat, and wish to pass to the case of continuous bodies, we must attribute to the coefficient k, which measures the velocity of transmission, a value proportional to the number of infinitely small masses which compose the given body.

3rd. If in the last equation which we obtained to express the value of v or $\phi(x, t)$, we suppose $t=0$, the equation necessarily represents the initial state, we have therefore in this way the equation (p), which we obtained formerly in Art. 233, namely,

$$\pi f(x) = \frac{1}{2}\int f(x)\,dx \begin{array}{l} + \sin x \int f(x) \sin x\, dx + \sin 2x \int f(x) \sin 2x\, dx + \&c. \\ + \cos x \int f(x) \cos x\, dx + \cos 2x \int f(x) \cos 2x\, dx + \&c. \end{array}$$

Thus the theórem which gives, between assigned limits, the development of an arbitrary function in a series of sines or cosines of multiple arcs is deduced from elementary rules of analysis. Here we find the origin of the process which we employed to make all the coefficients except one disappear by successive integrations from the equation

$$\phi(x) = a \begin{array}{l} + a_1 \sin x + a_2 \sin 2x + a_3 \sin 3x + \&c. \\ + b_1 \cos x + b_2 \cos 2x + b_3 \cos 3x + \&c. \end{array}$$

These integrations correspond to the elimination of the different unknowns in equations (m), Arts. 267 and 271, and we see clearly by the comparison of the two methods, that equation (B), Art. 279, holds for all values of x included between 0 and 2π, without its being established so as to apply to values of x which exceed those limits.

279. The function $\phi(x, t)$ which satisfies the conditions of the problem, and whose value is determined by equation (E), Art. 277, may be expressed as follows:

$$2\pi\phi(x, t) = \int d\alpha f(\alpha) + \left\{2 \sin x \int d\alpha f(\alpha) \sin \alpha + 2 \cos x \int d\alpha f(\alpha) \cos \alpha\right\} e^{-kt}$$

$$+ \left\{2 \sin 2x \int d\alpha f(\alpha) \sin 2\alpha + 2 \cos 2x \int d\alpha f(\alpha) \cos 2\alpha\right\} e^{-2^2kt}$$

$$+ \left\{2 \sin 3x \int d\alpha f(\alpha) \sin 3\alpha + 2 \cos 3x \int d\alpha f(\alpha) \cos 3\alpha\right\} e^{-3^2kt} + \&c.$$

$$\text{or } 2\pi\phi(x,t) = \int d\alpha f(\alpha)\,\{1 + (2\sin x\sin\alpha + 2\cos x\cos\alpha)e^{-kt}$$

$$+ (2\sin 2x\sin 2\alpha + 2\cos 2x\cos 2\alpha)e^{-2^2kt}$$

$$+ (2\sin 3x\sin 3\alpha + 2\cos 3x\cos 3\alpha)e^{-3^2kt} + \&\text{c.}\}$$

$$= \int d\alpha f(\alpha)\,[1 + 2\Sigma\cos i(\alpha - x)e^{-i^2kt}].$$

The sign Σ affects the number i, and indicates that the sum must be taken from $i=1$ to $i=\infty$. We can also include the first term under the sign Σ, and we have

$$2\pi\phi(x,t) = \int d\alpha f(\alpha)\,\Sigma_{-\infty}^{+\infty}\cos i(\alpha - x)\,e^{-i^2kt}.$$

We must then give to i all integral values from $-\infty$ to $+\infty$; which is indicated by writing the limits $-\infty$ and $+\infty$ next to the sign Σ, one of these values of i being 0. This is the most concise expression of the solution. To develope the second member of the equation, we suppose $i=0$, and then $i=1, 2, 3$, &c., and double each result except the first, which corresponds to $i=0$. When t is nothing, the function $\phi(x,t)$ necessarily represents the initial state in which the temperatures are equal to $f(x)$, we have therefore the identical equation,

$$f(x) = \frac{1}{2\pi}\int_0^{2\pi} d\alpha f(\alpha)\,\Sigma_{-\infty}^{+\infty}\cos i(\alpha - x)\ \ldots\ldots\ldots\ldots\ (\text{B}).$$

We have attached to the signs $\int$ and Σ the limits between which the integral sum must be taken. This theorem holds generally whatever be the form of the function $f(x)$ in the interval from $x=0$ to $x=2\pi$; the same is the case with that which is expressed by the equations which give the development of $F(x)$, Art. 235; and we shall see in the sequel that we can prove directly the truth of equation (B) independently of the foregoing considerations.

280. It is easy to see that the problem admits of no solution different from that given by equation (E), Art. 277. The function $\phi(x,t)$ in fact completely satisfied the conditions of the problem, and from the nature of the differential equation $\frac{dv}{dt} = k\frac{d^2v}{dx^2}$, no

other function can enjoy the same property. To convince ourselves of this we must consider that when the first state of the solid is represented by a given equation $v_1 = f(x)$, the fluxion $\frac{dv_1}{dt}$ is known, since it is equivalent to $k\frac{d^2 f(x)}{dx^2}$. Thus denoting by v_2 or $v_1 + k\frac{dv_1}{dt}dt$, the temperature at the commencement of the second instant, we can deduce the value of v_2 from the initial state and from the differential equation. We could ascertain in the same manner the values v_3, v_4, ... v_n of the temperature at any point whatever of the solid at the beginning of each instant. Now the function $\phi(x, t)$ satisfies the initial state, since we have $\phi(x, 0) = f(x)$. Further, it satisfies also the differential equation; consequently if it were differentiated, it would give the same values for $\frac{dv_1}{dt}$, $\frac{dv_2}{dt}$, $\frac{dv_3}{dt}$, &c., as would result from successive applications of the differential equation (a). Hence, if in the function $\phi(x, t)$ we give to t successively the values 0, ω, 2ω, 3ω, &c., ω denoting an element of time, we shall find the same values v_1, v_2, v_3, &c. as we could have derived from the initial state by continued application of the equation $\frac{dv}{dt} = k\frac{d^2v}{dx^2}$. Hence every function $\psi(x, t)$ which satisfies the differential equation and the initial state necessarily coincides with the function $\phi(x, t)$: for such functions each give the same function of x, when in them we suppose t successively equal to 0, ω, 2ω, 3ω ... $i\omega$, &c.

We see by this that there can be only one solution of the problem, and that if we discover in any manner a function $\psi(x, t)$ which satisfies the differential equation and the initial state, we are certain that it is the same as the former function given by equation (E).

281. The same remark applies to all investigations whose object is the varied movement of heat; it follows evidently from the very form of the general equation.

For the same reason the integral of the equation $\frac{dv}{dt} = k\frac{d^2v}{dx^2}$ can contain only one arbitrary function of x. In fact, when a value of v as a function of x is assigned for a certain value of the time t, it is evident that all the other values of v which correspond to any time whatever are determinate. We may therefore select arbitrarily the function of x, which corresponds to a certain state, and the general function of the two variables x and t then becomes determined. The same is not the case with the equation $\frac{d^2v}{dx^2}+\frac{d^2v}{dy^2}=0$, which was employed in the preceding chapter, and which belongs to the constant movement of heat; its integral contains two arbitrary functions of x and y: but we may reduce this investigation to that of the varied movement, by regarding the final and permanent state as derived from the states which precede it, and consequently from the initial state, which is given.

The integral which we have given

$$\frac{1}{2\pi}\int d\alpha f(\alpha)\,\Sigma e^{-i^2kt}\cos i(\alpha-x)$$

contains one arbitrary function $f(x)$, and has the same extent as the general integral, which also contains only one arbitrary function of x; or rather, it is this integral itself arranged in a form suitable to the problem. In fact, the equation $v_1=f(x)$ representing the initial state, and $v=\phi(x, t)$ representing the variable state which succeeds it, we see from the very form of the heated solid that the value of v does not change when $x\pm i2\pi$ is written instead of x, i being any positive integer. The function

$$\frac{1}{2\pi}\int d\alpha f(\alpha)\,\Sigma e^{-i^2kt}\cos i(\alpha-x)$$

satisfies this condition; it represents also the initial state when we suppose $t=0$, since we then have

$$f(x)=\frac{1}{2\pi}\int d\alpha f(\alpha)\,\Sigma\cos i(\alpha-x),$$

an equation which was proved above, Arts. 235 and 279, and is also easily verified. Lastly, the same function satisfies the differential equation $\frac{dv}{dt}=k\frac{d^2v}{dx^2}$. Whatever be the value of t, the temperature v is given by a very convergent series, and the different terms represent all the partial movements which combine to form the total movement. As the time increases, the partial states of higher orders alter rapidly, but their influence becomes inappreciable; so that the number of values which ought to be given to the exponent i diminishes continually. After a certain time the system of temperatures is represented sensibly by the terms which are found on giving to i the values 0, ± 1 and ± 2, or only 0 and ± 1, or lastly, by the first of those terms, namely, $\frac{1}{2\pi}\int d\alpha\, f(\alpha)$; there is therefore a manifest relation between the form of the solution and the progress of the physical phenomenon which has been submitted to analysis.

282. To arrive at the solution we considered first the simple values of the function v which satisfy the differential equation: we then formed a value which agrees with the initial state, and has consequently all the generality which belongs to the problem. We might follow a different course, and derive the same solution from another expression of the integral; when once the solution is known, the results are easily transformed. If we suppose the diameter of the mean section of the ring to increase infinitely, the function $\phi(x, t)$, as we shall see in the sequel, receives a different form, and coincides with an integral which contains a single arbitrary function under the sign of the definite integral. The latter integral might also be applied to the actual problem; but, if we were limited to this application, we should have but a very imperfect knowledge of the phenomenon; for the values of the temperatures would not be expressed by convergent series, and we could not discriminate between the states which succeed each other as the time increases. The periodic form which the problem supposes must therefore be attributed to the function which represents the initial state; but on modifying that integral in this manner, we should obtain no other result than

$$\phi(x, t) = \frac{1}{2\pi}\int d\alpha\, f(\alpha)\, \Sigma e^{-i^2kt} \cos i(\alpha - x).$$

From the last equation we pass easily to the integral in question, as was proved in the memoir which preceded this work. It is not less easy to obtain the equation from the integral itself. These transformations make the agreement of the analytical results more clearly evident; but they add nothing to the theory, and constitute no different analysis. In one of the following chapters we shall examine the different forms which may be assumed by the integral of the equation $\frac{dv}{dt} = k\frac{d^2v}{dx^2}$, the relations which they have to each other, and the cases in which they ought to be employed.

To form the integral which expresses the movement of heat in a ring, it was necessary to resolve an arbitrary function into a series of sines and cosines of multiple arcs; the numbers which affect the variable under the symbols *sine* and *cosine* are the natural numbers 1, 2, 3, 4, &c. In the following problem the arbitrary function is again reduced to a series of sines; but the coefficients of the variable under the symbol *sine* are no longer the numbers 1, 2, 3, 4, &c.: these coefficients satisfy a definite equation whose roots are all incommensurable and infinite in number.

Note on Sect. I, Chap. IV. Guglielmo Libri of Florence was the first to investigate the problem of the movement of heat in a ring on the hypothesis of the law of cooling established by Dulong and Petit. See his *Mémoire sur la théorie de la chaleur*, Crelle's Journal, Band VII., pp. 116—131, Berlin, 1831. (Read before the French Academy of Sciences, 1825.) M. Libri made the solution depend upon a series of partial differential equations, treating them as if they were linear. The equations have been discussed in a different manner by Mr Kelland, in his *Theory of Heat*, pp. 69—75, Cambridge, 1837. The principal result obtained is that the mean of the temperatures at opposite ends of any diameter of the ring is the same at the same instant. [A. F.]

CHAPTER V.

OF THE PROPAGATION OF HEAT IN A SOLID SPHERE.

SECTION I.

General solution.

283. THE problem of the propagation of heat in a sphere has been explained in Chapter II., Section 2, Article 117; it consists in integrating the equation

$$\frac{dv}{dt} = k\left(\frac{d^2v}{dx^2} + \frac{2}{x}\frac{dv}{dx}\right),$$

so that when $x = X$ the integral may satisfy the condition

$$\frac{dv}{dx} + hv = 0,$$

k denoting the ratio $\frac{K}{CD}$, and h the ratio $\frac{h}{K}$ of the two conducibilities; v is the temperature which is observed after the time t has elapsed in a spherical layer whose radius is x; X is the radius of the sphere; v is a function of x and t, which is equal to $F(x)$ when we suppose $t = 0$. The function $F(x)$ is given, and represents the initial and arbitrary state of the solid.

If we make $y = vx$, y being a new unknown, we have, after the substitutions, $\frac{dy}{dt} = k\frac{d^2y}{dx^2}$: thus we must integrate the last equation, and then take $v = \frac{y}{x}$. We shall examine, in the first place, what are the simplest values which can be attributed to y, and then form a general value which will satisfy at the same time the differential equation, the condition relative to the surface, and the initial state. It is easily seen that when these three conditions are fulfilled, the solution is complete, and no other can be found.

284. Let $y = e^{mt}u$, u being a function of x, we have

$$mu = k\frac{d^2u}{dx^2}.$$

First, we notice that when the value of t becomes infinite, the value of v must be nothing at all points, since the body is completely cooled. Negative values only can therefore be taken for m. Now k has a positive numerical value, hence we conclude that the value of u is a circular function, which follows from the known nature of the equation

$$mu = k\frac{d^2u}{dx^2}.$$

Let $u = A\cos nx + B\sin nx$; we have the condition $m = -kn^2$. Thus we can express a particular value of v by the equation

$$v = \frac{e^{-kn^2t}}{x}(A\cos nx + B\sin nx),$$

where n is any positive number, and A and B are constants. We may remark, first, that the constant A ought to be nothing; for the value of v which expresses the temperature at the centre, when we make $x = 0$, cannot be infinite; hence the term $A\cos nx$ should be omitted.

Further, the number n cannot be taken arbitrarily. In fact, if in the definite equation $\frac{dv}{dx} + hv = 0$ we substitute the value of v, we find

$$nx\cos nx + (hx - 1)\sin nx = 0.$$

As the equation ought to hold at the surface, we shall suppose in it $x = X$ the radius of the sphere, which gives

$$\frac{nX}{\tan nX} = 1 - hX.$$

Let λ be the number $1 - hX$, and $nX = \epsilon$, we have $\frac{\epsilon}{\tan\epsilon} = \lambda$. We must therefore find an arc ϵ, which divided by its tangent gives a known quotient λ, and afterwards take $n=\frac{\epsilon}{X}$. It is evident that there are an infinity of such arcs, which have a given ratio to their tangent; so that the equation of condition

$$\frac{nX}{\tan nX}=1-hX$$

has an infinite number of real roots.

285. Graphical constructions are very suitable for exhibiting the nature of this equation. Let $u=\tan\epsilon$ (fig. 12), be the equation

Fig. 12.

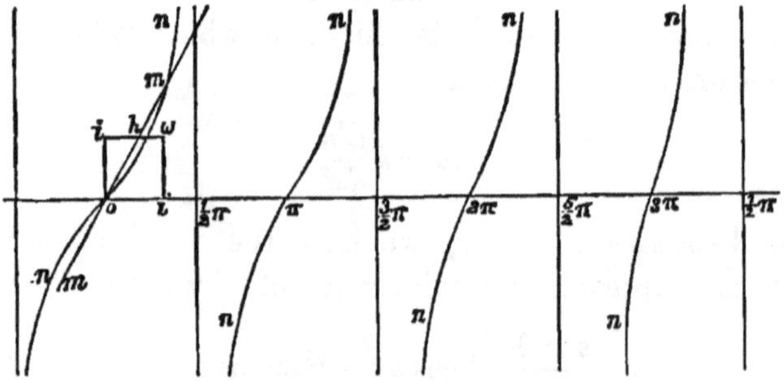

to a curve, of which the arc ϵ is the abscissa, and u the ordinate; and let $u=\frac{\epsilon}{\lambda}$ be the equation to a straight line, whose co-ordinates are also denoted by ϵ and u. If we eliminate u from these two equations, we have the proposed equation $\frac{\epsilon}{\lambda}=\tan\epsilon$. The unknown ϵ is therefore the abscissa of the point of intersection of the curve and the straight line. This curved line is composed of an infinity of arcs; all the ordinates corresponding to abscissæ

$$\frac{1}{2}\pi,\ \frac{3}{2}\pi,\ \frac{5}{2}\pi,\ \frac{7}{2}\pi,\ \&\text{c.}$$

are infinite, and all those which correspond to the points o, π, 2π, 3π, &c. are nothing. To trace the straight line whose equation is $u=\frac{\epsilon}{\lambda}=\frac{\epsilon}{1-hX}$, we form the square $oi\omega i$, and measuring the quantity hX from ω to h, join the point h with the origin o. The curve non whose equation is $u=\tan\epsilon$ has for tangent at the origin a line which divides the right angle into two equal parts, since the ultimate ratio of the arc to the tangent is 1. We conclude from this that if λ or $l-hX$ is a quantity less than unity, the straight line *mom* passes from the origin above the curve *non*, and there is a point of intersection of the straight line with the first branch. It is equally clear that the same straight line cuts all the further branches $n\pi n$, $n2\pi n$, &c. Hence the equation $\dfrac{\epsilon}{\tan\epsilon}=\lambda$ has an infinite number of real roots. The first is included between 0 and $\dfrac{\pi}{2}$, the second between π and $\dfrac{3\pi}{2}$, the third between 2π and $\dfrac{5\pi}{2}$, and so on. These roots approach very near to their upper limits when they are of a very advanced order.

286. If we wish to calculate the value of one of the roots, for example, of the first, we may employ the following rule: write down the two equations $\epsilon = \text{arc}\tan u$ and $u=\dfrac{\epsilon}{\lambda}$, arc tan u denoting the length of the arc whose tangent is u. Then taking any number for u, deduce from the first equation the value of ϵ; substitute this value in the second equation, and deduce another value of u; substitute the second value of u in the first equation; thence we deduce a value of ϵ, which, by means of the second equation, gives a third value of u. Substituting it in the first equation we have a new value of ϵ. Continue thus to determine u by the second equation, and ϵ by the first. The operation gives values more and more nearly approaching to the unknown ϵ, as is evident from the following construction.

In fact, if the point u correspond (see fig. 13) to the arbitrary value which is assigned to the ordinate u; and if we substitute this value in the first equation $\epsilon = \text{arc}\tan u$, the point ϵ will correspond to the abscissa which we have calculated by means of this equation. If this abscissa ϵ be substituted in the second equation $u=\dfrac{\epsilon}{\lambda}$, we shall find an ordinate u' which corresponds to the point u'. Substituting u' in the first equation, we find an abscissa ϵ' which corresponds to the point ϵ'; this abscissa being then substituted in the second equation gives rise to an ordinate u', which when substituted in the first, gives rise to a third abscissa ϵ'', and so on to infinity. That is to say, in order to represent the continued alternate employment of the two pre-

Fig. 13.

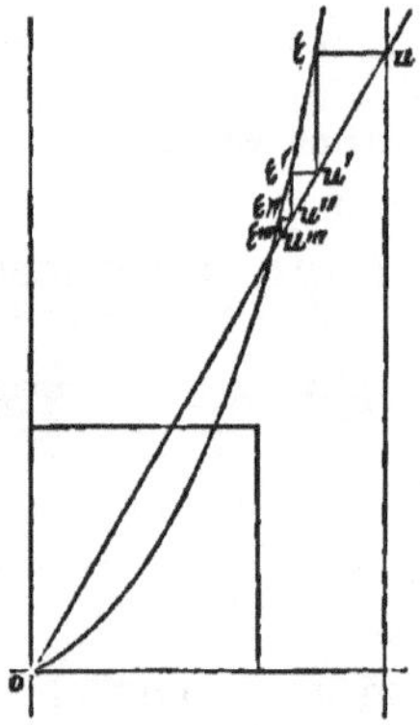

Fig. 14.

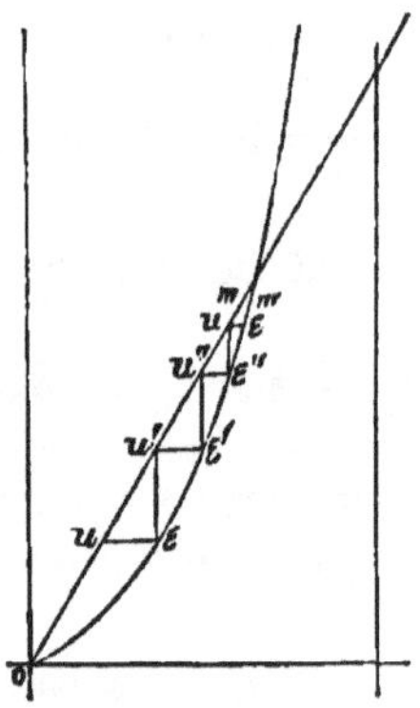

ceding equations, we must draw through the point u a horizontal line up to the curve, and through ϵ the point of intersection draw a vertical as far as the straight line, through the point of intersection u' draw a horizontal up to the curve, through the point of intersection ϵ' draw a vertical as far as the straight line, and so on to infinity, descending more and more towards the point sought.

287. The foregoing figure (13) represents the case in which the ordinate arbitrarily chosen for u is greater than that which corresponds to the point of intersection. If, on the other hand, we chose for the initial value of u a smaller quantity, and employed in the same manner the two equations $\epsilon = \text{arc} \tan u$, $u = \frac{\epsilon}{\lambda}$, we should again arrive at values successively closer to the unknown value. Figure 14 shews that in this case we rise continually towards the point of intersection by passing through the points $u\, \epsilon\, u'\, \epsilon'\, u''\, \epsilon''$, &c. which terminate the horizontal and vertical lines. Starting from a value of u which is too small, we obtain quantities $\epsilon\, \epsilon'\, \epsilon''\, \epsilon'''$, &c. which converge towards the unknown value, and are smaller than it; and starting from a value of u which is too great, we obtain quantities which also converge to the unknown value, and each of which is greater than it. We therefore ascertain

successively closer limits between the which magnitude sought is always included. Either approximation is represented by the formula

$$\epsilon = \ldots \text{arc tan}\left[\frac{1}{\lambda}\text{arc tan}\left\{\frac{1}{\lambda}\text{arc tan}\left(\frac{1}{\lambda}\text{arc tan}\frac{1}{\lambda}\right)\right\}\right].$$

When several of the operations indicated have been effected, the successive results differ less and less, and we have arrived at an approximate value of ϵ.

288. We might attempt to apply the two equations

$$\epsilon = \text{arc tan}\, u \text{ and } u = \frac{\epsilon}{\lambda}$$

in a different order, giving them the form $u = \tan\epsilon$ and $\epsilon = \lambda u$. We should then take an arbitrary value of ϵ, and, substituting it in the first equation, we should find a value of u, which being substituted in the second equation would give a second value of ϵ; this new value of ϵ could then be employed in the same manner as the first. But it is easy to see, by the constructions of the figures, that in following this course of operations we depart more and more from the point of intersection instead of approaching it, as in the former case. The successive values of ϵ which we should obtain would diminish continually to zero, or would increase without limit. We should pass successively from ϵ'' to u'', from u'' to ϵ', from ϵ' to u', from u' to ϵ, and so on to infinity.

The rule which we have just explained being applicable to the calculation of each of the roots of the equation

$$\frac{\epsilon}{\tan\epsilon} = 1 - hX,$$

which moreover have given limits, we must regard all these roots as known numbers. Otherwise, it was only necessary to be assured that the equation has an infinite number of real roots. We have explained this process of approximation because it is founded on a remarkable construction, which may be usefully employed in several cases, and which exhibits immediately the nature and limits of the roots; but the actual application of the process to the equation in question would be tedious; it would be easy to resort in practice to some other mode of approximation.

289. We now know a particular form which may be given to the function v so as to satisfy the two conditions of the problem. This solution is represented by the equation

$$v = \frac{Ae^{-kn^2t}\sin nx}{x} \quad \text{or} \quad v = ae^{-kn^2t}\frac{\sin nx}{nx}.$$

The coefficient a is any number whatever, and the number n is such that $\frac{nX}{\tan nX} = 1 - hX$. It follows from this that if the initial temperatures of the different layers were proportional to the quotient $\frac{\sin nx}{nx}$, they would all diminish together, retaining between themselves throughout the whole duration of the cooling the ratios which had been set up; and the temperature at each point would decrease as the ordinate of a logarithmic curve whose abscissa would denote the time passed. Suppose, then, the arc ϵ being divided into equal parts and taken as abscissa, we raise at each point of division an ordinate equal to the ratio of the sine to the arc. The system of ordinates will indicate the initial temperatures, which must be assigned to the different layers, from the centre to the surface, the whole radius X being divided into equal parts. The arc ϵ which, on this construction, represents the radius X, cannot be taken arbitrarily; it is necessary that the arc and its tangent should be in a given ratio. As there are an infinite number of arcs which satisfy this condition, we might thus form an infinite number of systems of initial temperatures, which could exist of themselves in the sphere, without the ratios of the temperatures changing during the cooling.

290. It remains only to form any initial state by means of a certain number, or of an infinite number of partial states, each of which represents one of the systems of temperatures which we have recently considered, in which the ordinate varies with the distance x, and is proportional to the quotient of the sine by the arc. The general movement of heat in the interior of a sphere will then be decomposed into so many particular movements, each of which is accomplished freely, as if it alone existed.

Denoting by n_1, n_2, n_3, &c., the quantities which satisfy the equation $\frac{nX}{\tan nX} = 1 - hX$, and supposing them to be arranged in order, beginning with the least, we form the general equation

$$vx = a_1 e^{-kn_1^2 t} \sin n_1 x + a_2 e^{-kn_2^2 t} \sin n_2 x + a_3 e^{-kn_3^2 t} \sin n_3 x + \&c.$$

If t be made equal to 0, we have as the expression of the initial state of temperatures

$$vx = a_1 \sin n_1 x + a_2 \sin n_2 x + a_3 \sin n_3 x + \&c.$$

The problem consists in determining the coefficients a_1, a_2, a_3 &c., whatever be the initial state. Suppose then that we know the values of v from $x = 0$ to $x = X$, and represent this system of values by $F(x)$; we have

$$F(x) = \frac{1}{x}(a_1 \sin n_1 x + a_2 \sin n_2 x + a_3 \sin n_3 x + a_4 \sin n_4 x + \&c.)^{1} \ldots (e).$$

291. To determine the coefficient a_1, multiply both members of the equation by $x \sin nx\, dx$, and integrate from $x = 0$ to $x = X$. The integral $\int \sin mx \sin nx\, dx$ taken between these limits is

$$\frac{1}{m^2 - n^2}(-m \sin nX \cos mX + n \sin mX \cos nX).$$

If m and n are numbers chosen from the roots n_1, n_2, n_3, &c., which satisfy the equation $\frac{nX}{\tan nX} = 1 - hX$, we have

$$\frac{mX}{\tan mX} = \frac{nX}{\tan nX},$$

or $$m \cos mX \sin nX - n \sin mX \cos nX = 0.$$

We see by this that the whole value of the integral is nothing; but a single case exists in which the integral does not vanish, namely, when $m = n$. It then becomes $\frac{0}{0}$; and, by application of known rules, is reduced to

$$\frac{1}{2} X - \frac{1}{4n} \sin 2nX.$$

[1] Of the possibility of representing an arbitrary function by a series of this form a demonstration has been given by Sir W. Thomson, *Camb. Math. Journal*, Vol. III. pp. 25—27. [A. F.]

It follows from this that in order to obtain the value of the coefficient a_1, in equation (e), we must write

$$2\int x \sin n_1x\, F(x)\, dx = a_1\left(X - \frac{1}{2n_1}\sin 2n_1X\right),$$

the integral being taken from $x=0$ to $x=X$. Similarly we have

$$2\int x \sin n_2x\, F(x)\, dx = a_2\left(X - \frac{1}{2n_2}\sin 2n_2X\right).$$

In the same manner all the following coefficients may be determined. It is easy to see that the definite integral $2\int x \sin nx\, F(x)\, dx$ always has a determinate value, whatever the arbitrary function $F(x)$ may be. If the function $F(x)$ be represented by the variable ordinate of a line traced in any manner, the function $xF(x)\sin nx$ corresponds to the ordinate of a second line which can easily be constructed by means of the first. The area bounded by the latter line between the abscissæ $x=0$ and $x=X$ determines the coefficient a_i, i being the index of the order of the root n.

The arbitrary function $F(x)$ enters each coefficient under the sign of integration, and gives to the value of v all the generality which the problem requires; thus we arrive at the following equation

$$\frac{xv}{2} = \frac{\sin n_1x \int x \sin n_1x\, F(x)\, dx}{X - \frac{1}{2n_1}\sin 2n_1X}\, e^{-kn_1^2t}$$

$$+ \frac{\sin n_2x \int x \sin n_2x\, F(x)\, dx}{X - \frac{1}{2n_2}\sin 2\,n_2X}\, e^{-kn_2^2t} + \&c.$$

This is the form which must be given to the general integral of the equation

$$\frac{dv}{dt} = k\left(\frac{d^2v}{dx^2} + \frac{2}{x}\frac{dv}{dx}\right),$$

in order that it may represent the movement of heat in a solid sphere. In fact, all the conditions of the problem are obeyed.

1st, The partial differential equation is satisfied; 2nd, the quantity of heat which escapes at the surface accords at the same time with the mutual action of the last layers and with the action of the air on the surface; that is to say, the equation $\frac{dv}{dx} + hx = 0$, which each part of the value of v satisfies when $x = X$, holds also when we take for v the sum of all these parts; 3rd, the given solution agrees with the initial state when we suppose the time nothing.

292. The roots n_1, n_2, n_3, &c. of the equation

$$\frac{nX}{\tan nX} = 1 - hX$$

are very unequal; whence we conclude that if the value of the time is considerable, each term of the value of v is very small, relatively to that which precedes it. As the time of cooling increases, the latter parts of the value of v cease to have any sensible influence; and those partial and elementary states, which at first compose the general movement, in order that the initial state may be represented by them, disappear almost entirely, one only excepted. In the ultimate state the temperatures of the different layers decrease from the centre to the surface in the same manner as in a circle the ratios of the sine to the arc decrease as the arc increases. This law governs naturally the distribution of heat in a solid sphere. When it begins to exist, it exists through the whole duration of the cooling. Whatever the function $F(x)$ may be which represents the initial state, the law in question tends continually to be established; and when the cooling has lasted some time, we may without sensible error suppose it to exist.

293. We shall apply the general solution to the case in which the sphere, having been for a long time immersed in a fluid, has acquired at all its points the same temperature. In this case the function $F(x)$ is 1, and the determination of the coefficients is reduced to integrating $x \sin nx\, dx$, from $x = 0$ to $x = X$: the integral is

$$\frac{\sin nX - nX \cos nX}{n^2}.$$

Hence the value of each coefficient is expressed thus:

$$a = \frac{2}{n}\frac{\sin nX - nX\cos nX}{nX - \sin nX \cos nX};$$

the order of the coefficient is determined by that of the root n, the equation which gives the values of n being

$$\frac{nX\cos nX}{\sin nX} = 1 - hX.$$

We therefore find

$$a = \frac{2}{n}\frac{hX}{nX\operatorname{cosec} nX - \cos nX}.$$

It is easy now to form the general value which is given by the equation

$$\frac{vx}{2Xh} = \frac{e^{-kn_1^2t}\sin n_1x}{n_1(n_1X\operatorname{cosec} n_1X - \cos n_1X)} + \frac{e^{-kn_2^2t}\sin n_2x}{n_2(n_2X\operatorname{cosec} n_2X - \cos n_2X)} + \&c.$$

Denoting by ϵ_1, ϵ_2, ϵ_3, &c. the roots of the equation

$$\frac{\epsilon}{\tan\epsilon} = 1 - hX,$$

and supposing them arranged in order beginning with the least; replacing n_1X, n_2X, n_3X, &c. by ϵ_1, ϵ_2, ϵ_3, &c., and writing instead of k and h their values $\frac{K}{CD}$ and $\frac{h}{K}$, we have for the expression of the variations of temperature during the cooling of a solid sphere, which was once uniformly heated, the equation

$$v = \frac{2h}{K}X\left\{\frac{\sin\frac{\epsilon_1x}{X}}{\frac{\epsilon_1x}{X}}\frac{e^{-\frac{K}{CD}\frac{\epsilon_1^2}{X^2}t}}{\epsilon_1\operatorname{cosec}\epsilon_1 - \cos\epsilon_1} + \frac{\sin\frac{\epsilon_2x}{X}}{\frac{\epsilon_2x}{X}}\frac{e^{-\frac{K}{CD}\frac{\epsilon_2^2}{X^2}t}}{\epsilon_2\operatorname{cosec}\epsilon_2 - \cos\epsilon_2} + \&c.\right\}.$$

Note. The problem of the sphere has been very completely discussed by Riemann, *Partielle Differentialgleichungen*, §§ 61—69. [A. F.]

SECTION II.

Different remarks on this solution.

294. We will now explain some of the results which may be derived from the foregoing solution. If we suppose the coefficient h, which measures the facility with which heat passes into the air, to have a very small value, or that the radius X of the sphere is very small, the least value of ϵ becomes very small; so that the equation $\frac{\epsilon}{\tan\epsilon} = 1 - \frac{h}{K}X$ is reduced to $\frac{\epsilon\left(1-\frac{1}{2}\epsilon^2\right)}{\epsilon - \frac{1}{2.3}\epsilon^3} = 1 - \frac{hX}{K}$, or, omitting the higher powers of ϵ, $\epsilon^2 = \frac{3hX}{K}$. On the other hand, the quantity $\frac{\epsilon}{\sin\epsilon} - \cos\epsilon$ becomes, on the same hypothesis, $\frac{2hX}{K}$. And the term $\frac{\sin\frac{\epsilon x}{X}}{\frac{\epsilon x}{X}}$ is reduced to 1. On making these substitutions in the general equation we have $v = e^{-\frac{3h}{CDX}t} + \&c.$ We may remark that the succeeding terms decrease very rapidly in comparison with the first, since the second root n_2 is very much greater than 0; so that if either of the quantities h or X has a small value, we may take, as the expression of the variations of temperature, the equation $v = e^{-\frac{3ht}{CDX}}$. Thus the different spherical envelopes of which the solid is composed retain a common temperature during the whole of the cooling. The temperature diminishes as the ordinate of a logarithmic curve, the time being taken for abscissa; the initial temperature 1 is reduced after the time t to $e^{-\frac{3ht}{CDX}}$. In order that the initial temperature may be reduced to the fraction $\frac{1}{m}$, the value of t must be $\frac{X}{3h}CD\log m$. Thus in spheres of the same material but

of different diameters, the times occupied in losing half or the same defined part of their actual heat, when the exterior conducibility is very small, are proportional to their diameters. The same is the case with solid spheres whose radius is very small; and we should also find the same result on attributing to the interior conducibility K a very great value. The statement holds generally when the quantity $\frac{hX}{K}$ is very small. We may regard the quantity $\frac{h}{K}$ as very small when the body which is being cooled is formed of a liquid continually agitated, and enclosed in a spherical vessel of small thickness. The hypothesis is in some measure the same as that of perfect conducibility; the temperature decreases then according to the law expressed by the equation $v = e^{-\frac{3ht}{CDX}}$.

295. By the preceding remarks we see that in a solid sphere which has been cooling for a long time, the temperature decreases from the centre to the surface as the quotient of the sine by the arc decreases from the origin where it is 1 to the end of a given arc ϵ, the radius of each layer being represented by the variable length of that arc. If the sphere has a small diameter, or if its interior conducibility is very much greater than the exterior conducibility, the temperatures of the successive layers differ very little from each other, since the whole arc ϵ which represents the radius X of the sphere is of small length. The variation of the temperature v common to all its points is then given by the equation $v = e^{-\frac{3ht}{CDX}}$. Thus, on comparing the respective times which two small spheres occupy in losing half or any aliquot part of their actual heat, we find those times to be proportional to the diameters.

296. The result expressed by the equation $v = e^{-\frac{3ht}{CDX}}$ belongs only to masses of similar form and small dimension. It has been known for a long time by physicists, and it offers itself as it were spontaneously. In fact, if any body is sufficiently small for the temperatures at its different points to be regarded as equal, it is easy to ascertain the law of cooling. Let 1 be the initial

temperature common to all points; it is evident that the quantity of heat which flows during the instant dt into the medium supposed to be maintained at temperature 0 is $hSvdt$, denoting by S the external surface of the body. On the other hand, if C is the heat required to raise unit of weight from the temperature 0 to the temperature 1, we shall have CDV for the expression of the quantity of heat which the volume V of the body whose density is D would take from temperature 0 to temperature 1. Hence $\frac{hSvdt}{CDV}$ is the quantity by which the temperature v is diminished when the body loses a quantity of heat equal to $hSvdt$. We ought therefore to have the equation

$$dv = -\frac{hSvdt}{CDV}, \text{ or } v = e^{\frac{-hSt}{CDV}}.$$

If the form of the body is a sphere whose radius is X, we shall have the equation $v = e^{\frac{-3ht}{CDX}}$.

297. Assuming that we observe during the cooling of the body in question two temperatures v_1 and v_2 corresponding to the times t_1 and t_2, we have

$$\frac{hS}{CDV} = \frac{\log v_1 - \log v_2}{t_2 - t_1}.$$

We can then easily ascertain by experiment the exponent $\frac{hS}{CDV}$. If the same observation be made on different bodies, and if we know in advance the ratio of their specific heats C and C', we can find that of their exterior conducibilities h and h'. Reciprocally, if we have reason to regard as equal the values h and h' of the exterior conducibilities of two different bodies, we can ascertain the ratio of their specific heats. We see by this that, by observing the times of cooling for different liquids and other substances enclosed successively in the same vessel whose thickness is small, we can determine exactly the specific heats of those substances.

We may further remark that the coefficient K which measures the interior conducibility does not enter into the equation

$$v = e^{\frac{-3ht}{CDX}}.$$

Thus the time of cooling in bodies of small dimension does not depend on the interior conducibility; and the observation of these times can teach us nothing about the latter property; but it could be determined by measuring the times of cooling in vessels of different thicknesses.

298. What we have said above on the cooling of a sphere of small dimension, applies to the movement of heat in a thermometer surrounded by air or fluid. We shall add the following remarks on the use of these instruments.

Suppose a mercurial thermometer to be dipped into a vessel filled with hot water, and that the vessel is being cooled freely in air at constant temperature. It is required to find the law of the successive falls of temperature of the thermometer.

If the temperature of the fluid were constant, and the thermometer dipped in it, its temperature would change, approaching very quickly that of the fluid. Let v be the variable temperature indicated by the thermometer, that is to say, its elevation above the temperature of the air; let u be the elevation of temperature of the fluid above that of the air, and t the time corresponding to these two values v and u. At the beginning of the instant dt which is about to elapse, the difference of the temperature of the thermometer from that of the fluid being $v-u$, the variable v tends to diminish and will lose in the instant dt a quantity proportional to $v-u$; so that we have the equation

$$dv = -h\,(v-u)\,dt.$$

During the same instant dt the variable u tends to diminish, and it loses a quantity proportional to u, so that we have the equation

$$du = -Hudt.$$

The coefficient H expresses the velocity of the cooling of the liquid in air, a quantity which may easily be discovered by experiment, and the coefficient h expresses the velocity with which the thermometer cools in the liquid. The latter velocity is very much greater than H. Similarly we may from experiment find the coefficient h by making the thermometer cool in fluid maintained at a constant temperature. The two equations

$$du = -Hudt \quad\text{and}\quad dv = -h\,(v-u)\,dt,$$

or $$u = Ae^{-Ht} \text{ and } \frac{dv}{dt} = -hv + hAe^{-Ht}$$

lead to the equation

$$v - u = be^{-ht} + aHe^{-Ht},$$

a and b being arbitrary constants. Suppose now the initial value of $v-u$ to be Δ, that is, that the height of the thermometer exceeds by Δ the true temperature of the fluid at the beginning of the immersion; and that the initial value of u is E. We can determine a and b, and we shall have

$$v - u = \Delta e^{-ht} + \frac{HE}{h-H}(e^{-Ht} - e^{-ht}).$$

The quantity $v-u$ is the error of the thermometer, that is to say, the difference which is found between the temperature indicated by the thermometer and the real temperature of the fluid at the same instant. This difference is variable, and the preceding equation informs us according to what law it tends to decrease. We see by the expression for the difference $v-u$ that two of its terms containing e^{-ht} diminish very rapidly, with the velocity which would be observed in the thermometer if it were dipped into fluid at constant temperature. With respect to the term which contains e^{-Ht}, its decrease is much slower, and is effected with the velocity of cooling of the vessel in air. It follows from this, that after a time of no great length the error of the thermometer is represented by the single term

$$\frac{HE}{h-H}e^{-Ht} \quad \text{or} \quad \frac{H}{h-H}u.$$

299. Consider now what experiment teaches as to the values of H and h. Into water at 8·5° (octogesimal scale) we dipped a thermometer which had first been heated, and it descended in the water from 40 to 20 degrees in six seconds. This experiment was repeated carefully several times. From this we find that the value of e^{-h} is 0·000042[1]; if the time is reckoned in minutes, that is to say, if the height of the thermometer be E at the beginning of a minute, it will be E(0·000042) at the end of the minute. Thus we find

$$h \log_{10} e = 4{\cdot}3761271.$$

[1] 0·00004206, strictly. [A. F.]

At the same time a vessel of porcelain filled with water heated to 60° was allowed to cool in air at 12°. The value of e^{-H} in this case was found to be 0·98514, hence that of $H\log_{10} e$ is 0·006500. We see by this how small the value of the fraction e^{-h} is, and that after a single minute each term multiplied by e^{-ht} is not half the ten-thousandth part of what it was at the beginning of the minute. We need not therefore take account of those terms in the value of $v-u$. The equation becomes

$$v-u=\frac{Hu}{h-H} \quad \text{or} \quad v-u=\frac{Hu}{h}+\frac{H}{H-h}\,\frac{Hu}{h}.$$

From the values found for H and h, we see that the latter quantity h is more than 673 times greater than H, that is to say, the thermometer cools in air more than 600 times faster than the vessel cools in air. Thus the term $\frac{Hu}{h}$ is certainly less than the 600th part of the elevation of temperature of the water above that of the air, and as the term $\frac{H}{h-H}\,\frac{Hu}{h}$ is less than the 600th part of the preceding term, which is already very small, it follows that the equation which we may employ to represent very exactly the error of the thermometer is

$$v-u=\frac{Hu}{h}.$$

In general if H is a quantity very great relatively to h, we have always the equation

$$v-u=\frac{Hu}{h}.$$

300. The investigation which we have just made furnishes very useful results for the comparison of thermometers.

The temperature marked by a thermometer dipped into a fluid which is cooling is always a little greater than that of the fluid. This excess or error of the thermometer differs with the height of the thermometer. The amount of the correction will be found by multiplying u the actual height of the thermometer by the ratio of H, the velocity of cooling of the vessel in air, to h the velocity of cooling of the thermometer in the fluid. We might suppose that the thermometer, when it was dipped into the fluid, marked a lower temperature. This is what almost always happens, but this state cannot last, the thermometer begins to approach to the temperature of the fluid; at the same time the fluid cools, so that the thermometer passes first to the same temperature as the fluid, and it then indicates a temperature very slightly different but always higher.

300*. We see by these results that if we dip different thermometers into the same vessel filled with fluid which is cooling slowly, they must all indicate very nearly the same temperature at the same instant. Calling h, h', h'', the velocities of cooling of the thermometers in the fluid, we shall have

$$\frac{Hu}{h}, \frac{Hu}{h'}, \frac{Hu}{h''},$$

as their respective errors. If two thermometers are equally sensitive, that is to say if the quantities h and h' are the same, their temperatures will differ equally from those of the fluid. The values of the coefficients h, h', h'' are very great, so that the errors of the thermometers are extremely small and often inappreciable quantities. We conclude from this that if a thermometer is constructed with care and can be regarded as exact, it will be easy to construct several other thermometers of equal exactness. It will be sufficient to place all the thermometers which we wish to graduate in a vessel filled with a fluid which cools slowly, and to place in it at the same time the thermometer which ought to serve as a model; we shall only have to observe all from degree to degree, or at greater intervals, and we must mark the points where the mercury is found at the same time in the different thermometers. These points will be at the divisions required. We have applied this process to the construction of the thermometers employed in our experiments, so that these instruments coincide always in similar circumstances.

This comparison of thermometers during the time of cooling not only establishes a perfect coincidence among them, and renders them all similar to a single model; but from it we derive also the means of exactly dividing the tube of the principal thermometer, by which all the others ought to be regulated. In this way we satisfy the fundamental condition of the instrument, which is, that any two intervals on the scale which include the same number of degrees should contain the same quantity of mercury. For the rest we omit here several details which do not directly belong to the object of our work.

301. We have determined in the preceding articles the temperature v received after the lapse of a time t by an interior spherical layer at a distance x from the centre. It is required now to calculate the value of the mean temperature of the sphere, or that which the solid would have if the whole quantity of heat which it contains were equally distributed throughout the whole mass. The volume of a sphere whose radius is x being $\frac{4\pi x^3}{3}$, the quantity of heat contained in a spherical envelope whose temperature is v, and radius x, will be $v\,d\left(\frac{4\pi x^3}{3}\right)$. Hence the mean temperature is

$$\int \frac{v\,.\,d\left(\frac{4\pi x^3}{3}\right)}{\frac{4\pi X^3}{3}} \quad \text{or} \quad \frac{3}{X^3}\int x^2 v\,dx,$$

the integral being taken from $x=0$ to $x=X$. Substitute for v its value

$$\frac{a_1}{x}e^{-kn_1^2t}\sin n_1x+\frac{a_2}{x}e^{-kn_2^2t}\sin n_2x+\frac{a_3}{x}e^{-kn_3^2t}\sin n_3x+\text{etc.}$$

and we shall have the equation

$$\frac{3}{X^3}\int x^2 v\,dx=\frac{3}{X^3}\left\{a_1\frac{\sin n_1X-n_1X\cos n_1X}{n_1^2}e^{-kn_1^2t}\right.$$
$$\left.+a_2\frac{\sin n_2X-n_2X\cos n_2X}{n_2^2}e^{-kn_2^2t}+\&\text{c.}\right\}.$$

We found formerly (Art. 293)

$$a_i=\frac{2}{n_i}\,\frac{\sin n_iX-n_iX\cos n_iX}{2n_iX-\frac{1}{2}\sin 2n_iX}.$$

We have, therefore, if we denote the mean temperature by z,

$$\frac{z}{3 \cdot 4} = \frac{(\sin \epsilon_1 - \epsilon_1 \cos \epsilon_1)^2}{\epsilon_1^3 (2\epsilon_1 - \sin 2\epsilon_1)} e^{-\frac{K\epsilon_1^2 t}{CDX^2}} + \frac{(\sin \epsilon_2 - \epsilon_2 \cos \epsilon_2)^2}{\epsilon_2^3 (2\epsilon_2 - \sin 2\epsilon_2)} e^{-\frac{K\epsilon_2^2 t}{CDX^2}} + \&c.,$$

an equation in which the coefficients of the exponentials are all positive.

302. Let us consider the case in which, all other conditions remaining the same, the value X of the radius of the sphere becomes infinitely great[1]. Taking up the construction described in Art. 285, we see that since the quantity $\frac{hX}{K}$ becomes infinite, the straight line drawn through the origin cutting the different branches of the curve coincides with the axis of x. We find then for the different values of ϵ the quantities π, 2π, 3π, etc.

Since the term in the value of z which contains $e^{-\frac{K}{CD}\frac{\epsilon_1^2}{X^2}t}$ becomes, as the time increases, very much greater than the following terms, the value of z after a certain time is expressed without sensible error by the first term only. The index $\frac{Kn^2}{CD}$ being equal to $\frac{K\pi^2}{CDX^2}$, we see that the final cooling is very slow in spheres of great diameter, and that the index of e which measures the velocity of cooling is inversely as the square of the diameter.

303. From the foregoing remarks we can form an exact idea of the variations to which the temperatures are subject during the cooling of a solid sphere. The initial values of the temperatures change successively as the heat is dissipated through the surface. If the temperatures of the different layers are at first equal, or if they diminish from the surface to the centre, they do not maintain their first ratios, and in all cases the system tends more and more towards a lasting state, which after no long delay is sensibly attained. In this final state the temperatures decrease

[1] Riemann has shewn, *Part. Diff. gleich.* § 69, that in the case of a very large sphere, uniformly heated initially, the surface temperature varies ultimately as the square root of the time inversely. [A. F.]

from the centre to the surface. If we represent the whole radius of the sphere by a certain arc ϵ less than a quarter of the circumference, and, after dividing this arc into equal parts, take for each point the quotient of the sine by the arc, this system of ratios will represent that which is of itself set up among the temperatures of layers of equal thickness. From the time when these ultimate ratios occur they continue to exist throughout the whole of the cooling. Each of the temperatures then diminishes as the ordinate of a logarithmic curve, the time being taken for abscissa. We can ascertain that this law is established by observing several successive values z, z', z'', z''', etc., which denote the mean temperature for the times t, $t+\Theta$, $t+2\Theta$, $t+3\Theta$, etc.; the series of these values converges always towards a geometrical progression, and when the successive quotients $\frac{z}{z'}$, $\frac{z'}{z''}$, $\frac{z''}{z'''}$, etc. no longer change, we conclude that the relations in question are established between the temperatures. When the diameter of the sphere is small, these quotients become sensibly equal as soon as the body begins to cool. The duration of the cooling for a given interval, that is to say the time required for the mean temperature z to be reduced to a definite part of itself $\frac{z}{m}$, increases as the diameter of the sphere is enlarged.

304. If two spheres of the same material and different dimensions have arrived at the final state in which whilst the temperatures are lowered their ratios are preserved, and if we wish to compare the durations of the same degree of cooling in both, that is to say, the time Θ which the mean temperature of the first occupies in being reduced to $\frac{z}{m}$, and the time Θ in which the temperature z' of the second becomes $\frac{z'}{m}$, we must consider three different cases. If the diameter of each sphere is small, the durations Θ and Θ' are in the same ratio as the diameters. If the diameter of each sphere is very great, the durations Θ and Θ' are in the ratio of the squares of the diameters; and if the diameters of the spheres are included between these two limits, the ratios of the times will be greater than that of the diameters, and less than that of their squares.

The exact value of the ratio has been already determined[1]. The problem of the movement of heat in a sphere includes that of the terrestrial temperatures. In order to treat of this problem at greater length, we have made it the object of a separate chapter[2].

305. The use which has been made above of the equation $\frac{\epsilon}{\tan\epsilon} = \lambda$ is founded on a geometrical construction which is very well adapted to explain the nature of these equations. The construction indeed shows clearly that all the roots are real; at the same time it ascertains their limits, and indicates methods for determining the numerical value of each root. The analytical investigation of equations of this kind would give the same results. First, we might ascertain that the equation $\epsilon - \lambda\tan\epsilon = 0$, in which λ is a known number less than unity, has no imaginary root of the form $m + n\sqrt{-1}$. It is sufficient to substitute this quantity for ϵ; and we see after the transformations that the first member cannot vanish when we give to m and n real values, unless n is nothing. It may be proved moreover that there can be no imaginary root of any form whatever in the equation

$$\epsilon - \lambda\tan\epsilon = 0, \text{ or } \frac{\epsilon\cos\epsilon - \lambda\sin\epsilon}{\cos\epsilon} = 0.$$

In fact, 1st, the imaginary roots of the factor $\frac{1}{\cos\epsilon} = 0$ do not belong to the equation $\epsilon - \lambda\tan\epsilon = 0$, since these roots are all of the form $m + n\sqrt{-1}$; 2nd, the equation $\sin\epsilon - \frac{\epsilon}{\lambda}\cos\epsilon = 0$ has necessarily all its roots real when λ is less than unity. To prove this proposition we must consider $\sin\epsilon$ as the product of the infinite number of factors

[1] It is $\Theta : \Theta' = \epsilon_1'^2 X^2 : \epsilon_1^2 X'^2$, as may be inferred from the exponent of the first term in the expression for z, Art. 301. [A. F.]

[2] The chapter referred to is not in this work. It forms part of the *Suite du mémorie sur la théorie du mouvement de la chaleur dans les corps solides*. See note, page 10.

The first memoir, entitled *Théorie du mouvement de la chaleur dans les corps solides*, is that which formed the basis of the *Théorie analytique du mouvement de la chaleur* published in 1822, but was considerably altered and enlarged in that work now translated. [A. F.]

$$\epsilon\left(1-\frac{\epsilon^2}{\pi^2}\right)\left(1-\frac{\epsilon^2}{2^2\pi^2}\right)\left(1-\frac{\epsilon^2}{3^2\pi^2}\right)\left(1-\frac{\epsilon^2}{4^2\pi^2}\right)\text{\&c.},$$

and consider $\cos\epsilon$ as derived from $\sin\epsilon$ by differentiation.

Suppose that instead of forming $\sin\epsilon$ from the product of an infinite number of factors, we employ only the m first, and denote the product by $\phi_m(\epsilon)$. To find the corresponding value of $\cos\epsilon$, we take

$$\frac{d}{d\epsilon}\phi_m(\epsilon) \text{ or } \phi_m'(\epsilon).$$

This done, we have the equation

$$\phi_m(\epsilon) - \epsilon\phi_m'(\epsilon) = 0.$$

Now, giving to the number m its successive values 1, 2, 3, 4, &c. from 1 to infinity, we ascertain by the ordinary principles of Algebra, the nature of the functions of ϵ which correspond to these different values of m. We see that, whatever m the number of factors may be, the equations in ϵ which proceed from them have the distinctive character of equations all of whose roots are real. Hence we conclude rigorously that the equation

$$\frac{\epsilon}{\tan\epsilon} = \lambda,$$

in which λ is less than unity, cannot have an imaginary root[1]. The same proposition could also be deduced by a different analysis which we shall employ in one of the following chapters.

Moreover the solution we have given is not founded on the property which the equation possesses of having all its roots real. It would not therefore have been necessary to prove this proposition by the principles of algebraical analysis. It is sufficient for the accuracy of the solution that the integral can be made to coincide with any initial state whatever; for it follows rigorously that it must then also represent all the subsequent states.

[1] The proof given by Riemann, *Part. Diff. Gleich.* § 67, is more simple. The method of proof is in part claimed by Poisson, *Bulletin de la Société Philomatique*, Paris, 1826, p. 147. [A. F.].

CHAPTER VI.

OF THE MOVEMENT OF HEAT IN A SOLID CYLINDER.

306. THE movement of heat in a solid cylinder of infinite length, is represented by the equations

$$\frac{dv}{dt} = \frac{K}{CD}\left(\frac{d^2v}{dx^2} + \frac{1}{x}\frac{dv}{dx}\right) \quad \text{and} \quad \frac{h}{K}V + \frac{dV}{dx} = 0,$$

which we have stated in Articles 118, 119, and 120. To integrate these equations we give to v the simple particular value expressed by the equation $v = ue^{-mt}$; m being any number, and u a function of x. We denote by k the coefficient $\frac{K}{CD}$ which enters the first equation, and by h the coefficient $\frac{h}{K}$ which enters the second equation. Substituting the value assigned to v, we find the following condition

$$\frac{m}{k}u + \frac{d^2u}{dx^2} + \frac{1}{x}\frac{du}{dx} = 0.$$

Next we choose for u a function of x which satisfies this differential equation. It is easy to see that the function may be expressed by the following series

$$u = 1 - \frac{gx^2}{2^2} + \frac{g^2x^4}{2^2 \cdot 4^2} - \frac{g^3x^6}{2^2 \cdot 4^2 \cdot 6^2} + \text{\&c.},$$

g denoting the constant $\frac{m}{k}$. We shall examine more particularly in the sequel the differential equation from which this series is derived; here we consider the function u to be known, and we have ue^{-gkt} as the particular value of v.

The state of the convex surface of the cylinder is subject to a condition expressed by the definite equation

$$hV + \frac{dV}{dx} = 0,$$

which must be satisfied when the radius x has its total value X; whence we obtain the definite equation

$$h\left(1 - g\frac{X^2}{2^2} + \frac{g^2X^4}{2^2 . 4^2} - \frac{g^3X^6}{2^2 . 4^2 . 6^2} + \&c.\right)$$
$$= \frac{2gX}{2^2} - \frac{4g^2X^3}{2^2 . 4^2} + \frac{6g^3X^5}{2^2 . 4^2 . 6^2} - \&c.:$$

thus the number g which enters into the particular value ue^{-gkt} is not arbitrary. The number must necessarily satisfy the preceding equation, which contains g and X.

We shall prove that this equation in g in which h and X are given quantities has an infinite number of roots, and that all these roots are real. It follows that we can give to the variable v an infinity of particular values of the form ue^{-gkt}, which differ only by the exponent g. We can then compose a more general value, by adding all these particular values multiplied by arbitrary coefficients. This integral which serves to resolve the proposed equation in all its extent is given by the following equation

$$v = a_1u_1e^{-g_1kt} + a_2u_2e^{-g_2kt} + a_3u_3e^{-gk_3t} + \&c.,$$

g_1, g_2, g_3, &c. denote all the values of g which satisfy the definite equation; u_1, u_2, u_3, &c. denote the values of u which correspond to these different roots; a_1, a_2, a_3, &c. are arbitrary coefficients which can only be determined by the initial state of the solid.

307. We must now examine the nature of the definite equation which gives the values of g, and prove that all the roots of this equation are real, an investigation which requires attentive examination.

In the series

$$1 - g\frac{X^2}{2^2} + \frac{g^2X^4}{2^2.4^2} - \frac{g^3X^6}{2^2.4^2.6^2} + \&c.,$$

which expresses the value which u receives when $x = X$, we shall replace $\frac{gX^2}{2^2}$ by the quantity θ, and denoting this function of θ by $f(\theta)$ or y, we have

$$y = f(\theta) = 1 - \theta + \frac{\theta^2}{2^2} - \frac{\theta^3}{2^2.3^2} + \frac{\theta^4}{2^2.3^2.4^2} + \&c.,$$

the definite equation becomes

$$\frac{hX}{2} = \frac{\theta - 2\frac{\theta^2}{2^2} + 3\frac{\theta^3}{2^2.3^2} - 4\frac{\theta^4}{2^2.3^2.4^2} + \&c.}{1 - \theta + \frac{\theta^2}{2^2} - \frac{\theta^3}{2^2.3^2} + \frac{\theta^4}{2^2.3^2.4^2} - \&c.},$$

or

$$\frac{hX}{2} + \theta\frac{f'(\theta)}{f(\theta)} = 0,$$

$f'(\theta)$ denoting the function $\frac{df(\theta)}{d\theta}$.

Each value of θ furnishes a value for g, by means of the equation

$$g\frac{X^2}{2^2} = \theta;$$

and we thus obtain the quantities g_1, g_2, g_3, &c. which enter in infinite number into the solution required.

The problem is then to prove that the equation

$$\frac{hX}{2} + \theta\frac{f'(\theta)}{f(\theta)} = 0$$

must have all its roots real. We shall prove in fact that the equation $f(\theta) = 0$ has all its roots real, that the same is the case consequently with the equation $f'(\theta) = 0$, and that it follows that the equation

$$A = \frac{\theta f'(\theta)}{f(\theta)}$$

has also all its roots real, A representing the known number

$$-\frac{hX}{2}.$$

308. The equation

$$y = 1 - \theta + \frac{\theta^2}{2^2} - \frac{\theta^3}{2^2 . 3^2} + \frac{\theta^4}{2^2 . 3^2 . 4^2} - \&c.,$$

on being differentiated twice, gives the following relation

$$y + \frac{dy}{d\theta} + \theta \frac{d^2y}{d\theta^2} = 0.$$

We write, as follows, this equation and all those which may be derived from it by differentiation,

$$y + \frac{dy}{d\theta} + \theta \frac{d^2y}{d\theta^2} = 0,$$

$$\frac{dy}{d\theta} + 2\frac{d^2y}{d\theta^2} + \theta \frac{d^3y}{d\theta^3} = 0,$$

$$\frac{d^2y}{d\theta^2} + 3\frac{d^3y}{d\theta^3} + \theta \frac{d^4y^4}{d\theta^4} = 0,$$

&c.,

and in general

$$\frac{d^iy}{d\theta^i} + (i+1)\frac{d^{i+1}y}{d\theta^{i+1}} + \theta \frac{d^{i+2}y}{d\theta^{i+2}} = 0.$$

Now if we write in the following order the algebraic equation $X = 0$, and all those which may be derived from it by differentiation,

$$X = 0, \quad \frac{dX}{dx} = 0, \quad \frac{d^2X}{dx^2} = 0, \quad \frac{d^3X}{dx^3} = 0, \text{ \&c.,}$$

and if we suppose that every real root of any one of these equations on being substituted in that which precedes and in that which follows it gives two results of opposite sign; it is certain that the proposed equation $X = 0$ has all its roots real, and that consequently the same is the case in all the subordinate equations

$$\frac{dX}{dx} = 0, \quad \frac{d^2X}{dx^2} = 0, \quad \frac{d^3X}{dx^3} = 0, \text{ \&c.}$$

These propositions are founded on the theory of algebraic equations, and have been proved long since. It is sufficient to prove that the equations

$$y = 0, \quad \frac{dy}{d\theta} = 0, \quad \frac{d^2y}{d\theta^2} = 0, \text{ \&c.}$$

fulfil the preceding condition. Now this follows from the general equation

$$\frac{d^i y}{d\theta^i} + (i+1)\frac{d^{i+1}y}{d\theta^{i+1}} + \theta\frac{d^{i+2}y}{d\theta^{i+2}} = 0:$$

for if we give to θ a positive value which makes the fluxion $\frac{d^{i+1}y}{d\theta^{i+1}}$ vanish, the other two terms $\frac{d^i y}{d\theta^i}$ and $\frac{d^{i+1}y}{d\theta^{i+1}}$ receive values of opposite sign. With respect to the negative values of θ it is evident, from the nature of the function $f(\theta)$, that no negative value substituted for θ can reduce to nothing, either that function, or any of the others which are derived from it by differentiation: for the substitution of any negative quantity gives the same sign to all the terms. Hence we are assured that the equation $y=0$ has all its roots real and positive.

309. It follows from this that the equation $f'(\theta)=0$ or $y'=0$ also has all its roots real; which is a known consequence from the principles of algebra. Let us examine now what are the successive values which the term $\theta\frac{f'(\theta)}{f(\theta)}$ or $\theta\frac{y'}{y}$ receives when we give to θ values which continually increase from $\theta=0$ to $\theta=\infty$. If a value of θ makes y' nothing, the quantity $\theta\frac{y'}{y}$ becomes nothing also; it becomes infinite when θ makes y nothing. Now it follows from the theory of equations that in the case in question, every root of $y'=0$ lies between two consecutive roots of $y=0$, and reciprocally. Hence denoting by θ_1 and θ_3 two consecutive roots of the equation $y'=0$, and by θ_2 that root of the equation $y=0$ which lies between θ_1 and θ_3, every value of θ included between θ_1 and θ_2 gives to y a sign different from that which the function y would receive if θ had a value included between θ_2 and θ_3. Thus the quantity $\theta\frac{y'}{y}$ is nothing when $\theta=\theta_1$; it is infinite when $\theta=\theta_2$, and nothing when $\theta=\theta_3$. The quantity $\theta\frac{y'}{y}$ must therefore necessarily take all possible values, from θ to infinity, in the interval from θ to θ_2, and must also take all possible values of the opposite sign, from infinity to zero, in the interval from θ_2 to θ_3. Hence the equation $A=\theta\frac{y'}{y}$ necessarily has one

real root between θ_1 and θ_3 and since the equation $y'=0$ has all its roots real in infinite number, it follows that the equation $A=\theta\frac{y'}{y}$ has the same property. In this manner we have achieved the proof that the definite equation

$$\frac{hX}{2}=\frac{\frac{gX^2}{2^2}-2\frac{g^2X^4}{2^2.4^2}+3\frac{g^3X^6}{2^2.4^2.6^2}-\&c.}{1-\frac{gX^2}{2^2}+\frac{g^2X^4}{2^2.4^2}-\frac{g^3X^6}{2^2.4^2.6^2}+\&c.},$$

in which the unknown is g, has all its roots real and positive. We proceed to continue the investigation of the function u and of the differential equation which it satisfies.

310. From the equation $y+\frac{dy}{d\theta}+\theta\frac{d^2y}{d\theta^2}=0$, we derive the general equation $\frac{d^iy}{d\theta^i}+(i+1)\frac{d^{i+1}y}{d\theta^{i+1}}+\theta\frac{d^{i+2}y}{d\theta^{i+2}}=0$, and if we suppose $\theta=0$ we have the equation

$$\frac{d^{i+1}y}{d\theta^{i+1}}=-\frac{1}{i+1}\frac{d^iy}{d\theta^i},$$

which serves to determine the coefficients of the different terms of the development of the function $f(\theta)$, since these coefficients depend on the values which the differential coefficients receive when the variable in them is made to vanish. Supposing the first term to be known and to be equal to 1, we have the series

$$y=1-\theta+\frac{\theta^2}{2^2}-\frac{\theta^3}{2^2.3^2}+\frac{\theta^4}{2^2.3^2.4^2}-\&c.$$

If now in the equation proposed

$$gu+\frac{d^2u}{dx^2}+\frac{1}{x}\frac{du}{dx}=0$$

we make $g\frac{x^2}{2^2}=\theta$, and seek for the new equation in u and θ, regarding u as a function of θ, we find

$$u+\frac{du}{d\theta}+\theta\frac{d^2u}{d\theta^2}=0.$$

Whence we conclude

$$u = 1 - \theta + \frac{\theta^2}{2^2} - \frac{\theta^3}{2^2 . 3^2} + \frac{\theta^4}{2^2 . 3^2 . 4^2} - \&c.,$$

or

$$u = 1 - \frac{gx^2}{2^2} + \frac{g^2x^4}{2^2 . 4^2} - \frac{g^3x^6}{2^2 . 4^2 . 6^2} + \&c.$$

It is easy to express the sum of this series. To obtain the result, develope as follows the function $\cos(\alpha \sin x)$ in cosines of multiple arcs. We have by known transformations (x)

$$2\cos(\alpha \sin x) = e^{\frac{1}{2}\alpha e^{x\sqrt{-1}}} e^{-\frac{1}{2}\alpha e^{-x\sqrt{-1}}} + e^{-\frac{1}{2}\alpha e^{x\sqrt{-1}}} e^{\frac{1}{2}\alpha e^{-x\sqrt{-1}}},$$

and denoting $e^{x\sqrt{-1}}$ by ω,

$$2\cos(\alpha \sin x) = e^{\frac{\alpha\omega}{2}} e^{-\frac{\alpha\omega^{-1}}{2}} + e^{-\frac{\alpha\omega}{2}} e^{\frac{\alpha\omega^{-1}}{2}}.$$

Developing the second member according to powers of ω, we find the term which does not contain ω in the development of $2\cos(\alpha \sin x)$ to be

$$2\left(1 - \frac{\alpha^2}{2^2} + \frac{\alpha^4}{2^2 . 4^2} - \frac{\alpha^6}{2^2 . 4^2 . 6^2} + \&c.\right).$$

The coefficients of ω^1, ω^3, ω^5, &c. are nothing, the same is the case with the coefficients of the terms which contain ω^{-1}, ω^{-3}, ω^{-5}, &c.; the coefficient of ω^{-2} is the same as that of ω^2; the coefficient of ω^4 is

$$2\left(\frac{\alpha^4}{2 . 4 . 6 . 8} - \frac{\alpha^6}{2^2 . 4 . 6 . 8 . 10} + \&c.\right);$$

the coefficient of ω^{-4} is the same as that of ω^4. It is easy to express the law according to which the coefficients succeed; but without stating it, let us write $2\cos 2x$ instead of $(\omega^2 + \omega^{-2})$, or $2\cos 4x$ instead of $(\omega^4 + \omega^{-4})$, and so on: hence the quantity $2\cos(\alpha \sin x)$ is easily developed in a series of the form

$$A + B\cos 2x + C\cos 4x + D\cos 6x + \&c.,$$

and the first coefficient A is equal to

$$2\left(1 - \frac{\alpha^2}{2^2} + \frac{\alpha^4}{2^2 . 4^2} - \frac{\alpha^6}{2^2 . 4^2 . 6^2} + \&c.\right);$$

if we now compare the general equation which we gave formerly

$$\frac{1}{2}\pi\phi(x) = \frac{1}{2}\int\phi(x)dx + \cos x\int\phi(x)\cos x dx + \&c.$$

2 cos(α sin x) = $e^{i\alpha \sin x} + e^{-i\alpha \sin x}$

with the equation

$$2\cos(\alpha \sin x) = A + B\cos 2x + C\cos 4x + \&c.,$$

we shall find the values of the coefficients A, B, C expressed by definite integrals. It is sufficient here to find that of the first coefficient A. We have then

$$\frac{1}{2}A = \frac{1}{\pi}\int \cos(\alpha \sin x)\,dx,$$

the integral should be taken from $x = 0$ to $x = \pi$. Hence the value of the series $1 - \frac{\alpha^2}{2^2} + \frac{\alpha^4}{2^2 . 4^2} - \frac{\alpha^6}{2^2 . 4^2 . 6^2} + \&c.$ is that of the definite integral $\int_0^\pi dx \cos(\alpha \sin x)$. We should find in the same manner by comparison of two equations the values of the successive coefficients B, C, &c.; we have indicated these results because they are useful in other researches which depend on the same theory. It follows from this that the particular value of u which satisfies the equation

$$gu + \frac{d^2u}{dx^2} + \frac{1}{x}\frac{du}{dx} = 0 \text{ is } \frac{1}{\pi}\int \cos(x\sqrt{g}\sin r)\,dr,$$

the integral being taken from $r = 0$ to $r = \pi$. Denoting by q this value of u, and making $u = qS$, we find $S = a + b\int \frac{dx}{xq^2}$, and we have as the complete integral of the equation $gu + \frac{d^2u}{dx^2} + \frac{1}{x}\frac{du}{dr} = 0$,

$$u = \left[a + b\int \frac{dx}{x\left\{\int \cos(x\sqrt{g}\sin r)\,dr\right\}^2}\right]\int \cos(x\sqrt{g}\sin r)\,dr.$$

a and b are arbitrary constants. If we suppose $b = 0$, we have, as formerly,

$$u = \int \cos(x\sqrt{g}\sin r)\,dr.$$

With respect to this expression we add the following remarks.

311. The equation

$$\frac{1}{\pi}\int_0^\pi \cos(\theta \sin u)\,du = 1 - \frac{\theta^2}{2^2} + \frac{\theta^4}{2^2 . 4^2} - \frac{\theta^6}{2^2 . 4^2 . 6^2} + \&c.$$

verifies itself. We have in fact

$$\int \cos(\theta \sin u)\, du = \int du \left(1 - \frac{\theta^2 \sin^2 u}{\lfloor 2} + \frac{\theta^4 \sin^4 u}{\lfloor 4} - \frac{\theta^6 \sin^6 u}{\lfloor 6} + \&c.\right);$$

and integrating from $u = 0$ to $u = \pi$, denoting by S_2, S_4, S_6, &c. the definite integrals

$$\int \sin^2 u\, du,\ \int \sin^4 u\, du,\ \int \sin^6 u\, du,\ \&c.,$$

we have

$$\int \cos(\theta \sin u)\, du = \pi - \frac{\theta^2}{\lfloor 2} S_2 + \frac{\theta^4}{\lfloor 4} S_4 - \frac{\theta^6}{\lfloor 6} S_6 + \&c.,$$

it remains to determine S_2, S_4, S_6, &c. The term $\sin^n u$, n being an even number, may be developed thus

$$\sin^n u = A_n + B_n \cos 2u + C_n \cos 4u + \&c.$$

Multiplying by du and integrating between the limits $u = 0$ and $u = \pi$, we have simply $\int \sin^n u\, du = A_n \pi$, the other terms vanish. From the known formula for the development of the integral powers of sines, we have

$$A_2 = \frac{1}{2^2} \cdot \frac{2}{1},\quad A_4 = \frac{1}{2^4} \cdot \frac{3.4}{1.2},\quad A_6 = \frac{1}{2^6} \cdot \frac{4.5.6}{1.2.3},\ \&c.$$

Substituting these values of S_2, S_4, S_6, &c., we find

$$\frac{1}{\pi} \int \cos(\theta \sin u)\, du = 1 - \frac{\theta^2}{2^2} + \frac{\theta^4}{2^2 . 4^2} - \frac{\theta^6}{2^2 . 4^2 . 6^2} + \&c.$$

We can make this result more general by taking, instead of $\cos(t \sin u)$, any function whatever ϕ of $t \sin u$.

Suppose then that we have a function $\phi(z)$ which may be developed thus

$$\phi(z) = \phi + z\phi' + \frac{z^2}{\lfloor 2} \phi'' + \frac{z^3}{\lfloor 3} \phi''' + \&c.;$$

we shall have

$$\phi(t \sin u) = \phi + t\phi' \sin u + \frac{t^2}{\lfloor 2} \phi'' \sin^2 u + \frac{t^3}{\lfloor 3} \phi''' \sin^3 u + \&c.$$

and $$\frac{1}{\pi} \int du\, \phi(t \sin u) = \phi + t S_1 \phi' + \frac{t^2}{\lfloor 2} S_2 \phi'' + \frac{t^3}{\lfloor 3} S_3 \phi''' + \&c. \quad (e).$$

Now, it is easy to see that the values of S_1, S_3, S_5, &c. are nothing. With respect to S_2, S_4, S_6, &c. their values are the quantities which we previously denoted by A_2, A_4, A_6, &c. For this reason, substituting these values in the equation (e) we have generally, whatever the function ϕ may be,

$$\frac{1}{\pi}\int \phi\,(t\sin u)\,du = \phi + \frac{t^2}{2}\phi'' + \frac{t^4}{2^2.4^2}\phi^{\text{iv}} + \frac{t^6}{2^2.4^2.6^2}\phi^{\text{vi}} + \&c.,$$

in the case in question, the function $\phi(z)$ represents $\cos z$, and we have $\phi = 1$, $\phi'' = -1$, $\phi^{\text{iv}} = 1$, $\phi^{\text{vi}} = -1$, and so on.

312. To ascertain completely the nature of the function $f(\theta)$, and of the equation which gives the values of g, it would be necessary to consider the form of the line whose equation is

$$y = 1 - \theta + \frac{\theta^2}{2^2} - \frac{\theta^3}{2^2.3^2} + \&c.,$$

which forms with the axis of abscissæ areas alternately positive and negative which cancel each other; the preceding remarks, also, on the expression of the values of series by means of definite integrals, might be made more general. When a function of the variable x is developed according to powers of x, it is easy to deduce the function which would represent the same series, if the powers x, x^2, x^3, &c. were replaced by $\cos x$, $\cos 2x$, $\cos 3x$, &c. By making use of this reduction and of the process employed in the second paragraph of Article 235, we obtain the definite integrals which are equivalent to given series; but we could not enter upon this investigation, without departing too far from our main object.

It is sufficient to have indicated the methods which have enabled us to express the values of series by definite integrals.

We will add only the development of the quantity $\theta\frac{f'(\theta)}{f(\theta)}$ in a continued fraction.

313. The undetermined y or $f(\theta)$ satisfies the equation

$$y + \frac{dy}{d\theta} + \theta\frac{d^2y}{d\theta^2} = 0,$$

whence we derive, denoting the functions

$$\frac{dy}{d\theta},\ \frac{d^2y}{d\theta^2},\ \frac{d^3y}{d\theta^3},\ \&c.$$

by y', y'', y''', &c.,

$$-y = y' + \theta y'' \quad \text{or} \quad \frac{y'}{y} = \frac{-y'}{y' + \theta y''} = \frac{-1}{1 + \theta\dfrac{y''}{y'}},$$

$$-y' = 2y'' + \theta y''', \quad \frac{y''}{y'} = \frac{-y''}{2y'' + \theta y'''} = \frac{-1}{2 + \theta\dfrac{y'''}{y''}},$$

$$-y'' = 3y''' + \theta y^{\text{iv}}, \quad \frac{y'''}{y''} = \frac{-y'''}{3y''' + \theta y^{\text{iv}}} = \frac{-1}{3 + \theta\dfrac{y^{\text{iv}}}{y'''}},$$

&c.;

whence we conclude

$$\frac{y'}{y} = \frac{-1}{1-}\ \frac{\theta}{2-}\ \frac{\theta}{3-}\ \frac{\theta}{4-}\ \frac{\theta}{5-\&c.}.$$

Thus the value of the function $-\dfrac{\theta f'(\theta)}{f(\theta)}$ which enters into the definite equation, when expressed as an infinite continued fraction, is

$$\frac{\theta}{1-}\ \frac{\theta}{2-}\ \frac{\theta}{3-}\ \frac{\theta}{4-}\ \frac{\theta}{5-\&c.}.$$

314. We shall now state the results at which we have up to this point arrived.

If the variable radius of the cylindrical layer be denoted by x, and the temperature of the layer by v, a function of x and the time t; the required function v must satisfy the partial differential equation

$$\frac{dv}{dt} = k\left(\frac{d^2v}{dx^2} + \frac{1}{x}\frac{dv}{dx}\right);$$

for v we may assume the following value

$$v = ue^{-mt};$$

u is a function of x, which satisfies the equation

$$\frac{m}{k}u + \frac{d^2u}{dx^2} + \frac{1}{x}\frac{du}{dx} = 0.$$

If we make $\theta = \frac{m}{k}\frac{x^2}{2^2}$, and consider u as a function of x, we have

$$u + \frac{du}{d\theta} + \theta\frac{d^2u}{d\theta^2} = 0.$$

The following value

$$u = 1 - \theta + \frac{\theta^2}{2^2} - \frac{\theta^3}{2^2.3^2} + \frac{\theta^4}{2^2.3^2.4^2} - \&c.$$

satisfies the equation in u and θ. We therefore assume the value of u in terms of x to be

$$u = 1 - \frac{m}{k}\frac{x^2}{2^2} + \frac{m^2}{k^2}\frac{x^2}{2^2.4^2} - \frac{m^3}{k^3}\frac{x^3}{2^2.4^2.6^2} + \&c.,$$

the sum of this series is

$$\frac{1}{\pi}\int \cos\left(x\sqrt{\frac{m}{k}}\sin r\right)dr;$$

the integral being taken from $r = 0$ to $r = \pi$. This value of v in terms of x and m satisfies the differential equation, and retains a finite value when x is nothing. Further, the equation $hu + \frac{du}{dx} = 0$ must be satisfied when $x = X$ the radius of the cylinder. This condition would not hold if we assigned to the quantity m any value whatever; we must necessarily have the equation

$$\frac{hX}{2} = \frac{\theta}{1-}\,\frac{\theta}{2-}\,\frac{\theta}{3-}\,\frac{\theta}{4-}\,\frac{\theta}{5-\&c.},$$

in which θ denotes $\frac{m}{k}\frac{X^2}{2^2}$.

This definite equation, which is equivalent to the following,

$$\frac{hX}{2}\left(1 - \theta + \frac{\theta^2}{2^2} - \frac{\theta^3}{2^2.3^2} + \&c.\right) = \theta - \frac{2\theta^2}{2^2} + \frac{3\theta^3}{2^2.3^2} - \&c.,$$

gives to θ an infinity of real values denoted by θ_1, θ_2, θ_3, &c.; the corresponding values of m are

$$\frac{2^2k\theta_1}{X^2},\ \frac{2^2k\theta_2}{X^2},\ \frac{2^2k\theta_3}{X^2},\ \&c.;$$

thus a particular value of v is expressed by

$$\pi v = e^{-\frac{2^2kt\theta_1}{X^2}}\int \cos\left(2\frac{x}{X}\sqrt{\theta_1}\sin q\right)dq.$$

We can write, instead of θ_1, one of the roots θ_1, θ_2, θ_3, &c., and compose by means of them a more general value expressed by the equation

$$\pi v = a_1 e^{-\frac{2^2 k t \theta_1}{X^2}} \int \cos\left(2\frac{x}{X}\sqrt{\theta_1}\sin q\right) dq$$

$$+ a_2 e^{-\frac{2^2 k t \theta_2}{X^2}} \int \cos\left(2\frac{x}{X}\sqrt{\theta_2}\sin q\right) dq$$

$$+ a_3 e^{-\frac{2^2 k t \theta_3}{X^2}} \int \cos\left(2\frac{x}{X}\sqrt{\theta_3}\sin q\right) dq + \&c.$$

a_1, a_2, a_3, &c. are arbitrary coefficients: the variable q disappears after the integrations, which should be taken from $q=0$ to $q=\pi$.

315. To prove that this value of v satisfies all the conditions of the problem and contains the general solution, it remains only to determine the coefficients a_1, a_2, a_3, &c. from the initial state. Take the equation

$$v = a_1 e^{-m_1 t} u_1 + a_2 e^{-m_2 t} u_2 + a_3 e^{-m_3 t} u_3 + \&c.,$$

in which u_1, u_2, u_3, &c. are the different values assumed by the function u, or

$$1 - \frac{m}{k}\frac{x^2}{2^2} + \frac{m^2}{k^2}\frac{x^4}{2^2 4^2} - \&c.$$

when, instead of $\frac{m}{k}$, the values g_1, g_2, g_3, &c. are successively substituted. Making in it $t=0$, we have the equation

$$V = a_1 u_1 + a_2 u_2 + a_3 u_3 + \&c.,$$

in which V is a given function of x. Let $\phi(x)$ be this function; if we represent the function u_i whose index is i by $\psi(x\sqrt{g_i})$, we have

$$\phi(x) = a_1\psi(x\sqrt{g_1}) + a_2\psi(x\sqrt{g_2}) + a_3\psi(x\sqrt{g_3}) + \&c.$$

To determine the first coefficient, multiply each member of the equation by $\sigma_1 dx$, σ_1 being a function of x, and integrate from $x=0$ to $x=X$. We then determine the function σ_1, so that after the integrations the second member may reduce to the first term only, and the coefficient a_1 may be found, all the other integrals having nul values. Similarly to determine the second coefficient a_2, we multiply both terms of the equation

$$\phi(x) = a_2 u_1 + a_2 u_2 + a_3 u_3 + \&c.$$

by another factor $\sigma_2\, dx$, and integrate from $x=0$ to $x=X$. The factor σ_2 must be such that all the integrals of the second member vanish, except one, namely that which is affected by the coefficient a_2. In general, we employ a series of functions of x denoted by σ_1, σ_2, σ_3, &c. which correspond to the functions u_1, u_2, u_3, &c.; each of the factors σ has the property of making all the terms which contain definite integrals disappear in integration except one; in this manner we obtain the value of each of the coefficients a_1, a_2, a_3, &c. We must now examine what functions enjoy the property in question.

316. Each of the terms of the second member of the equation is a definite integral of the form $a\int \sigma u dx$; u being a function of x which satisfies the equation

$$\frac{m}{k}u + \frac{d^2u}{dx^2} + \frac{1}{x}\frac{du}{dx} = 0;$$

we have therefore $a\int \sigma u dx = -a\frac{k}{m}\int\left(\frac{\sigma}{x}\frac{du}{dx} + \sigma\frac{d^2u}{dx}\right)$.

Developing, by the method of integration by parts, the terms

$$\int \frac{\sigma}{x}\frac{du}{dx}dx \text{ and } \int \sigma \frac{d^2u}{dx^2}dx,$$

we have
$$\int \frac{\sigma}{x}\frac{du}{dx}dx = C + u\frac{\sigma}{x} - \int u d\left(\frac{\sigma}{x}\right)$$

and
$$\int \sigma\frac{d^2u}{dx^2}dx = D + \frac{du}{dx}\sigma - u\frac{d\sigma}{dx} + \int u\frac{d^2\sigma}{dx^2}dx.$$

The integrals must be taken between the limits $x=0$ and $x=X$, by this condition we determine the quantities which enter into the development, and are not under the integral signs. To indicate that we suppose $x=0$ in any expression in x, we shall affect that expression with the suffix α; and we shall give it the suffix ω to indicate the value which the function of x takes, when we give to the variable x its last value X.

Supposing $x=0$ in the two preceding equations we have

$$0=C+\left(u\frac{\sigma}{x}\right)_\alpha \text{ and } 0=D+\left(\frac{du}{dx}\sigma-u\frac{d\sigma}{dx}\right)_\alpha,$$

thus we determine the constants C and D. Making then $x=X$ in the same equations, and supposing the integral to be taken from $x=0$ to $x=X$, we have

$$\int\frac{\sigma}{x}\frac{du}{dx}dx=\left(u\frac{\sigma}{x}\right)_\omega-\left(u\frac{\sigma}{x}\right)_\alpha-\int ud\left(\frac{\sigma}{x}\right)$$

and $$\int\sigma\frac{d^2u}{dx^2}dx=\left(\frac{du}{dx}\sigma-u\frac{d\sigma}{dx}\right)_\omega-\left(\frac{du}{dx}\sigma-u\frac{d\sigma}{dx}\right)_\alpha+\int u\frac{d^2\sigma}{dx^2}dx,$$

thus we obtain the equation

$$-\frac{m}{k}\int\sigma udx=\int\left\{u\frac{d^2\sigma}{dx^2}-u\frac{d\left(\frac{\sigma}{x}\right)}{dx}\right\}dx+\left(\frac{du}{dx}\sigma-u\frac{d\sigma}{dx}+u\frac{\sigma}{x}\right)_\omega$$
$$-\left(\frac{du}{dx}\sigma-u\frac{d\sigma}{dx}+u\frac{\sigma}{x}\right)_\alpha.$$

317. If the quantity $\frac{d^2\sigma}{dx^2}-\frac{d\left(\frac{\sigma}{x}\right)}{dx}$ which multiplies u under the sign of integration in the second member were equal to the product of σ by a constant coefficient, the terms

$$\int\left\{u\frac{d^2\sigma}{dx^2}-\frac{d\left(\frac{\sigma}{x}\right)}{dx}dx\right\} \text{ and } \int\sigma udx$$

would be collected into one, and we should obtain for the required integral $\int\sigma udx$ a value which would contain only determined quantities, with no sign of integration. It remains only to equate that value to zero.

Suppose then the factor σ to satisfy the differential equation of the second order $\frac{n}{k}\sigma+\frac{d^2\sigma}{dx^2}-\frac{d\left(\frac{\sigma}{x}\right)}{dx}=0$ in the same manner as the function u satisfies the equation

$$\frac{m}{k}u+\frac{d^2u}{dx^2}+\frac{1}{x}\frac{du}{dx}=0,$$

m and n being constant coefficients, we have

$$\frac{n-m}{k}\int \sigma u dx = \left(\frac{du}{dx}\sigma - u\frac{d\sigma}{dx} + u\frac{\sigma}{x}\right)_\omega - \left(\frac{du}{dx}\sigma - u\frac{d\sigma}{dx} + u\frac{\sigma}{x}\right)_a.$$

Between u and σ a very simple relation exists, which is discovered when in the equation $\frac{n}{k}\sigma + \frac{d^2\sigma}{dx^2} - \frac{d\left(\frac{\sigma}{x}\right)}{dx} = 0$ we suppose $\sigma = xs$; as the result of this substitution we have the equation

$$\frac{n}{k}s + \frac{d^2s}{dx^2} + \frac{1}{x}\frac{ds}{dx} = 0,$$

which shews that the function s depends on the function u given by the equation

$$\frac{m}{k}u + \frac{d^2u}{dx^2} + \frac{1}{x}\frac{du}{dx} = 0.$$

To find s it is sufficient to change m into n in the value of u; the value of u has been denoted by $\psi\left(x\sqrt{\frac{m}{k}}\right)$, that of σ will therefore be $x\psi\left(x\sqrt{\frac{n}{k}}\right)$.

We have then

$$\frac{du}{dx}\sigma - u\frac{d\sigma}{dx} + u\frac{\sigma}{x}$$

$$= x\sqrt{\frac{m}{k}}\psi'\left(x\sqrt{\frac{m}{k}}\right)\psi\left(x\sqrt{\frac{n}{k}}\right) - x\sqrt{\frac{n}{k}}\psi'\left(x\sqrt{\frac{n}{k}}\right)\psi\left(x\sqrt{\frac{m}{k}}\right)$$

$$-\psi\left(x\sqrt{\frac{m}{k}}\right)\psi\left(x\sqrt{\frac{n}{k}}\right) + \psi\left(x\sqrt{\frac{m}{k}}\right)\psi\left(x\sqrt{\frac{n}{k}}\right);$$

the two last terms destroy each other, it follows that on making $x = 0$, which corresponds to the suffix a, the second member vanishes completely. We conclude from this the following equation

$$\frac{n-m}{k}\int \sigma u dx = X\sqrt{\frac{m}{k}}\psi'\left(X\sqrt{\frac{m}{k}}\right)\psi\left(X\sqrt{\frac{n}{k}}\right)$$

$$- X\sqrt{\frac{n}{k}}\psi'\left(X\sqrt{\frac{n}{k}}\right)\psi\left(X\sqrt{\frac{m}{k}}\right) \ldots\ldots (f).$$

It is easy to see that the second member of this equation is always nothing when the quantities m and n are selected from those which we formerly denoted by m_1, m_2, m_3, &c.

We have in fact

$$hX = -X\sqrt{\frac{m}{k}}\frac{\psi'\left(X\sqrt{\frac{m}{k}}\right)}{\psi\left(X\sqrt{\frac{m}{k}}\right)} \text{ and } hX = -X\sqrt{\frac{n}{k}}\frac{\psi'\left(X\sqrt{\frac{n}{k}}\right)}{\psi\left(X\sqrt{\frac{n}{k}}\right)},$$

comparing the values of hX we see that the second member of the equation (f) vanishes.

It follows from this that after we have multiplied by σdx the two terms of the equation

$$\phi(x) = a_1u_1 + a_2u_2 + a_3u_3 + \&c.,$$

and integrated each side from $x = 0$ to $x = X$, in order that each of the terms of the second member may vanish, it suffices to take for σ the quantity xu or $x\psi\left(x\sqrt{\frac{m}{k}}\right)$.

We must except only the case in which $n = m$, when the value of $\int \sigma u dx$ derived from the equation (f) is reduced to the form $\frac{0}{0}$, and is determined by known rules.

318. If $\sqrt{\frac{m}{k}} = \mu$ and $\sqrt{\frac{n}{k}} = \nu$, we have

$$\int x\psi(\mu)\psi(\nu)\,dx = \frac{\mu X\psi'(\mu X)\psi(\nu X) - \nu X\psi'(\nu X)\psi(\mu X)}{\nu^2 - \mu^2}.$$

If the numerator and denominator of the second member are separately differentiated with respect to ν, the factor becomes, on making $\mu = \nu$,

$$\frac{\mu X^2\psi'^2 - X\psi\psi' - \mu X^2\psi\psi''}{2\mu}.$$

We have on the other hand the equation

$$\mu^2 u + \frac{d^2u}{dx^2} + \frac{1}{x}\frac{du}{dx} = 0, \text{ or } \mu^2\psi + \frac{\mu}{x}\psi' + \mu^2\psi'' = 0,$$

and also $$hx\psi + \mu x\psi' = 0,$$

or, $$h\psi + \mu\psi' = 0;$$

hence we have

$$\left(\mu^2 - \frac{h}{x}\right)\psi + \mu^2\psi'' = 0,$$

we can therefore eliminate the quantities ψ' and ψ'' from the integral which is required to be evaluated, and we shall find as the value of the integral sought

$$\tfrac{1}{2}X^2\psi^2\left(\frac{\mu^2+h^2}{\mu^2}\right) \quad \text{or} \quad \frac{X^2U_i^2}{2}\left(1+\frac{kh^2}{m_i}\right),$$

putting for μ its value, and denoting by U_i the value which the function u or $\psi\left(x\sqrt{\frac{m_i}{k}}\right)$ takes when we suppose $x = X$. The index i denotes the order of the root m of the definite equation which gives an infinity of values of m. If we substitute m_i or $\frac{2^2k\theta_i}{X^2}$ in $\frac{X^2U_i^2}{2}\left(1+\frac{kh^2}{m_i}\right)$, we have

$$\tfrac{1}{2}X^2U_i^2\left\{1+\left(\frac{hX}{2\sqrt{\theta_i}}\right)^2\right\}.$$

319. It follows from the foregoing analysis that we have the two equations

$$\int_0^X xu_ju_i\,dx = 0 \quad \text{and} \quad \int_0^X xu_i^2dx = \left\{1+\left(\frac{hX}{2\sqrt{\theta_i}}\right)^2\right\}\frac{X^2U_i^2}{2},$$

the first holds whenever the number i and j are different, and the second when these numbers are equal.

Taking then the equation $\phi(x) = a_1u_1 + a_2u_2 + a_3u_3 + \&c.$, in which the coefficients a_1, a_2, a_3, &c. are to be determined, we shall find the coefficient denoted by a_i by multiplying the two members of the equation by xu_idx, and integrating from $x = 0$ to $x = X$; the second member is reduced by this integration to one term only, and we have the equation

$$2\int x\phi(x)\,u_idx = a_iX^2U_i^2\left\{1+\left(\frac{hX}{2\sqrt{\theta_i}}\right)^2\right\},$$

which gives the value of a_i. The coefficients a_1, a_2, a_3, ... a_i, being thus determined, the condition relative to the initial state expressed by the equation $\phi(x) = a_1 u_1 + a_2 u_2 + a_3 u_3 + \&c.$, is fulfilled.

We can now give the complete solution of the proposed problem; it is expressed by the following equation:

$$\frac{vX^2}{2} = \frac{\int_0^X x\phi(x)\, u_1 dx}{U_1^2\left(1+\frac{h^2X^2}{2^2\theta_1}\right)} u_1 e^{-\frac{2^2kt}{X^2}\theta_1} + \frac{\int_0^X x\phi(x)\, u_2 dx}{U_2^2\left(1+\frac{h^2X^2}{2^2\theta_2}\right)} u_2 e^{-\frac{2^2kt}{X^2}\theta_2}$$

$+ \&c.$

The function of x denoted by u in the preceding equation is expressed by

$$\tfrac{1}{\pi}\int \cos\left(\frac{2x}{X}\sqrt{\theta_i}\sin q\right) dq;$$

all the integrals with respect to x must be taken from $x = 0$ to $x = X$, and to find the function u we must integrate from $q = 0$ to $q = \pi$; $\phi(x)$ is the initial value of the temperature, taken in the interior of the cylinder at a distance x from the axis, which function is arbitrary, and θ_1, θ_2, θ_3, &c. are the real and positive roots of the equation

$$\frac{hX}{2} = \frac{\theta}{1-}\,\frac{\theta}{2-}\,\frac{\theta}{3-}\,\frac{\theta}{4-}\,\frac{\theta}{5-\&c.}.$$

320. If we suppose the cylinder to have been immersed for an infinite time in a liquid maintained at a constant temperature, the whole mass becomes equally heated, and the function $\phi(x)$ which represents the initial state is represented by unity. After this substitution, the general equation represents exactly the gradual progress of the cooling.

If t the time elapsed is infinite, the second member contains only one term, namely, that which involves the least of all the roots θ_1, θ_2, θ_3, &c.; for this reason, supposing the roots to be arranged according to their magnitude, and θ to be the least, the final state of the solid is expressed by the equation

$$\frac{vX^2}{2} = \frac{\int x\phi(x)\, u_1 dx}{U_1^2\left(1+\frac{h^2X^2}{2^2\theta_1}\right)} u_1 e^{-\frac{2^2kt}{X^2}\theta}.$$

From the general solution we might deduce consequences similar to those offered by the movement of heat in a spherical mass. We notice first that there are an infinite number of particular states, in each of which the ratios established between the initial temperatures are preserved up to the end of the cooling. When the initial state does not coincide with one of these simple states, it is always composed of several of them, and the ratios of the temperatures change continually, according as the time increases. In general the solid arrives very soon at the state in which the temperatures of the different layers decrease continually preserving the same ratios. When the radius X is very small[1], we find that the temperatures decrease in proportion to the fraction $e^{-\frac{2h}{CDX}}$.

If on the contrary the radius X is very large[2], the exponent of e in the term which represents the final system of temperatures contains the square of the whole radius. We see by this what influence the dimension of the solid has upon the final velocity of cooling. If the temperature[3] of the cylinder whose radius is X, passes from the value A to the lesser value B, in the time T, the temperature of a second cylinder of radius equal to X' will pass from A to B in a different time T'. If the two sides are thin, the ratio of the times T and T' will be that of the diameters. If, on the contrary, the diameters of the cylinders are very great, the ratio of the times T and T' will be that of the squares of the diameters.

[1] When X is very small, $\theta = \frac{hX}{2}$, from the equation in Art. 314. Hence

$$e^{-\frac{2^2 kt}{X^2}\theta} \text{ becomes } e^{-\frac{2hkt}{X}}.$$

In the text, h is the surface conducibility.

[2] When X is very large, a value of θ nearly equal to one of the roots of the quadratic equation $1 = \frac{\theta}{2-}\frac{\theta}{3-}\frac{\theta}{4-}\frac{\theta}{5}$ will make the continued fraction in Art. 314 assume its proper magnitude. Hence $\theta = 1{\cdot}446$ nearly, and

$$e^{-\frac{2^2 kt}{X^2}\theta} \text{ becomes } e^{-\frac{5{\cdot}78kt}{X^2}}.$$

The least root of $f(\theta) = 0$ is $1{\cdot}4467$, neglecting terms after θ^4.

[3] The temperature intended is the mean temperature, which is equal to

$$\frac{1}{\pi X^2}\int_0^X v\,d(\pi x^2), \text{ or } \frac{2}{X^2}\int_0^X vx\,dx. \qquad \text{[A. F.]}$$

CHAPTER VII.

PROPAGATION OF HEAT IN A RECTANGULAR PRISM.

321. THE equation $\frac{d^2v}{dx^2}+\frac{d^2v}{dy^2}+\frac{d^2v}{dz^2}=0$, which we have stated in Chapter II., Section IV., Article 125, expresses the uniform movement of heat in the interior of a prism of infinite length, submitted at one end to a constant temperature, its initial temperatures being supposed nul. To integrate this equation we shall, in the first place, investigate a particular value of v, remarking that this function v must remain the same, when y changes sign or when z changes sign; and that its value must become infinitely small, when the distance x is infinitely great. From this it is easy to see that we can select as a particular value of v the function $ae^{-mx}\cos ny \cos pz$; and making the substitution we find $m^2-n^2-p^2=0$. Substituting for n and p any quantities whatever, we have $m=\sqrt{n^2+p^2}$. The value of v must also satisfy the definite equation $\frac{h}{k}v+\frac{dv}{dy}=0$, when $y=l$ or $-l$, and the equation $\frac{h}{k}v+\frac{du}{dz}=0$ when $z=l$ or $-l$ (Chapter II., Section IV., Article 125). If we give to v the foregoing value, we have

$$-n\sin ny+\frac{h}{k}\cos ny=0 \quad\text{and}\quad -p\sin pz+\frac{h}{k}\cos pz=0,$$

or

$$\frac{hl}{k}=pl\tan pl,\quad \frac{hl}{k}=nl\tan nl.$$

We see by this that if we find an arc ϵ, such that $\epsilon\tan\epsilon$ is equal to the whole known quantity $\frac{h}{k}l$, we can take for n or p the quantity $\frac{\epsilon}{l}$. Now, it is easy to see that there are an infinite number of arcs which, multiplied respectively by their tangents, give the same definite product $\frac{hl}{k}$, whence it follows that we can find for n or p an infinite number of different values.

322. If we denote by ϵ_1, ϵ_2, ϵ_3, &c. the infinite number of arcs which satisfy the definite equation $\epsilon \tan \epsilon = \frac{hl}{k}$, we can take for n any one of these arcs divided by l. The same would be the case with the quantity p; we must then take $m^2 = n^2 + p^2$. If we gave to n and p other values, we could satisfy the differential equation, but not the condition relative to the surface. We can then find in this manner an infinite number of particular values of v, and as the sum of any collection of these values still satisfies the equation, we can form a more general value of v.

Take successively for n and p all the possible values, namely, $\frac{\epsilon_1}{l}$, $\frac{\epsilon_2}{l}$, $\frac{\epsilon_3}{l}$, &c. Denoting by a_1, a_2, a_3, &c., b_1, b_2, b_3, &c., constant coefficients, the value of v may be expressed by the following equation:

$$\begin{aligned} v = &(a_1 e^{-x\sqrt{n_1^2+n_1^2}} \cos n_1 y + a_2 e^{-x\sqrt{n_2^2+n_1^2}} \cos n_2 y + \&c.)\, b_1 \cos n_1 z \\ &+ (a_1 e^{-x\sqrt{n_1^2+n_2^2}} \cos n_1 y + a_2 e^{-x\sqrt{n_2^2+n_2^2}} \cos n_2 y + \&c.)\, b_2 \cos n_2 z \\ &+ (a_1 e^{-x\sqrt{n_1^2+n_3^2}} \cos n_1 y + a_2 e^{-x\sqrt{n_2^2+n_3^2}} \cos n_2 y + \&c.)\, b_3 \cos n_3 z \\ &+ \&c. \end{aligned}$$

323. If we now suppose the distance x nothing, every point of the section A must preserve a constant temperature. It is therefore necessary that, on making $x = 0$, the value of v should be always the same, whatever value we may give to y or to z; provided these values are included between 0 and l. Now, on making $x = 0$, we find

$$\begin{aligned} v = &(a_1 \cos n_1 y + a_2 \cos n_2 y + a_3 \cos n_3 y + \&c.) \\ &\times (b_1 \cos n_1 z + b_2 \cos n_2 y + b_3 \cos n_3 y + \&c.). \end{aligned}$$

Denoting by 1 the constant temperature of the end A, assume the two equations

$$1 = a_1 \cos n_1 y + a_2 \cos n_2 y + a_3 \cos n_3 y + \&c,$$

$$1 = b_1 \cos n_1 y + b_2 \cos n_2 y + b_3 \cos n_3 y + \&c.$$

It is sufficient then to determine the coefficients a_1, a_2, a_3, &c., whose number is infinite, so that the second member of the equation may be always equal to unity. This problem has already been solved in the case where the numbers n_1, n_2, n_3, &c. form the series of odd numbers (Chap. III., Sec. II., Art. 177). Here n_1, n_2, n_3, &c. are incommensurable quantities given by an equation of infinitely high degree.

324. Writing down the equation

$$1 = a_1 \cos n_1 y + a_2 \cos n_2 y + a_3 \cos n_3 y + \&c.,$$

multiply the two members of the equation by $\cos n_1 y\, dy$, and take the integral from $y=0$ to $y=l$. We thus determine the first coefficient a_1. The remaining coefficients may be determined in a similar manner.

In general, if we multiply the two members of the equation by $\cos \nu y$, and integrate it, we have corresponding to a single term of the second member, represented by $a \cos ny$, the integral

$$a\int \cos ny \cos \nu y\, dy \text{ or } \frac{1}{2}a\int \cos (n-\nu)\, y\, dy + \frac{1}{2}a\int \cos (n+\nu)\, y\, dy,$$

or,
$$\frac{a}{2}\left\{\frac{1}{n-\nu}\sin (n-\nu)\, y + \frac{1}{n+\nu}\sin (n+\nu)\, y\right\},$$

and making $y=l$,

$$\frac{a}{2}\left\{\frac{(n+\nu)\sin (n-\nu)\, l + (n-\nu)\sin (n+\nu)\, l}{n^2-\nu^2}\right\}.$$

Now, every value of n satisfies the equation $n \tan nl = \frac{h}{k}$; the same is the case with ν, we have therefore

$$n \tan \nu l = \nu \tan \nu l;$$

or
$$n \sin nl \cos \nu l - \nu \sin \nu l \cos nl = 0.$$

Thus the foregoing integral, which reduces to

$$\frac{a}{n^2 - \nu^2}(n \sin nl \cos \nu l - \nu \cos nl \sin \nu l),$$

is nothing, except only in the case where $n = \nu$. Taking then the integral

$$\frac{a}{2}\left\{\frac{\sin(n-\nu)l}{n-\nu} + \frac{\sin(n+\nu)l}{n+\nu}\right\},$$

we see that if we have $n = \nu$, it is equal to the quantity

$$\frac{1}{2}a\left(l + \frac{\sin 2nl}{2n}\right).$$

It follows from this that if in the equation

$$1 = a_1 \cos n_1 y + a_2 \cos n_2 y + a_3 \cos n_3 y + \&c.$$

we wish to determine the coefficient of a term of the second member denoted by $a \cos ny$, we must multiply the two members by $\cos ny\, dy$, and integrate from $y = 0$ to $y = l$. We have the resulting equation

$$\int_0^l \cos ny\, dy = \frac{1}{2}a\left(l + \frac{\sin 2nl}{2n}\right) = \frac{1}{n}\sin nl,$$

whence we deduce $\dfrac{\sin nl}{2nl + \sin 2nl} = \dfrac{1}{4}a$. In this manner the coefficients a_1, a_2, a_3, &c. may be determined; the same is the case with b_1, b_2, b_3, &c., which are respectively the same as the former coefficients.

325. It is easy now to form the general value of v. 1st, it satisfies the equation $\dfrac{d^2v}{dx^2} + \dfrac{d^2v}{dy^2} + \dfrac{d^2v}{dz^2} = 0$; 2nd, it satisfies the two conditions $k\dfrac{dv}{dy} + hv = 0$, and $k\dfrac{dv}{dz} + hv = 0$; 3rd, it gives a constant value to v when we make $x = 0$, whatever else the values of y and z may be, included between 0 and l; hence it is the complete solution of the proposed problem.

We have thus arrived at the equation

$$\frac{1}{4} = \frac{\sin n_1 l \cos n_1 y}{2n_1 l + \sin 2n_1 l} + \frac{\sin n_2 l \cos n_2 y}{2n_2 l + \sin 2n_2 l} + \frac{\sin n_3 l \cos n_3 y}{2n_3 l + \sin 2n_3 l} + \&c.,$$

or denoting by ϵ_1, ϵ_2, ϵ_3, &c. the arcs $n_1 l$, $n_2 l$, $n_3 l$, &c.

$$\frac{1}{4} = \frac{\sin\epsilon_1 \cos\frac{\epsilon_1 y}{l}}{2\epsilon_1 + \sin\epsilon_1} + \frac{\sin\epsilon_2 \cos\frac{\epsilon_2 y}{l}}{2\epsilon_2 + \sin\epsilon_2} + \frac{\sin\epsilon_3 \cos\frac{\epsilon_3 y}{l}}{2\epsilon_3 + \sin\epsilon_3} + \&\text{c.},$$

an equation which holds for all values of y included between 0 and l, and consequently for all those which are included between 0 and $-l$, when $x = 0$.

Substituting the known values of a_1, b_1, a_2, b_2, a_3, b_3, &c. in the general value of v, we have the following equation, which contains the solution of the proposed problem,

$$\begin{aligned}\frac{v}{4.4} = &\frac{\sin n_1 l \cos n_1 z}{2n_1 l + \sin 2n_1 l}\left(\frac{\sin n_1 l \cos n_1 y}{2n_1 l + \sin 2n_1 l}\, e^{-x\sqrt{n_1^2+n_1^2}} + \&\text{c.}\right)\\
&+ \frac{\sin n_2 l \cos n_2 z}{2n_2 l + \sin 2n_2 l}\left(\frac{\sin n_1 l \cos n_1 y}{2n_1 l + \sin 2n_1 l}\, e^{-x\sqrt{n_2^2+n_1^2}} + \&\text{c.}\right)\\
&+ \frac{\sin n_3 l \cos n_3 z}{2n_3 l + \sin 2n_3 l}\left(\frac{\sin n_1 l \cos n_1 y}{2n_1 l + \sin 2n_1 l}\, e^{-x\sqrt{n_3^2+n_1^2}} + \&\text{c.}\right)\\
&+ \&\text{c.} \ldots\ldots\ldots\ldots\ldots\ldots\ldots\ldots \text{(E)}.\end{aligned}$$

The quantities denoted by n_1, n_2, n_3, &c. are infinite in number, and respectively equal to the quantities $\frac{\epsilon_1}{l}$, $\frac{\epsilon_2}{l}$, $\frac{\epsilon_3}{l}$, &c.; the arcs, ϵ_1, ϵ_2, ϵ_3, &c., are the roots of the definite equation $\epsilon \tan \epsilon = \frac{hl}{k}$.

326. The solution expressed by the foregoing equation E is the only solution which belongs to the problem; it represents the general integral of the equation $\frac{d^2v}{dx^2} + \frac{d^2v}{dy^2} + \frac{d^2v}{dz^2} = 0$, in which the arbitrary functions have been determined from the given conditions. It is easy to see that there can be no different solution. In fact, let us denote by $\psi(x, y, z)$ the value of v derived from the equation (E), it is evident that if we gave to the solid initial temperatures expressed by $\psi(x, y, z)$, no change could happen in the system of temperatures, provided that the section at the origin were retained at the constant temperature 1: for the equation $\frac{d^2v}{dx^2} + \frac{d^2v}{dy^2} + \frac{d^2v}{dz^2} = 0$ being satisfied, the instantaneous variation of

the temperature is necessarily nothing. The same would not be the case, if after having given to each point within the solid whose co-ordinates are x, y, z the initial temperature $\psi(x, y, z)$, we gave to all points of the section at the origin the temperature 0. We see clearly, and without calculation, that in the latter case the state of the solid would change continually, and that the original heat which it contains would be dissipated little by little into the air, and into the cold mass which maintains the end at the temperature 0. This result depends on the form of the function $\psi(x, y, z)$, which becomes nothing when x has an infinite value as the problem supposes.

A similar effect would exist if the initial temperatures instead of being $+\psi(x, y, z)$ were $-\psi(x, y, z)$ at all the internal points of the prism; provided the section at the origin be maintained always at the temperature 0. In each case, the initial temperatures would continually approach the constant temperature of the medium, which is 0; and the final temperatures would all be nul.

327. These preliminaries arranged, consider the movement of heat in two prisms exactly equal to that which was the subject of the problem. For the first solid suppose the initial temperatures to be $+\psi(x, y, z)$, and that the section at origin A is maintained at the fixed temperature 1. For the second solid suppose the initial temperatures to be $-\psi(x, y, z)$, and that at the origin A all points of the section are maintained at the temperature 0. It is evident that in the first prism the system of temperatures cannot change, and that in the second this system varies continually up to that at which all the temperatures become nul.

If now we make the two different states coincide in the same solid, the movement of heat is effected freely, as if each system alone existed. In the initial state formed of the two united systems, each point of the solid has zero temperature, except the points of the section A, in accordance with the hypothesis. Now the temperatures of the second system change more and more, and vanish entirely, whilst those of the first remain unchanged. Hence after an infinite time, the permanent system of temperatures becomes that represented by equation E, or $v = \psi(x, y, z)$. It must be remarked that this result depends on the condition relative to the initial state; it occurs whenever the initial heat contained in the prism is so distributed, that it would vanish entirely, if the end A were maintained at the temperature 0.

328. We may add several remarks to the preceding solution. 1st, it is easy to see the nature of the equation $\epsilon \tan \epsilon = \frac{hl}{k}$; we need only suppose (see fig. 15) that we have constructed the curve $u = \epsilon \tan \epsilon$, the arc ϵ being taken for abscissa, and u for ordinate. The curve consists of asymptotic branches.

Fig. 15.

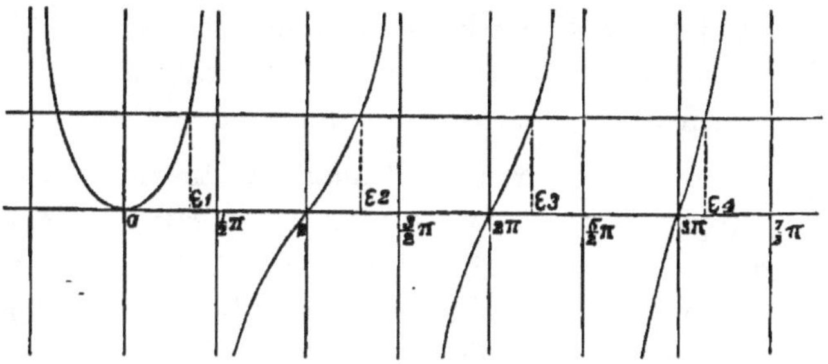

The abscissæ which correspond to the asymptotes are $\frac{1}{2}\pi$, $\frac{3}{2}\pi$, $\frac{5}{2}\pi$, $\frac{7}{2}\pi$, &c.; those which correspond to points of intersection are π, 2π, 3π, &c. If now we raise at the origin an ordinate equal to the known quantity $\frac{hl}{k}$, and through its extremity draw a parallel to the axis of abscissæ, the points of intersection will give the roots of the proposed equation $\epsilon \tan \epsilon = \frac{hl}{k}$. The construction indicates the limits between which each root lies. We shall not stop to indicate the process of calculation which must be employed to determine the values of the roots. Researches of this kind present no difficulty.

329. 2nd. We easily conclude from the general equation (E) that the greater the value of x becomes, the greater that term of the value of v becomes, in which we find the fraction $e^{-x\sqrt{n_1^2+n_1^2}}$, with respect to each of the following terms. In fact, n_1, n_2, n_3, &c. being increasing positive quantities, the fraction $e^{-x\sqrt{2n_1^2}}$ is

greater than any of the analogous fractions which enter into the subsequent terms.

Suppose now that we can observe the temperature of a point on the axis of the prism situated at a very great distance x, and the temperature of a point on this axis situated at the distance $x+1$, 1 being the unit of measure; we have then $y=0$, $z=0$, and the ratio of the second temperature to the first is sensibly equal to the fraction $e^{-\sqrt{2n_1^2}}$. This value of the ratio of the temperatures at the two points on the axis becomes more exact as the distance x increases.

It follows from this that if we mark on the axis points each of which is at a distance equal to the unit of measure from the preceding, the ratio of the temperature of a point to that of the point which precedes it, converges continually to the fraction $e^{-\sqrt{2n_1^2}}$; thus the temperatures of points situated at equal distances end by decreasing in geometrical progression. This law always holds, whatever be the thickness of the bar, provided we consider points situated at a great distance from the source of heat.

It is easy to see, by means of the construction, that if the quantity called l, which is half the thickness of the prism, is very small, n_1 has a value very much smaller than n_2, or n_3, &c.; it follows from this that the first fraction $e^{-x\sqrt{2n_1^2}}$ is very much greater than any of the analogous fractions. Thus, in the case in which the thickness of the bar is very small, it is unnecessary to be very far distant from the source of heat, in order that the temperatures of points equally distant may decrease in geometrical progression. The law holds through the whole extent of the bar.

330. If the half thickness l is a very small quantity, the general value of v is reduced to the first term which contains $e^{-x\sqrt{2n_1^2}}$. Thus the function v which expresses the temperature of a point whose co-ordinates are x, y, and z, is given in this case by the equation

$$v=\left(\frac{4\sin nl}{2nl+\sin 2nl}\right)^2 \cos ny \cos nz\, e^{-x\sqrt{2n^2}},$$

the arc ϵ or nl becomes very small, as we see by the construction. The equation $\epsilon\tan\epsilon=\frac{hl}{k}$ reduces then to $\epsilon^2=\frac{hl}{k}$; the first value of

ϵ, or ϵ_1, is $\sqrt{\frac{hl}{k}}$; by inspection of the figure we know the values of the other roots, so that the quantities $\epsilon_1, \epsilon_2, \epsilon_3, \epsilon_4, \epsilon_5$, &c. are the following $\sqrt{\frac{hl}{k}}$, π, 2π, 3π, 4π, &c. The values of n_1, n_2, n_3, n_4, n_5, &c. are, therefore,

$$\frac{1}{\sqrt{l}}\sqrt{\frac{h}{k}},\ \frac{\pi}{l},\ \frac{2\pi}{l},\ \frac{3\pi}{l},\ \&c.;$$

whence we conclude, as was said above, that if l is a very small quantity, the first value n is incomparably greater than all the others, and that we must omit from the general value of v all the terms which follow the first. If now we substitute in the first term the value found for n, remarking that the arcs nl and $2nl$ are equal to their sines, we have

$$v = \cos\left(\frac{y}{l}\sqrt{\frac{hl}{k}}\right)\cos\left(\frac{z}{l}\sqrt{\frac{hl}{k}}\right)e^{-x\sqrt{\frac{2h}{kl}}},$$

the factor $\sqrt{\frac{hl}{k}}$ which enters under the symbol cosine being very small, it follows that the temperature varies very little, for different points of the same section, when the half thickness l is very small. This result is so to speak self-evident, but it is useful to remark how it is explained by analysis. The general solution reduces in fact to a single term, by reason of the thinness of the bar, and we have on replacing by unity the cosines of very small arcs $v = e^{-x\sqrt{\frac{2h}{kl}}}$, an equation which expresses the stationary temperatures in the case in question.

We found the same equation formerly in Article 76; it is obtained here by an entirely different analysis.

331. The foregoing solution indicates the character of the movement of heat in the interior of the solid. It is easy to see that when the prism has acquired at all its points the stationary temperatures which we are considering, a constant flow of heat passes through each section perpendicular to the axis towards the end which was not heated. To determine the quantity of flow which corresponds to an abscissa x, we must consider that the quantity which flows during unit of time, across one element of

the section, is equal to the product of the coefficient k, of the area $dydz$, of the element dt, and of the ratio $\frac{dv}{dx}$ taken with the negative sign. We must therefore take the integral $-k\int dy\int dz\frac{dv}{dx}$, from $z=0$ to $z=l$, the half thickness of the bar, and then from $y=0$ to $y=l$. We thus have the fourth part of the whole flow.

The result of this calculation discloses the law according to which the quantity of heat which crosses a section of the bar decreases; and we see that the distant parts receive very little heat from the source, since that which emanates directly from it is directed partly towards the surface to be dissipated into the air. That which crosses any section whatever of the prism forms, if we may so say, a sheet of heat whose density varies from one point of the section to another. It is continually employed to replace the heat which escapes at the surface, through the whole end of the prism situated to the right of the section: it follows therefore that the whole heat which escapes during a certain time from this part of the prism is exactly compensated by that which penetrates it by virtue of the interior conducibility of the solid.

To verify this result, we must calculate the produce of the flow established at the surface. The element of surface is $dxdy$, and v being its temperature, $hvdxdy$ is the quantity of heat which escapes from this element during the unit of time. Hence the integral $h\int dx\int dy\, v$ expresses the whole heat which has escaped from a finite portion of the surface. We must now employ the known value of v in y, supposing $z=l$, then integrate once from $y=0$ to $y=l$, and a second time from $x=x$ up to $x=\infty$. We thus find half the heat which escapes from the upper surface of the prism; and taking four times the result, we have the heat lost through the upper and lower surfaces.

If we now make use of the expression $h\int dx\int dz\, v$, and give to y in v its value l, and integrate once from $z=0$ to $z=l$, and a second time from $x=0$ to $x=\infty$; we have one quarter of the heat which escapes at the lateral surfaces.

The integral $h\int dx\int dy\, v$, taken between the limits indicated gives

$$\frac{ha}{m\sqrt{m^2+n^2}}\sin ml\cos nl\, e^{-x\sqrt{m^2+n^2}},$$

and the integral $h\int dx\int dz\, v$ gives

$$\frac{ha}{n\sqrt{m^2+n^2}}\cos ml\sin nl\, e^{-x\sqrt{m^2+n^2}}.$$

Hence the quantity of heat which the prism loses at its surface, throughout the part situated to the right of the section whose abscissa is x, is composed of terms all analogous to

$$\frac{4ha}{\sqrt{m^2+n^2}}e^{-x\sqrt{m^2+n^2}}\left\{\frac{1}{m}\sin ml\cos nl+\frac{1}{n}\cos ml\sin nl\right\}.$$

On the other hand the quantity of heat which during the same time penetrates the section whose abscissa is x is composed of terms analogous to

$$\frac{4ka\sqrt{m^2+n^2}}{mn}e^{-x\sqrt{m^2+n^2}}\sin ml\sin nl;$$

the following equation must therefore necessarily hold

$$\frac{k\sqrt{m^2+n^2}}{mn}\sin ml\sin nl=\frac{h}{m\sqrt{m^2+n^2}}\sin ml\cos nl$$
$$+\frac{h}{n\sqrt{m^2+n^2}}\cos ml\sin nl,$$

or $\quad k(m^2+n^2)\sin ml\sin nl=hm\cos ml\sin nl+hn\sin ml\cos nl$:

now we have separately,

$$km^2\sin ml\cos nl=hm\cos ml\sin nl,$$

or
$$\frac{m\sin ml}{\cos ml}=\frac{h}{k};$$

we have also

$$kn^2\sin nl\sin ml=hn\cos nl\sin ml,$$

or
$$\frac{n\sin nl}{\cos nl}=\frac{h}{k}.$$

Hence the equation is satisfied. This compensation which is incessantly established between the heat dissipated and the heat transmitted, is a manifest consequence of the hypothesis; and analysis reproduces here the condition which has already been expressed; but it was useful to notice this conformity in a new problem, which had not yet been submitted to analysis.

332. Suppose the half side l of the square which serves as the base of the prism to be very long, and that we wish to ascertain the law according to which the temperatures at the different points of the axis decrease; we must give to y and z nul values in the general equation, and to l a very great value. Now the construction shews in this case that the first value of ϵ is $\frac{\pi}{2}$, the second $\frac{3\pi}{2}$, the third $\frac{5\pi}{2}$, &c. Let us make these substitutions in the general equation, and replace $n_1 l$, $n_2 l$, $n_3 l$, $n_4 l$, &c. by their values $\frac{\pi}{2}, \frac{3\pi}{2}, \frac{5\pi}{2}, \frac{7\pi}{2}$, and also substitute the fraction α for $e^{-\frac{x}{l}\frac{\pi}{2}}$; we then find

$$\begin{aligned} v\left(\frac{\pi}{4}\right)^2 = {} & 1\left(\alpha^{\sqrt{1^2+1^2}} - \frac{1}{3}\alpha^{\sqrt{1^2+3^2}} + \frac{1}{5}\alpha^{\sqrt{1^2+5^2}} - \&c.\right) \\ & - \frac{1}{3}\left(\alpha^{\sqrt{3^2+1^2}} - \frac{1}{3}\alpha^{\sqrt{3^2+3^2}} + \frac{1}{5}\alpha^{\sqrt{3^2+5^2}} - \&c.\right) \\ & + \frac{1}{5}\left(\alpha^{\sqrt{5^2+1^2}} - \frac{1}{3}\alpha^{\sqrt{5^2+3^2}} + \frac{1}{5}\alpha^{\sqrt{5^2+5^2}} - \&c.\right) \\ & - \&c. \end{aligned}$$

We see by this result that the temperature at different points of the axis decreases rapidly according as their distance from the origin increases. If then we placed on a support heated and maintained at a permanent temperature, a prism of infinite height, having as base a square whose half side l is very great; heat would be propagated through the interior of the prism, and would be dissipated at the surface into the surrounding air which is supposed to be at temperature 0. When the solid had arrived at a fixed state, the points of the axis would have very unequal temperatures, and at a height equal to half the side of the base the temperature of the hottest point would be less than one fifth part of the temperature of the base.

CHAPTER VIII.

OF THE MOVEMENT OF HEAT IN A SOLID CUBE.

333. It still remains for us to make use of the equation

$$\frac{dv}{dt}=\frac{K}{CD}\left(\frac{d^2v}{dx^2}+\frac{d^2v}{dy^2}+\frac{d^2v}{dz^2}\right) \quad\text{.................}(a),$$

which represents the movement of heat in a solid cube exposed to the action of the air (Chapter II., Section V.). Assuming, in the first place, for v the very simple value $e^{-mt}\cos nx \cos py \cos qz$, if we substitute it in the proposed equation, we have the equation of condition $m=k(n^2+p^2+q^2)$, the letter k denoting the coefficient $\frac{K}{CD}$. It follows from this that if we substitute for n, p, q any quantities whatever, and take for m the quantity $k(n^2+p^2+q^2)$, the preceding value of v will always satisfy the partial differential equation. We have therefore the equation $v=e^{-k(n^2+p^2+q^2)t}\cos nx \cos py \cos qz$. The nature of the problem requires also that if x changes sign, and if y and z remain the same, the function should not change; and that this should also hold with respect to y or z: now the value of v evidently satisfies these conditions.

334. To express the state of the surface, we must employ the following equations:

$$\left.\begin{aligned} \pm K\frac{dv}{dx}+hv=0 \\ \pm K\frac{dv}{dy}+hv=0 \\ \pm K\frac{dv}{dz}+hv=0 \end{aligned}\right\} \quad\text{.............................}(b).$$

These ought to be satisfied when $x = \pm a$, or $y = \pm a$, or $z = \pm a$. The centre of the cube is taken to be the origin of co-ordinates: and the side is denoted by a.

The first of the equations (b) gives

$$\mp e^{-mt} n \sin nx \cos py \cos qz + \frac{h}{K} \cos nx \cos py \cos qz = 0,$$

or

$$\mp n \tan nx + \frac{h}{K} = 0,$$

an equation which must hold when $x = \pm a$.

It follows from this that we cannot take any value whatever for n, but that this quantity must satisfy the condition $na \tan na = \frac{h}{K} a$. We must therefore solve the definite equation $\epsilon \tan \epsilon = \frac{h}{K} a$, which gives the value of ϵ, and take $n = \frac{\epsilon}{a}$. Now the equation in ϵ has an infinity of real roots; hence we can find for n an infinity of different values. We can ascertain in the same manner the values which may be given to p and to q; they are all represented by the construction which was employed in the preceding problem (Art. 321). Denoting these roots by n_1, n_2, n_3, &c.; we can then give to v the particular value expressed by the equation

$$v = e^{-kt(n^2+p^2+q^2)} \cos nx \cos py \cos qz,$$

provided we substitute for n one of the roots n_1, n_2, n_3, &c., and select p and q in the same manner.

335. We can thus form an infinity of particular values of v, and it evident that the sum of several of these values will also satisfy the differential equation (a), and the definite equations (b). In order to give to v the general form which the problem requires, we may unite an indefinite number of terms similar to the term

$$ae^{-kt(n^2+p^2+q^2)} \cos nx \cos py \cos qz.$$

The value of v may be expressed by the following equation:

$$v = (a_1 \cos n_1x\, e^{-kn_1^2t} + a_2 \cos n_2x\, e^{-kn_2^2t} + a_3 \cos n_3x\, e^{-kn_3^2t} + \&c.),$$

$$(b_1 \cos n_1y\, e^{-kn_1^2t} + b_2 \cos n_2y\, e^{-kn_2^2t} + b_3 \cos n_3y\, e^{-kn_3^2t} + \&c.),$$

$$(c_1 \cos n_1z\, e^{-kn_1^2t} + c_2 \cos n_2z\, e^{-kn_2^2t} + c_3 \cos n_3y\, e^{-kn_3^2t} + \&c.).$$

The second member is formed of the product of the three factors written in the three horizontal lines, and the quantities a_1, a_2, a_3, &c. are unknown coefficients. Now, according to the hypothesis, if t be made $=0$, the temperature must be the same at all points of the cube. We must therefore determine a_1, a_2, a_3, &c., so that the value of v may be constant, whatever be the values of x, y, and z, provided that each of these values is included between a and $-a$. Denoting by 1 the initial temperature at all points of the solid, we shall write down the equations (Art. 323)

$$1 = a_1 \cos n_1 x + a_2 \cos n_2 x + a_3 \cos n_3 x + \&c.,$$

$$1 = b_1 \cos n_1 y + b_2 \cos n_2 y + b_3 \cos n_3 y + \&c.,$$

$$1 = c_1 \cos n_1 z + c_2 \cos n_2 z + c_3 \cos n_3 z + \&c.,$$

in which it is required to determine a_1, a_2, a_3, &c. After multiplying each member of the first equation by $\cos nx$, integrate from $x=0$ to $x=a$: it follows then from the analysis formerly employed (Art. 324) that we have the equation

$$1 = \frac{\sin n_1 a \cos n_1 x}{\frac{1}{2} n_1 a \left(1 + \frac{\sin 2n_1 a}{2n_1 a}\right)} + \frac{\sin n_2 a \cos n_2 x}{\frac{1}{2} n_2 a \left(1 + \frac{\sin 2n_2 a}{2n_2 a}\right)} + \frac{\sin n_3 a \cos n_3 x}{\frac{1}{2} n_3 a \left(1 + \frac{\sin 2n_3 a}{2n_3 a}\right)}$$
$$+ \&c.$$

Denoting by μ_i the quantity $\frac{1}{2}\left(1 + \frac{\sin 2n_i a}{2n_i a}\right)$, we have

$$1 = \frac{\sin n_1 a}{n_1 a \mu_1} \cos n_1 x + \frac{\sin n_2 a}{n_2 a \mu_2} \cos n_2 x + \frac{\sin n_3 a}{n_3 a \mu_3} \cos n_3 x + \&c.$$

This equation holds always when we give to x a value included between a and $-a$.

From it we conclude the general value of v, which is given by the following equation

$$v = \left(\frac{\sin n_1 a}{n_1 a \mu_1} \cos n_1 x \, e^{-kn_1^2 t} + \frac{\sin n_2 a}{n_2 a \mu_2} \cos n_2 x \; e^{-kn_2^2 t} + \&c.\right),$$

$$\left(\frac{\sin n_1 a}{n_1 a \mu_1} \cos n_1 y \, e^{-kn_1^2 t} + \frac{\sin n_2 a}{n_2 a \mu_2} \cos n_2 y \; e^{-kn_2^2 t} + \&c.\right),$$

$$\left(\frac{\sin n_1 a}{n_1 a \mu_1} \cos n_1 z \, e^{-kn_1^2 t} + \frac{\sin n_2 a}{n_2 a \mu_2} \cos n_2 z \; e^{-kn_2^2 t} + \&c.\right).$$

336. The expression for v is therefore formed of three similar functions, one of x, the other of y, and the third of z, which is easily verified directly.

In fact, if in the equation

$$\frac{dv}{dt}=k\left(\frac{d^2v}{dx^2}+\frac{d^2v}{dy^2}+\frac{d^2v}{dz^2}\right),$$

we suppose $v=XYZ$; denoting by X a function of x and t, by Y a function of y and t, and by Z a function of z and t, we have

$$YZ=\frac{dX}{dt}+ZX\frac{dY}{dt}+XY\frac{dZ}{dt}=k\left(YZ\frac{d^2X}{dx^2}+ZX\frac{d^2Y}{dy^2}+XY\frac{d^2Z}{dz^2}\right),$$

or $$\frac{1}{X}\frac{dX}{dt}+\frac{1}{Y}\frac{dY}{dt}+\frac{1}{Z}\frac{dZ}{dt}=k\left(\frac{1}{X}\frac{d^2X}{dx^2}+\frac{1}{Y}\frac{d^2Y}{dy^2}+\frac{1}{Z}\frac{d^2Z}{dz^2}\right),$$

which implies the three separate equations

$$\frac{dX}{dt}=k\frac{d^2X}{dx^2},\quad \frac{dY}{dt}=k\frac{d^2Y}{dy^2},\quad \frac{dZ}{dt}=k\frac{d^2Z}{dz^2}.$$

We must also have as conditions relative to the surface,

$$\frac{dV}{dx}+\frac{k}{K}V=0,\quad \frac{dV}{dy}+\frac{k}{K}V=0,\quad \frac{dV}{dz}+\frac{k}{K}V=0,$$

whence we deduce

$$\frac{dX}{dx}+\frac{h}{K}X=0,\quad \frac{dY}{dy}+\frac{h}{K}Y=0,\quad \frac{dZ}{dz}+\frac{h}{K}Z=0.$$

It follows from this, that, to solve the problem completely, it is enough to take the equation $\frac{du}{dt}=k\frac{d^2u}{dx^2}$, and to add to it the equation of condition $\frac{du}{dx}+\frac{h}{K}u=0$, which must hold when $x=a$. We must then put in the place of x, either y or z, and we shall have the three functions X, Y, Z, whose product is the general value of v.

Thus the problem proposed is solved as follows:

$$v=\phi(x,t)\,\phi(y,t)\,\phi(z,t);$$

$$\phi(x,t)=\frac{\sin n_1a}{n_1a\mu_1}\cos n_1x\,e^{-kn_1^2t}+\frac{\sin n_2a}{n_2a\mu_2}\cos n_2x\,e^{-kn_2^2t}$$
$$+\frac{\sin n_3a}{n_3a\mu_3}\cos n_3x\,e^{-kn_3^2t}+\&c.;$$

n_1, n_2, n_3, &c. being given by the following equation

$$\epsilon \tan \epsilon = \frac{ha}{K},$$

in which ϵ represents na and the value of μ_i is

$$\frac{1}{2}\left(1+\frac{\sin 2n_i a}{2n_i a}\right).$$

In the same manner the functions $\phi(y, t)$, $\phi(z, t)$ are found.

337. We may be assured that this value of v solves the problem in all its extent, and that the complete integral of the partial differential equation (a) must necessarily take this form in order to express the variable temperatures of the solid.

In fact, the expression for v satisfies the equation (a) and the conditions relative to the surface. Hence the variations of temperature which result in one instant from the action of the molecules and from the action of the air on the surface, are those which we should find by differentiating the value of v with respect to the time t. It follows that if, at the beginning of any instant, the function v represents the system of temperatures, it will still represent those which hold at the commencement of the following instant, and it may be proved in the same manner that the variable state of the solid is always expressed by the function v, in which the value of t continually increases. Now this function agrees with the initial state: hence it represents all the later states of the solid. Thus it is certain that any solution which gives for v a function different from the preceding must be wrong.

338. If we suppose the time t, which has elapsed, to have become very great, we no longer have to consider any but the first term of the expression for v; for the values n_1, n_2, n_3, &c. are arranged in order beginning with the least. This term is given by the equation

$$v = \left(\frac{\sin n_1 a}{n_1 a \mu_1}\right)^3 \cos n_1 x \cos n_1 y \cos n_1 z \, e^{-3kn_1^2 t};$$

this then is the principal state towards which the system of temperatures continually tends, and with which it coincides without sensible error after a certain value of t. In this state the temperature at every point decreases proportionally to the powers of the fraction $e^{-3kn_1^2}$; the successive states are then all similar, or rather they differ only in the magnitudes of the temperatures which all diminish as the terms of a geometrical progression, preserving their ratios. We may easily find, by means of the preceding equation, the law by which the temperatures decrease from one point to another in direction of the diagonals or the edges of the cube, or lastly of a line given in position. We might ascertain also what is the nature of the surfaces which determine the layers of the same temperature. We see that in the final and regular state which we are here considering, points of the same layer preserve always equal temperatures, which would not hold in the initial state and in those which immediately follow it. During the infinite continuance of the ultimate state the mass is divided into an infinity of layers all of whose points have a common temperature.

339. It is easy to determine for a given instant the mean temperature of the mass, that is to say, that which is obtained by taking the sum of the products of the volume of each molecule by its temperature, and dividing this sum by the whole volume. We thus form the expression $\iiint \frac{v\,dx\,dy\,dz}{2^3 a^3}$, which is that of the mean temperature V. The integral must be taken successively with respect to x, y, and z, between the limits a and $-a$: v being equal to the product XYZ, we have

$$V = \int X dx \int Y dy \int Z dz;$$

thus the mean temperature is $\left(\int \frac{X dx}{2a}\right)^3$, since the three complete integrals have a common value, hence

$$\sqrt[3]{V} = \left(\frac{\sin n_1 a}{n_1 a}\right)^2 \frac{1}{\mu_1} e^{-kn_1^2 t} + \left(\frac{\sin n_2 a}{n_2 a}\right)^2 \frac{1}{\mu_2} e^{-kn_2^2 t} + \&c.$$

The quantity na is equal to ϵ, a root of the equation $\epsilon \tan \epsilon = \frac{ha}{K}$, and μ is equal to $\frac{1}{2}\left(1 + \frac{\sin 2\epsilon}{2\epsilon}\right)$. We have then, denoting the different roots of this equation by $\epsilon_1, \epsilon_2, \epsilon_3$, &c.,

$$2\sqrt[3]{V} = \left(\frac{\sin \epsilon_1}{\epsilon_1}\right)^2 \frac{e^{-k\frac{\epsilon_1^2}{a^2}t}}{1+\frac{\sin 2\epsilon_1}{2\epsilon_1}} + \left(\frac{\sin \epsilon_2}{\epsilon_2}\right)^2 \frac{e^{-k\frac{\epsilon_2^2}{a^2}t}}{1+\frac{\sin 2\epsilon_2}{2\epsilon_2}} + \&c.$$

ϵ_1 is between 0 and $\frac{1}{2}\pi$, ϵ_2 is between π and $\frac{3\pi}{2}$, ϵ_3 between 2π and $\frac{5}{2}\pi$, the roots ϵ_2, ϵ_3, ϵ_4, &c. approach more and more nearly to the inferior limits π, 2π, 3π, &c., and end by coinciding with them when the index i is very great. The double arcs $2\epsilon_1$, $2\epsilon_2$, $2\epsilon_3$, &c., are included between 0 and π, between 2π and 3π, between 4π and 5π; for which reason the sines of these arcs are all positive: the quantities $1+\frac{\sin 2\epsilon_1}{2\epsilon_1}$, $1+\frac{\sin 2\epsilon_2}{2\epsilon_2}$, &c., are positive and included between 1 and 2. It follows from this that all the terms which enter into the value of $\sqrt[3]{V}$ are positive.

340. We propose now to compare the velocity of cooling in the cube, with that which we have found for a spherical mass. We have seen that for either of these bodies, the system of temperatures converges to a permanent state which is sensibly attained after a certain time; the temperatures at the different points of the cube then diminish all together preserving the same ratios, and the temperatures of one of these points decrease as the terms of a geometric progression whose ratio is not the same in the two bodies. It follows from the two solutions that the ratio for the sphere is e^{-kn^2} and for the cube $e^{-3\frac{\epsilon^2}{a^2}k}$. The quantity n is given by the equation

$$na\frac{\cos na}{\sin na} = 1 - \frac{h}{K}a,$$

a being the semi-diameter of the sphere, and the quantity ϵ is given by the equation $\epsilon \tan \epsilon = \frac{h}{K}a$, a being the half side of the cube.

This arranged, let us consider two different cases; that in which the radius of the sphere and the half side of the cube are each equal to a, a very small quantity; and that in which the value of a is very great. Suppose then that the two bodies are of small dimensions; $\frac{ha}{K}$ having a very small value, the same is the case with ϵ, we have therefore $\frac{ha}{K}=\epsilon^2$, hence the fraction

$$e^{-3\frac{\epsilon^2}{a^2}k} \text{ is equal to } e^{-\frac{3h}{CDa}}.$$

Thus the ultimate temperatures which we observe are expressed in the form $Ae^{-\frac{3ht}{CDa}}$. If now in the equation $\frac{na\cos na}{\sin na}=1-\frac{h}{K}a$, we suppose the second member to differ very little from unity, we find $\frac{h}{K}=\frac{n^2a}{3}$, hence the fraction e^{-kn^2} is $e^{-\frac{3h}{CDa}}$.

We conclude from this that if the radius of the sphere is very small, the final velocities of cooling are the same in that solid and in the circumscribed cube, and that each is in inverse ratio of the radius; that is to say, if the temperature of a cube whose half side is a passes from the value A to the value B in the time t, a sphere whose semi-diameter is a will also pass from the temperature A to the temperature B in the same time. If the quantity a were changed for each body so as to become a', the time required for the passage from A to B would have another value t', and the ratio of the times t and t' would be that of the half sides a and a'. The same would not be the case when the radius a is very great: for ϵ is then equal to $\frac{1}{2}\pi$, and the values of na are the quantities π, 2π, 3π, 4π, &c.

We may then easily find, in this case, the values of the fractions $e^{-3\frac{\epsilon^2}{a^2}k}$, e^{-kn^2}; they are $e^{-\frac{3k\pi^2}{4a^2}}$ and $e^{-\frac{k\pi^2}{a^2}}$.

From this we may derive two remarkable consequences: 1st, when two cubes are of great dimensions, and a and a' are their half-sides; if the first occupies a time t in passing from the temperature A to the temperature B, and the second the time t' for the same interval; the times t and t' will be proportional to the squares a^2 and a'^2 of the half-sides. We found a similar result for spheres of great dimensions. 2nd, If the length a of the half-side of a cube is considerable, and a sphere has the same magnitude a for radius, and during the time t the temperature of the cube falls from A to B, a different time t' will elapse whilst the temperature of the sphere is falling from A to B, and the times t and t' are in the ratio of 4 to 3.

Thus the cube and the inscribed sphere cool equally quickly when their dimension is small; and in this case the duration of the cooling is for each body proportional to its thickness. If the dimension of the cube and the inscribed sphere is great, the final duration of the cooling is not the same for the two solids. This duration is greater for the cube than for the sphere, in the ratio of 4 to 3, and for each of the two bodies severally the duration of the cooling increases as the square of the diameter.

341. We have supposed the body to be cooling slowly in atmospheric air whose temperature is constant. We might submit the surface to any other condition, and imagine, for example, that all its points preserve, by virtue of some external cause, the fixed temperature 0. The quantities n, p, q, which enter into the value of v under the symbol cosine, must in this case be such that $\cos nx$ becomes nothing when x has its complete value a, and that the same is the case with $\cos py$ and $\cos qz$. If $2a$ the side of the cube is represented by π, 2π being the length of the circumference whose radius is 1; we can express a particular value of v by the following equation, which satisfies at the same time the general equation of movement of heat, and the state of the surface,

$$v = e^{-3\frac{Kt}{CD}} \cos x \,.\, \cos y \,.\, \cos z.$$

This function is nothing, whatever be the time t, when x or y or z receive their extreme values $+\frac{\pi}{2}$ or $-\frac{\pi}{2}$: but the expression for the temperature cannot have this simple form until after a considerable time has elapsed, unless the given initial state is itself represented by $\cos x \cos y \cos z$. This is what we have supposed in Art. 100, Sect. VIII. Chap. I. The foregoing analysis proves the truth of the equation employed in the Article we have just cited.

Up to this point we have discussed the fundamental problems in the theory of heat, and have considered the action of that element in the principal bodies. Problems of such kind and order have been chosen, that each presents a new difficulty of a higher degree. We have designedly omitted a numerous variety of intermediate problems, such as the problem of the linear movement of heat in a prism whose ends are maintained at fixed temperatures, or exposed to the atmospheric air. The expression for the varied movement of heat in a cube or rectangular prism which is cooling in an aëriform medium might be generalised, and any initial state whatever supposed. These investigations require no other principles than those which have been explained in this work.

A memoir was published by M. Fourier in the *Mémoires de l'Académie des Sciences*, Tome VII. Paris, 1827, pp. 605—624, entitled, *Mémoire sur la distinction des racines imaginaires, et sur l'application des théorèmes d'analyse algébrique aux équations transcendantes qui dependent de la théorie de la chaleur.* It contains a proof of two propositions in the theory of heat. If there be two solid bodies of similar convex forms, such that corresponding elements have the same density, specific capacity for heat, and conductivity, and the same initial distribution of temperature, the condition of the two bodies will always be the same after times which are as the squares of the dimensions, when, 1st, corresponding elements of the surfaces are maintained at constant temperatures, or 2nd, when the temperatures of the exterior medium at corresponding points of the surface remain constant.

For the velocities of flow along lines of flow across the terminal areas s, s' of corresponding prismatic elements are as $u-v : u'-v'$, where (u, v), (u', v') are temperatures at pairs of points at the same distance $\frac{1}{2}\Delta$ on opposite sides of s and s'; and if $n : n'$ is the ratio of the dimensions, $u-v : u'-v' = n' : n$. If then, dt, dt' be corresponding times, the quantities of heat received by the prismatic elements are as $sk(u-v)\,dt : s'k(u'-v')\,dt'$, or as $n^2n'dt : n'^2ndt'$. But the volumes being as $n^3 : n'^3$, if the corresponding changes of temperature are always equal we must have

$$\frac{n^2n'dt}{n^3} = \frac{n'^2ndt'}{n'^3}, \text{ or } \frac{dt}{dt'} = \frac{n^2}{n'^2}.$$

In the second case we must suppose $H : H' = n' : n$. [A. F.]

CHAPTER IX.

OF THE DIFFUSION OF HEAT.

FIRST SECTION.

Of the free movement of heat in an infinite line.

342. HERE we consider the movement of heat in a solid homogeneous mass, all of whose dimensions are infinite. The solid is divided by planes infinitely near and perpendicular to a common axis; and it is first supposed that one part only of the solid has been heated, that, namely, which is enclosed between two parallel planes A and B, whose distance is g; all other parts have the initial temperature 0; but any plane included between A and B has a given initial temperature, regarded as arbitrary, and common to every point of the plane; the temperature is different for different planes. The initial state of the mass being thus defined, it is required to determine by analysis all the succeeding states. The movement in question is simply linear, and in direction of the axis of the plane; for it is evident that there can be no transfer of heat in any plane perpendicular to the axis, since the initial temperature at every point in the plane is the same.

Instead of the infinite solid we may suppose a prism of very small thickness, whose lateral surface is wholly impenetrable to heat. The movement is then considered only in the infinite line which is the common axis of all the sectional planes of the prism.

The problem is more general, when we attribute temperatures entirely arbitrary to all points of the part of the solid which has been heated, all other points of the solid having the initial temperature 0. The laws of the distribution of heat in an infinite solid mass ought to have a simple and remarkable character; since the movement is not disturbed by the obstacle of surfaces, or by the action of a medium.

343. The position of each point being referred to three rectangular axes, on which we measure the co-ordinates x, y, z, the temperature sought is a function of the variables x, y, z, and of the time t. This function v or $\phi(x, y, z, t)$ satisfies the general equation

$$\frac{dv}{dt} = \frac{K}{CD}\left(\frac{d^2v}{dx^2} + \frac{d^2v}{dy^2} + \frac{d^2v}{dz^2}\right) \qquad\qquad (a).$$

Further, it must necessarily represent the initial state which is arbitrary; thus, denoting by $F(x, y, z)$ the given value of the temperature at any point, taken when the time is nothing, that is to say, at the moment when the diffusion begins, we must have

$$\phi(x, y, z, 0) = F(x, y, z) \qquad\qquad (b).$$

Hence we must find a function v of the four variables x, y, z, t, which satisfies the differential equation (a) and the definite equation (b).

In the problems which we previously discussed, the integral is subject to a third condition which depends on the state of the surface: for which reason the analysis is more complex, and the solution requires the employment of exponential terms. The form of the integral is very much more simple, when it need only satisfy the initial state; and it would be easy to determine at once the movement of heat in three dimensions. But in order to explain this part of the theory, and to ascertain according to what law the diffusion is effected, it is preferable to consider first the linear movement, resolving it into the two following problems: we shall see in the sequel how they are applied to the case of three dimensions.

344. First problem: a part ab of an infinite line is raised at all points to the temperature 1; the other points of the line are at the actual temperature 0; it is assumed that the heat cannot be dispersed into the surrounding medium; we have to determine

what is the state of the line after a given time. This problem may be made more general, by supposing, 1st, that the initial temperatures of the points included between a and b are unequal and represented by the ordinates of any line whatever, which we shall regard first as composed of two symmetrical parts (see fig. 16);

Fig. 16.

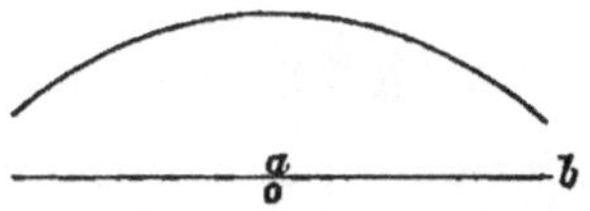

2nd, that part of the heat is dispersed through the surface of the solid, which is a prism of very small thickness, and of infinite length.

The second problem consists in determining the successive states of a prismatic bar, infinite in length, one extremity of which is submitted to a constant temperature. The solution of these two problems depends on the integration of the equation

$$\frac{dv}{dt} = \frac{K}{CD}\frac{d^2v}{dx^2} - \frac{HL}{CDS}v,$$

(Article 105), which expresses the linear movement of heat. v is the temperature which the point at distance x from the origin must have after the lapse of the time t; K, H, C, D, L, S, denote the internal and surface conducibilities, the specific capacity for heat, the density, the contour of the perpendicular section, and the area of this section.

345. Consider in the first instance the case in which heat is propagated freely in an infinite line, one part of which ab has received any initial temperatures; all other points having the initial temperature 0. If at each point of the bar we raise the ordinate of a plane curve so as to represent the actual temperature at that point, we see that after a certain value of the time t, the state of the solid is expressed by the form of the curve. Denote by $v = F(x)$ the equation which corresponds to the given initial state, and first, for the sake of making the investigation

more simple, suppose the initial form of the curve to be composed of two symmetrical parts, so that we have the condition

$$F(x) = F(-x).$$

Let $$\frac{K}{CD} = k, \quad \frac{HL}{CDS} = h;$$

in the equation $\frac{dv}{dt} = k\frac{d^2v}{dx^2} - hv$, make $v = e^{-ht}u$, and we have

$$\frac{du}{dt} = k\frac{d^2u}{dz^2}.$$

Assume a particular value of u, namely, $a\cos qx\, e^{-kq^2t}$; a and q being arbitrary constants. Let q_1, q_2, q_3, &c. be a series of any values whatever, and a_1, a_2, a_3, &c. a series of corresponding values of the coefficient Q, we have

$$u = a_1\cos(q_1x)\, e^{-kq_1^2t} + a_2\cos(q_2x)\, e^{-kq_2^2t} + a_3\cos(q_3x)\, e^{-kq_3^2t} + \&c.$$

Suppose first that the values q_1, q_2, q_3, &c. increase by infinitely small degrees, as the abscissæ q of a certain curve; so that they become equal to $dq, 2dq, 3dq$, &c.; dq being the constant differential of the abscissa; next that the values a_1, a_2, a_3, &c. are proportional to the ordinates Q of the same curve, and that they become equal to Q_1dq, Q_2dq, Q_3dq, &c., Q being a certain function of q. It follows from this that the value of u may be expressed thus:

$$u = \int dq\, Q\cos qx\, e^{-kq^2t},$$

Q is an arbitrary function $f(q)$, and the integral may be taken from $q = 0$ to $q = \infty$. The difficulty is reduced to determining suitably the function Q.

346. To determine Q, we must suppose $t = 0$ in the expression for u, and equate u to $F(x)$. We have therefore the equation of condition

$$F(x) = \int_0^\infty dq\, Q\cos qx.$$

If we substituted for Q any function of q, and conducted the integration from $q = 0$ to $q = \infty$, we should find a function of x: it is required to solve the inverse problem, that is to say, to ascertain what function of q, after being substituted for Q, gives as the result the function $F(x)$, a remarkable problem whose solution demands attentive examination.

Developing the sign of the integral, we write as follows, the equation from which the value of Q must be derived:

$$F(x) = dq\,Q_1 \cos q_1 x + dqQ_2 \cos q_2 x + dqQ_3 \cos q_3 x + \&c.$$

In order to make all the terms of the second member disappear, except one, multiply each side by $dx \cos rx$, and then integrate with respect to x from $x = 0$ to $x = n\pi$, where n is an infinite number, and r represents a magnitude equal to any one of q_1, q_2, q_3, &c., or which is the same thing dq, $2dq$, $3dq$, &c. Let q_i be any value whatever of the variable q, and q_j another value, namely, that which we have taken for r; we shall have $r = jdq$, and $q = idq$. Consider then the infinite number n to express how many times unit of length contains the element dq, so that we have $n = \dfrac{1}{dq}$. Proceeding to the integration we find that the value of the integral $\int dx \cos qx \cos rx$ is nothing, whenever r and q have different magnitudes; but its value is $\frac{1}{2} n\pi$, when $q = r$. This follows from the fact that integration eliminates from the second member all the terms, except one; namely, that which contains q_j or r. The function which affects the same term is Q_j, we have therefore

$$\int dx\,F(x) \cos qx = dq\;Q_j \frac{1}{2} n\pi,$$

and substituting for ndq its value 1, we have

$$\frac{\pi Q_j}{2} = \int dx\,F(x) \cos qx.$$

We find then, in general, $\dfrac{\pi Q}{2} = \displaystyle\int_0^\infty dx\,F(x) \cos qx$. Thus, to determine the function Q which satisfies the proposed condition, we must multiply the given function $F(x)$ by $dx \cos qx$, and integrate from x nothing to x infinite, multiplying the result by $\dfrac{2}{\pi}$; that is to say, from the equation $F(x) = \int dq f(q) \cos qx$, we deduce $f(q) = \dfrac{2}{\pi} \int dx\,F(x) \cos qx$, the function $F(x)$ representing the initial temperatures of an infinite prism, of which an intermediate part only is heated. Substituting the value of $f(q)$ in the expression for $F(x)$, we obtain the general equation

$$\frac{\pi}{2}F(x)=\int_0^\infty dq\cos qx\int_0^\infty dx\,F(x)\cos qx\ldots\ldots\ldots(\epsilon).$$

347. If we substitute in the expression for v the value which we have found for the function Q, we have the following integral, which contains the complete solution of the proposed problem,

$$\frac{\pi v}{2}=e^{-ht}\int_0^\infty dq\cos qx\,e^{-kq^2t}\int_0^\infty dx\,F(x)\cos qx.$$

The integral, with respect to x, being taken from x nothing to x infinite, the result is a function of q; and taking then the integral with respect to q from $q=0$ to $q=\infty$, we obtain for v a function of x and t, which represents the successive states of the solid. Since the integration with respect to x makes this variable disappear, it may be replaced in the expression of v by any variable α, the integral being taken between the same limits, namely from $\alpha=0$ to $\alpha=\infty$. We have then

$$\frac{\pi v}{2}=e^{-ht}\int_0^\infty dq\cos qx\,e^{-kq^2t}\int_0^\infty d\alpha\,F(\alpha)\cos q\alpha,$$

or

$$\frac{\pi v}{2}=e^{-ht}\int_0^\infty d\alpha\,F(\alpha)\int_0^\infty dq\,e^{-kq^2t}\cos qx\cos q\alpha.$$

The integration with respect to q will give a function of x, t and α, and taking the integral with respect to α we find a function of x and t only. In the last equation it would be easy to effect the integration with respect to q, and thus the expression of v would be changed. We can in general give different forms to the integral of the equation

$$\frac{dv}{dt}=k\frac{d^2v}{dx^2}-hv,$$

they all represent the same function of x and t.

348. Suppose in the first place that all the initial temperatures of points included between a and b, from $x=-1$, to $x=1$, have the common value 1, and that the temperatures of all the

other points are nothing, the function $F(x)$ will be given by this condition. It will then be necessary to integrate, with respect to x, from $x=0$ to $x=1$, for the rest of the integral is nothing according to the hypothesis. We shall thus find

$$Q=\frac{2}{\pi}\frac{\sin q}{q} \text{ and } \frac{\pi v}{2}=e^{-ht}\int_0^\infty \frac{dq}{q}e^{-kq^2t}\cos qx \sin q.$$

The second member may easily be converted into a convergent series, as will be seen presently; it represents exactly the state of the solid at a given instant, and if we make in it $t=0$, it expresses the initial state.

Thus the function $\frac{2}{\pi}\int_0^\infty \frac{dq}{q}\sin q \cos qx$ is equivalent to unity, if we give to x any value included between -1 and 1: but this function is nothing if to x any other value be given not included between -1 and 1. We see by this that discontinuous functions also may be expressed by definite integrals.

349. In order to give a second application of the preceding formula, let us suppose the bar to have been heated at one of its points by the constant action of the same source of heat, and that it has arrived at its permanent state which is known to be represented by a logarithmic curve.

It is required to ascertain according to what law the diffusion of heat is effected after the source of heat is withdrawn. Denoting by $F(x)$ the initial value of the temperature, we shall have $F(x)=Ae^{-x\sqrt{\frac{HL}{KS}}}$; A is the initial temperature of the point most heated. To simplify the investigation let us make $A=1$, and $\frac{HL}{KS}=1$. We have then $F(x)=e^{-x}$, whence we deduce $Q=\int dx\, e^{-x}\cos qx$, and taking the integral from x nothing to x infinite, $Q=\frac{1}{1+q^2}$. Thus the value of v in x and t is given by the following equation:

$$\frac{\pi v}{2}=e^{-ht}\int_0^\infty \frac{dq\cos qx}{1+q^2}e^{-q^2kt}.$$

350. If we make $t=0$, we have $\frac{\pi v}{2}=\int_0^\infty \frac{dq\cos qx}{1+q^2}$; which corresponds to the initial state. Hence the expression $\frac{2}{\pi}\int_0^\infty \frac{dq\cos qx}{1+q^2}$ is equal to e^{-x}. It must be remarked that the function $F(x)$, which represents the initial state, does not change its value according to hypothesis when x becomes negative. The heat communicated by the source before the initial state was formed, is propagated equally to the right and the left of the point 0, which directly receives it: it follows that the line whose equation is $y=\frac{2}{\pi}\int_0^\infty \frac{dq\cos qx}{1+q^2}$ is composed of two symmetrical branches which are formed by repeating to right and left of the axis of y the part of the logarithmic curve which is on the right of the axis of y, and whose equation is $y=e^{-x}$. We see here a second example of a discontinuous function expressed by a definite integral. This function $\frac{2}{\pi}\int_0^\infty \frac{dq\cos qx}{1+q^2}$ is equivalent to e^{-x} when x is positive, but it is e^x when x is negative[1].

351. The problem of the propagation of heat in an infinite bar, one end of which is subject to a constant temperature, is reducible, as we shall see presently, to that of the diffusion of heat in an infinite line; but it must be supposed that the initial heat, instead of affecting equally the two contiguous halves of the solid, is distributed in it in contrary manner; that is to say that representing by $F(x)$ the temperature of a point whose distance from the middle of the line is x, the initial temperature of the opposite point for which the distance is $-x$, has for value $-F(x)$.

This second problem differs very little from the preceding, and might be solved by a similar method: but the solution may also be derived from the analysis which has served to determine for us the movement of heat in solids of finite dimensions.

Suppose that a part ab of the infinite prismatic bar has been heated in any manner, see fig. (16*), and that the opposite part $a\beta$ is in like state, but of contrary sign; all the rest of the solid having the initial temperature 0. Suppose also that the surround-

[1] Cf. Riemann, *Part. Diff. Gleich.* § 16, p. 34. [A. F.]

ing medium is maintained at the constant temperature 0, and that it receives heat from the bar or communicates heat to it through

Fig. 16*.

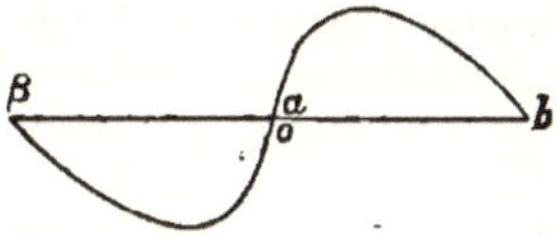

the external surface. It is required to find, after a given time t, what will be the temperature v of a point whose distance from the origin is x.

We shall consider first the heated bar as having a finite length $2X$, and as being submitted to some external cause which maintains its two ends at the constant temperature 0; we shall then make $X = \infty$.

352. We first employ the equation

$$\frac{dv}{dt} = \frac{K}{CD}\frac{d^2v}{dx^2} - \frac{HL}{CDS}v; \quad \text{or} \quad \frac{dv}{dt} = k\frac{d^2v}{dx^2} - hv;$$

and making $v = e^{-ht}u$ we have

$$\frac{du}{dt} = k\frac{d^2u}{dx^2},$$

the general value of u may be expressed as follows:

$$u = a_1 e^{-kg_1^2 t}\sin g_1 x + a_2 e^{-kg_2^2 t}\sin g_2 x + a_3 e^{-kg_3^2 t}\sin g_3 x + \&c.$$

Making then $x = X$, which ought to make the value of v nothing, we have, to determine the series of exponents g, the condition $\sin gX = 0$, or $gX = i\pi$, i being an integer.

Hence

$$u = a_1 e^{-k\frac{\pi^2}{X^2}t}\sin\frac{\pi x}{X} + a_2 e^{-k\frac{\pi^2 2^2 t}{X^2}}\sin\frac{2\pi x}{X} + \&c.$$

It remains only to find the series of constants a_1, a_2, a_3, &c. Making $t = 0$ we have

$$u = F(x) = a_1\sin\frac{\pi x}{X} + a_2\sin\frac{2\pi x}{X} + a_3\sin\frac{3\pi x}{X} + \&c.$$

Let $\frac{\pi x}{X} = r$, and denote $F(x)$ or $F\left(\frac{rX}{\pi}\right)$ by $f(r)$; we have

$$f(r) = a_1 \sin r + a_2 \sin 2r + a_3 \sin 3r + \&c.$$

Now, we have previously found $a_i = \frac{2}{\pi}\int dr f(r) \sin ir$, the integral being taken from $r = 0$ to $r = \pi$. Hence

$$\frac{X}{2} a_i = \int dx\, F(x) \sin \frac{i\pi x}{X}.$$

The integral with respect to x must be taken from $x = 0$ to $x = X$. Making these substitutions, we form the equation

$$v = \frac{2}{X} e^{-ht} \left\{ e^{-k\frac{\pi^2}{X^2}t} \sin \frac{\pi x}{X} \int_0^X dx\, F(x) \sin \frac{\pi x}{X} \right.$$
$$\left. + e^{-k\frac{\pi^2}{X^2}2^2t} \sin \frac{2\pi x}{X} \int dx\, F(x) \sin \frac{2\pi x}{X} + \&c. \right\} \ldots (a).$$

353. Such would be the solution if the prism had a finite length represented by $2X$. It is an evident consequence of the principles which we have laid down up to this point; it remains only to suppose the dimension X infinite. Let $X = n\pi$, n being an infinite number; also let q be a variable whose infinitely small increments dq are all equal; we write $\frac{1}{dq}$ instead of n. The general term of the series which enters into equation (a) being

$$e^{-k\frac{\pi^2}{X^2}i^2t} \sin \frac{i\pi x}{X} \int dx\, F(x) \sin \frac{i\pi x}{X},$$

we represent by $\frac{q}{dq}$ the number i, which is variable and becomes infinite. Thus we have

$$X = \frac{\pi}{dq}, \quad n = \frac{1}{dq}, \quad i = \frac{q}{dq}.$$

Making these substitutions in the term in question we find $e^{-kq^2t} \sin qx \int dx\, F(x) \sin qx$. Each of these terms must be divided by X or $\frac{\pi}{dq}$, becoming thereby an infinitely small quantity, and the sum of the series is simply an integral, which must be taken with respect to q from $q=0$ to $q=\infty$. Hence

$$v=\frac{2}{\pi}e^{-ht}\int dq\, e^{-kq^2t}\sin qx\int dx\, F(x)\sin qx \ldots\ldots\ldots(a),$$

the integral with respect to x must be taken from $x=0$ to $x=\infty$. We may also write

$$\frac{\pi v}{2}=e^{-ht}\int_0^\infty dq\, e^{-kq^2t}\sin qx\int_0^\infty d\alpha F(\alpha)\sin q\alpha,$$

or $$\frac{\pi v}{2}=e^{-ht}\int_0^\infty d\alpha\, F(\alpha)\int_0^\infty dq\, e^{-kq^2t}\sin qx\sin q\alpha.$$

Equation (a) contains the general solution of the problem; and, substituting for $F(x)$ any function whatever, subject or not to a continuous law, we shall always be able to express the value of the temperature in terms of x and t: only it must be remarked that the function $F(x)$ corresponds to a line formed of two equal and alternate parts[1].

354. If the initial heat is distributed in the prism in such a manner that the line *FFFF* (fig. 17), which represents the initial

Fig. 17.

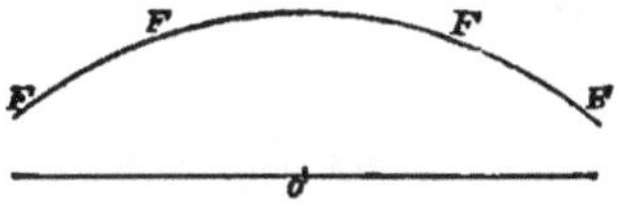

state, is formed of two equal arcs situated right and left of the fixed point 0, the variable movement of the heat is expressed by the equation

$$\frac{\pi v}{2}=e^{-ht}\int_0^\infty d\alpha\, F(\alpha)\int_0^\infty dq\, e^{-kq^2t}\cos qx\cos q\alpha.$$

Fig. 18.

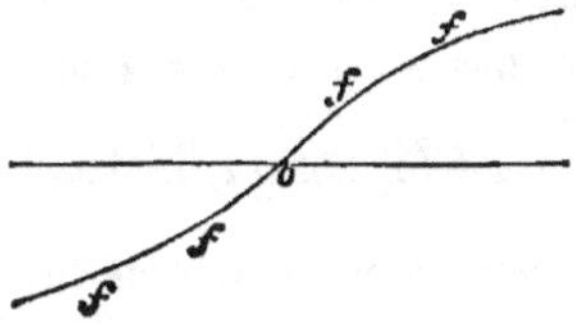

If the line *ffff* (fig. 18), which represents the initial state, is

[1] That is to say, $F(x)=-F(-x)$. [A. F.]

formed of two similar and alternate arcs, the integral which gives the value of the temperature is

$$\frac{\pi v}{2} = e^{-ht}\int_0^\infty d\alpha f(\alpha)\int_0^\infty dq\, e^{-kq^2t}\sin qx \sin q\alpha.$$

If we suppose the initial heat to be distributed in any manner, it will be easy to derive the expression for v from the two preceding solutions. In fact, whatever the function $\phi(x)$ may be, which represents the given initial temperature, it can always be decomposed into two others $F(x)+f(x)$, one of which corresponds to the line *FFFF*, and the other to the line *ffff*, so that we have these three conditions

$$F(x) = F(-x), f(x) = -f(-x), \phi(x) = F(x) + f(x).$$

We have already made use of this remark in Articles 233 and 234. We know also that each initial state gives rise to a variable partial state which is formed as if it alone existed. The composition of these different states introduces no change into the temperatures which would have occurred separately from each of them. It follows from this that denoting by v the variable temperature produced by the initial state which represents the total function $\phi(x)$, we must have

$$\frac{\pi v}{2} = e^{-ht}\left(\int_0^\infty dq\, e^{-kq^2t}\cos qx\int_0^\infty d\alpha\, F(\alpha)\cos q\alpha \right.$$
$$\left. + \int_0^\infty dq\, e^{-kq^2t}\sin qx\int_0^\infty d\alpha f(\alpha)\sin q\alpha\right).$$

If we took the integrals with respect to α between the limits $-\infty$ and $+\infty$, it is evident that we should double the results. We may then, in the preceding equation, omit from the first member the denominator 2, and take the integrals with respect to α in the second form $\alpha=-\infty$ to $\alpha=+\infty$. We easily see also that we could write $\int_{-\infty}^{+\infty} d\alpha\, \phi(\alpha)\cos q\alpha$, instead of $\int_{-\infty}^{+\infty} d\alpha\, F(\alpha)\cos q\alpha$; for it follows from the condition to which the function $f(\alpha)$ is subject, that we must have

$$0 = \int_{-\infty}^{+\infty} d\alpha f(\alpha)\cos q\alpha.$$

We can also write

$$\int_{-\infty}^{+\infty} d\alpha\, \phi(\alpha) \sin q\alpha \text{ instead of } \int_{-\infty}^{+\infty} d\alpha\, f(\alpha) \cos q\alpha,$$

for we evidently have

$$0 = \int_{-\infty}^{+\infty} d\alpha\, F(\alpha) \sin q\alpha.$$

We conclude from this

$$\pi v = e^{-ht} \int_0^{\infty} dq\, e^{-kq^2t} \left(\int_{-\infty}^{+\infty} d\alpha\, \phi(\alpha) \cos q\alpha \cos qx + \int_{-\infty}^{+\infty} d\alpha\, \phi(\alpha) \sin q\alpha \sin qx \right),$$

or,

$$\pi v = e^{-ht} \int_0^{\infty} dq\, e^{-kq^2t} \int_{-\infty}^{+\infty} d\alpha\, \phi(\alpha) \cos q(x-\alpha),$$

or,

$$\pi v = e^{-ht} \int_{-\infty}^{+\infty} d\alpha\, \phi(\alpha) \int_0^{\infty} dq\, e^{-kq^2t} \cos q(x-\alpha).$$

355. The solution of this second problem indicates clearly what the relation is between the definite integrals which we have just employed, and the results of the analysis which we have applied to solids of a definite form. When, in the convergent series which this analysis furnishes, we give to the quantities which denote the dimensions infinite values; each of the terms becomes infinitely small, and the sum of the series is nothing but an integral. We might pass directly in the same manner and without any physical considerations from the different trigonometrical series which we have employed in Chapter III. to definite integrals; it will be sufficient to give some examples of these transformations in which the results are remarkable.

356. In the equation

$$\frac{1}{4}\pi = \sin u + \frac{1}{3}\sin 3u + \frac{1}{5}\sin 5u + \&c.$$

we shall write instead of u the quantity $\frac{x}{n}$; x is a new variable, and n is an infinite number equal to $\frac{1}{dq}$; q is a quantity formed by the successive addition of infinitely small parts equal to dq. We shall represent the variable number i by $\frac{q}{dq}$. If in the general term $\frac{1}{2i+1}\sin(2i+1)\frac{x}{n}$ we put for i and n their values, the term becomes $\frac{dq}{2q}\sin 2qx$. Hence the sum of the series is $\frac{1}{2}\int\frac{dq}{q}\sin 2qx$, the integral being taken from $q=0$ to $q=\infty$; we have therefore the equation $\frac{1}{4}\pi=\frac{1}{2}\int_0^\infty\frac{dq}{q}\sin 2qx$ which is always true whatever be the positive value of x. Let $2qx=r$, r being a new variable, we have $\frac{dq}{q}=\frac{dr}{r}$ and $\frac{1}{2}\pi=\int_0^\infty\frac{dr}{r}\sin r$; this value of the definite integral $\int\frac{dr}{r}\sin r$ has been known for some time. If on supposing r negative we took the same integral from $r=0$ to $r=-\infty$, we should evidently have a result of contrary sign $-\frac{1}{2}\pi$.

357. The remark which we have just made on the value of the integral $\int\frac{dr}{r}\sin r$, which is $\frac{1}{2}\pi$ or $-\frac{1}{2}\pi$, serves to make known the nature of the expression

$$\frac{2}{\pi}\int_0^\infty\frac{dq\sin q}{q}\cos qx,$$

whose value we have already found (Article 348) to be equal to 1 or 0 according as x is or is not included between 1 and -1.

We have in fact

$$\int\frac{dq}{q}\cos qx\sin q=\frac{1}{2}\int\frac{dq}{q}\sin q(x+1)-\frac{1}{2}\int\frac{dq}{q}\sin q(x-1);$$

the first term is equal to $\frac{1}{4}\pi$ or $-\frac{1}{4}\pi$ according as $x+1$ is a positive or negative quantity; the second $\frac{1}{2}\int\frac{dq}{q}\sin q(x-1)$ is equal to $\frac{1}{4}\pi$ or $-\frac{1}{4}\pi$, according as $x-1$ is a positive or negative quantity. Hence the whole integral is nothing if $x+1$ and $x-1$ have the same sign; for, in this case, the two terms cancel each other. But if these quantities are of different sign, that is to say if we have at the same time

$$x+1>0 \text{ and } x-1<0,$$

the two terms add together and the value of the integral is $\frac{1}{2}\pi$. Hence the definite integral[1] $\dfrac{2}{\pi}\displaystyle\int_0^\infty \dfrac{dq}{q}\sin q\cos qx$ is a function of x equal to 1 if the variable x has any value included between 1 and -1; and the same function is nothing for every other value of x not included between the limits 1 and -1.

358. We might deduce also from the transformation of series into integrals the properties of the two expressions[2]

$$\frac{2}{\pi}\int_0^\infty \frac{dq\cos qx}{1+q^2}, \text{ and } \frac{2}{\pi}\int_0^\infty \frac{qdq\sin qx}{1+q^2};$$

the first (Art. 350) is equivalent to e^{-x} when x is positive, and to e^x when x is negative. The second is equivalent to e^{-x} if x is positive, and to $-e^x$ if x is negative, so that the two integrals have the same value, when x is positive, and have values of contrary sign when x is negative. One is represented by the line $eeee$ (fig. 19), the other by the line $\epsilon\epsilon\epsilon\epsilon$ (fig. 20).

Fig. 19.

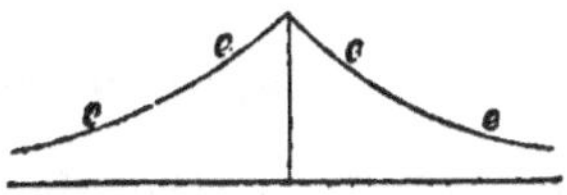

Fig. 20.

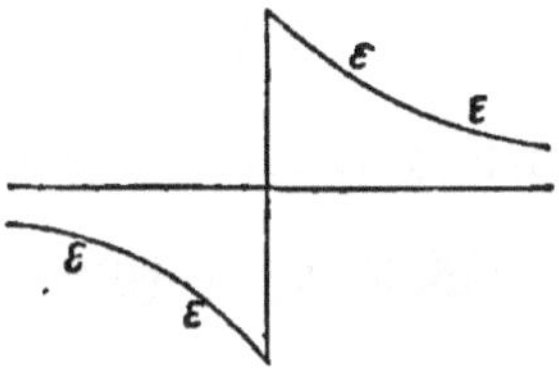

The equation

$$\frac{1}{2\alpha}\sin\frac{\pi x}{\alpha} = \frac{\sin\alpha\sin x}{\pi^2-\alpha^2} + \frac{\sin 2\alpha\sin 2x}{\pi^2-2^2\alpha^2} + \frac{\sin 3\alpha\sin 3x}{\pi^2-3^2\alpha^2} + \&c.,$$

which we have arrived at (Art. 226), gives immediately the integral $\dfrac{2}{\pi}\displaystyle\int_0^\infty \dfrac{dq\sin q\pi\sin qx}{1-q^2}$; which expression[3] is equivalent to $\sin x$, if x is included between 0 and π, and its value is 0 whenever x exceeds π.

---

[1] At the limiting values of x the value of this integral is $\frac{1}{2}$; Riemann, § 15.

[2] Cf. Riemann, § 16.

[3] The substitutions required in the equation are $\dfrac{\alpha x}{\pi}$ for x, dq for $\dfrac{\alpha}{\pi}$, q for $i\dfrac{\alpha}{\pi}$.

We then have $\sin x$ equal to a series equivalent to the above integral for values of x between 0 and π, the original equation being true for values of x between 0 and α.

[A. F.]

359. The same transformation applies to the general equation

$$\tfrac{1}{2}\pi\phi(u) = \sin u \int du\,\phi(u)\sin u + \sin 2u \int du\,\phi(u)\sin 2u + \&c.$$

Making $u = \frac{x}{n}$, denote $\phi(u)$ or $\phi\left(\frac{x}{n}\right)$ by $f(x)$, and introduce into the analysis a quantity q which receives infinitely small increments equal to dq, n will be equal to $\frac{1}{dq}$ and i to $\frac{q}{dq}$; substituting these values in the general term

$$\sin\frac{ix}{n}\int\frac{dx}{n}\phi\left(\frac{x}{n}\right)\sin\frac{ix}{n},$$

we find $dq \sin qx \int dx f(x) \sin qx$. The integral with respect to u is taken from $u = 0$ to $u = \pi$, hence the integration with respect to x must be taken from $x = 0$ to $x = n\pi$, or from x nothing to x infinite.

We thus obtain a general result expressed by the equation

$$\tfrac{1}{2}\pi f(x) = \int_0^\infty dq \sin qx \int_0^\infty dx f(x) \sin qx \;\ldots\ldots\ldots\; (e),$$

for which reason, denoting by Q a function of q such that we have $f(u) = \int dq Q \sin qu$ an equation in which $f(u)$ is a given function, we shall have $Q = \frac{2}{\pi}\int du f(u) \sin qu$, the integral being taken from u nothing to u infinite. We have already solved a similar problem (Art. 346) and proved the general equation

$$\tfrac{1}{2}\pi F(x) = \int_0^\infty dq \cos qx \int_0^\infty dx\, F(x) \cos qx \ldots\ldots\ldots (\epsilon),$$

which is analogous to the preceding.

360. To give an application of these theorems, let us suppose $f(x) = x^r$, the second member of equation (e) by this substitution becomes $\int dq \sin qx \int dx \sin qx\, x^r$.

The integral

$$\int dx \sin qx\, x^r \text{ or } \frac{1}{q^{r+1}}\int q dx \sin qx\,(qx)^r$$

is equivalent to $\frac{1}{q^{r+1}}\int du \sin u\, u^r$, the integral being taken from u nothing to u infinite.

Let μ be the integral

$$\int_0^\infty du \sin u\, u^r;$$

it remains to form the integral

$$\int dq \sin qx \frac{1}{q^{r+1}}\mu, \text{ or } \mu x^r \int du \sin u\, u^{-(r+1)};$$

denoting the last integral by ν, taken from u nothing to u infinite, we have as the result of two successive integrations the term $x^r \mu\nu$. We must then have, according to the condition expressed by the equation (e),

$$\tfrac{1}{2}\pi x^r = \mu\nu x^r \text{ or } \mu\nu = \tfrac{1}{2}\pi;$$

thus the product of the two transcendants

$$\int_0^\infty du\, u^r \sin u \text{ and } \int_0^\infty \frac{du}{u} u^{-r} \sin u \text{ is } \tfrac{1}{2}\pi.$$

For example, if $r = -\frac{1}{2}$, we find the known result

$$\int_0^\infty \frac{du \sin u}{\sqrt{u}} = \sqrt{\frac{\pi}{2}} \quad\ldots\ldots\ldots\ldots(a);$$

in the same manner we find

$$\int_0^\infty \frac{du \cos u}{\sqrt{u}} = \sqrt{\frac{\pi}{2}} \quad\ldots\ldots\ldots\ldots(b):$$

and from these two equations we might also conclude the following[1], $\int_0^\infty dq e^{-q^2} = \frac{1}{2}\sqrt{\pi}$, which has been employed for some time.

361. By means of the equations (e) and (ϵ) we may solve the following problem, which belongs also to partial differential analysis. What function Q of the variable q must be placed under

[1] The way is simply to use the expressions $e^{-z} = +\cos\sqrt{-1}\,z + \sqrt{-1}\sin\sqrt{-1}\,z$, transforming a and b by writing y^2 for u and recollecting that $\sqrt{\sqrt{-1}} = \frac{1+\sqrt{-1}}{\sqrt{2}}$. Cf. § 407. [R. L. E.]

the integral sign in order that the expression $\int dq Q\, e^{-qx}$ may be equal to a given function, the integral being taken from q nothing to q infinite[1]? But without stopping for different consequences, the examination of which would remove us from our chief object, we shall limit ourselves to the following result, which is obtained by combining the two equations (e) and (ϵ).

They may be put under the form

$$\frac{1}{2}\pi f(x) = \int_0^\infty dq \sin qx \int_0^\infty d\alpha f(\alpha) \sin q\alpha,$$

and

$$\frac{1}{2}\pi F(x) = \int_0^\infty dq \cos qx \int_0^\infty d\alpha F(\alpha) \cos q\alpha.$$

If we took the integrals with respect to α from $-\infty$ to $+\infty$, the result of each integration would be doubled, which is a necessary consequence of the two conditions

$$f(\alpha) = -f(-\alpha) \text{ and } F(\alpha) = F(-\alpha).$$

We have therefore the two equations

$$\pi f(x) = \int_0^\infty dq \sin qx \int_{-\infty}^\infty d\alpha f(\alpha) \sin q\alpha,$$

and

$$\pi F(x) = \int_0^\infty dq \cos qx \int_{-\infty}^\infty d\alpha F(\alpha) \cos q\alpha.$$

We have remarked previously that any function $\phi(x)$ can always be decomposed into two others, one of which $F(x)$ satisfies the condition $F(x) = F(-x)$, and the other $f(x)$ satisfies the condition $f(x) = -f(-x)$. We have thus the two equations

$$0 = \int_{-\infty}^{+\infty} d\alpha F(\alpha) \sin q\alpha, \quad \text{and } 0 = \int_{-\infty}^{+\infty} d\alpha f(\alpha) \cos q\alpha,$$

[1] To do this write $\pm x\sqrt{-1}$ in $f(x)$ and add, therefore

$$2\int Q \cos qx\, dq = f(x\sqrt{-1}) + f(-x\sqrt{-1}),$$

which remains the same on writing $-x$ for x,

$$\text{therefore } Q = \frac{1}{\pi}\int dx\, [f(x\sqrt{-1}) + f(-x\sqrt{-1})] \cos qx\, dx.$$

Again we may subtract and use the sine but the difficulty of dealing with imaginary quantities recurs continually. [R. L. E.]

whence we conclude

$$\pi\,[F(x)+f(x)]=\pi\phi(x)=\int_0^\infty dq\sin qx\int_{-\infty}^{+\infty}d\alpha f(\alpha)\sin q\alpha$$

$$+\int_0^\infty dq\cos qx\int_{-\infty}^{+\infty}d\alpha F(\alpha)\cos q\alpha,$$

and $\pi\phi(x)=\int_0^\infty dq\sin qx\int_{-\infty}^{+\infty}d\alpha\phi(\alpha)\sin q\alpha$

$$+\int_0^\infty dq\cos qx\int_{-\infty}^{+\infty}d\alpha\phi(\alpha)\cos q\alpha,$$

or $$\pi\phi(x)=\int_{-\infty}^{+}d\alpha\phi(\alpha)\int_0^\infty dq\,(\sin qx\sin q\alpha+\cos qx\cos q\alpha);$$

or lastly[1], $$\phi(x)=\frac{1}{\pi}\int_{-\infty}^{+\infty}d\alpha\phi(\alpha)\int_0^\infty dq\cos q(x-\alpha)\quad\ldots\ldots\ldots(E).$$

The integration with respect to q gives a function of x and α, and the second integration makes the variable α disappear. Thus the function represented by the definite integral $\int dq\cos q(x-\alpha)$ has the singular property, that if we multiply it by any function $\phi(\alpha)$ and by $d\alpha$, and integrate it with respect to α between infinite limits, the result is equal to $\pi\phi(x)$; so that the effect of the integration is to change α into x, and to multiply by the number π.

362. We might deduce equation (E) directly from the theorem

[1] Poisson, in his *Mémoire sur la Théorie des Ondes*, in the *Mémoires de l'Académie des Sciences*, Tome I., Paris, 1818, pp. 85—87, first gave a direct proof of the theorem

$$f(x)=\frac{1}{\pi}\int_0^\infty dq\int_{-\infty}^{+\infty}d\alpha\, e^{-kq}\cos(qx-q\alpha)f(\alpha),$$

in which k is supposed to be a small positive quantity which is made equal to 0 after the integrations.

Boole, *On the Analysis of Discontinuous Functions*, in the *Transactions of the Royal Irish Academy*, Vol. XXI., Dublin, 1848, pp. 126—130, introduces some analytical representations of discontinuity, and regards Fourier's Theorem as unproved unless equivalent to the above proposition.

Deflers, at the end of a *Note sur quelques intégrales définies &c.*, in the *Bulletin des Sciences, Société Philomatique*, Paris, 1819, pp. 161—166, indicates a proof of Fourier's Theorem, which Poisson repeats in a modified form in the *Journal Polytechnique*, Cahier 19, p. 454. The special difficulties of this proof have been noticed by De Morgan, *Differential and Integral Calculus*, pp. 619, 628.

An excellent discussion of the class of proofs here alluded to is given by Mr J. W. L. Glaisher in an article *On sin ∞ and cos ∞*, *Messenger of Mathematics*, Ser. I., Vol. V., pp. 232—244, Cambridge, 1871. [A. F.]

stated in Article 234, which gives the development of any function $F(x)$ in a series of sines and cosines of multiple arcs. We pass from the last proposition to those which we have just demonstrated, by giving an infinite value to the dimensions. Each term of the series becomes in this case a differential quantity[1]. Transformations of functions into trigonometrical series are some of the elements of the analytical theory of heat; it is indispensable to make use of them to solve the problems which depend on this theory.

The reduction of arbitrary functions into definite integrals, such as are expressed by equation (E), and the two elementary equations from which it is derived, give rise to different consequences which are omitted here since they have a less direct relation with the physical problem. We shall only remark that the same equations present themselves sometimes in analysis under other forms. We obtain for example this result

$$\phi(x) = \frac{1}{\pi}\int_0^\infty d\alpha\,\phi(\alpha)\int_0^\infty dq \cos q(x-\alpha) \quad\ldots\ldots\ldots(E')$$

which differs from equation (E) in that the limits taken with respect to α are 0 and ∞ instead of being $-\infty$ and $+\infty$.

In this case it must be remarked that the two equations (E) and (E') give equal values for the second member when the variable x is positive. If this variable is negative, equation (E') always gives a nul value for the second member. The same is not the case with equation (E), whose second member is equivalent to $\pi\phi(x)$, whether we give to x a positive or negative value. As to equation (E') it solves the following problem. To find a function of x such that if x is positive, the value of the function may be $\phi(x)$, and if x is negative the value of the function may be always nothing[2].

363. The problem of the propagation of heat in an infinite line may besides be solved by giving to the integral of the partial differential equation a different form which we shall indicate in

[1] Riemann, *Part. Diff. Gleich.* § 32, gives the proof, and deduces the formulæ corresponding to the cases $F(x) = \pm F(-x)$.

[2] These remarks are essential to clearness of view. The equations from which (E') and its cognate form may be derived will be found in Todhunter's *Integral Calculus*, Cambridge, 1862, § 316, Equations (3) and (4). [A. F.]

the following article. We shall first examine the case in which the source of heat is constant.

Suppose that, the initial heat being distributed in any manner throughout the infinite bar, we maintain the section A at a constant temperature, whilst part of the heat communicated is dispersed through the external surface. It is required to determine the state of the prism after a given time, which is the object of the second problem that we have proposed to ourselves. Denoting by 1 the constant temperature of the end A, by 0 that of the medium, we have $e^{-x\sqrt{\frac{HL}{KS}}}$ as the expression of the final temperature of a point situated at the distance x from this extremity, or simply e^{-x}, assuming for simplicity the quantity $\frac{HL}{KS}$ to be equal to unity. Denoting by v the variable temperature of the same point after the time t has elapsed, we have, to determine v, the equation

$$\frac{dv}{dt} = \frac{K}{CD}\frac{d^2v}{dx^2} - \frac{HL}{CDS}v,$$

let now $$v = e^{-x\sqrt{\frac{HL}{KS}}} + u',$$

we have $$\frac{du'}{dt} = \frac{K}{CD}\frac{d^2u'}{dx^2} - \frac{HL}{CDS}u';$$

or $$\frac{du'}{dt} = k\frac{d^2u'}{dx^2} - hu',$$

replacing $\frac{K}{CD}$ by k and $\frac{HL}{CDS}$ by h.

If we make $u' = e^{-ht}u$ we have $\frac{du}{dt} = k\frac{d^2u}{dx^2}$; the value of u' or $v - e^{-x\sqrt{\frac{HL}{KS}}}$ is that of the difference between the actual and the final temperatures; this difference u', which tends more and more to vanish, and whose final value is nothing, is equivalent at first to

$$F(x) - e^{-x\sqrt{\frac{h}{k}}},$$

denoting by $F(x)$ the initial temperature of a point situated at the distance x. Let $f(x)$ be the excess of the initial temperature over the final temperature, we must find for u a function which satisfies the equation $\frac{du}{dt}=k\frac{d^2u}{dx^2}-hu$, and whose initial value is $f(x)$, and final value 0. At the point A, or $x=0$, the quantity $v-e^{-x\sqrt{\frac{HL}{KS}}}$ has, by hypothesis, a constant value equal to 0. We see by this that u represents an excess of heat which is at first accumulated in the prism, and which then escapes, either by being propagated to infinity, or by being scattered into the medium. Thus to represent the effect which results from the uniform heating of the end A of a line infinitely prolonged, we must imagine, 1st, that the line is also prolonged to the left of the point A, and that each point situated to the right is now affected with the initial excess of temperature; 2nd, that the other half of the line to the left of the point A is in a contrary state; so that a point situated at the distance $-x$ from the point A has the initial temperature $-f(x)$: the heat then begins to move freely through the interior of the bar, and to be scattered at the surface.

The point A preserves the temperature 0, and all the other points arrive insensibly at the same state. In this manner we are able to refer the case in which the external source incessantly communicates new heat, to that in which the primitive heat is propagated through the interior of the solid. We might therefore solve the proposed problem in the same manner as that of the diffusion of heat, Articles 347 and 353; but in order to multiply methods of solution in a matter thus new, we shall employ the integral under a different form from that which we have considered up to this point.

364. The equation $\frac{du}{dt}=k\frac{d^2u}{dx^2}$ is satisfied by supposing u equal to $e^{-x}e^{kt}$. This function of x and t may also be put under the form of a definite integral, which is very easily deduced from the known value of $\int dqe^{-q^2}$. We have in fact $\sqrt{\pi}=\int dqe^{-q^2}$, when the integral is taken from $q=-\infty$ to $q=+\infty$. We have therefore also

$$\sqrt{\pi}=\int dqe^{-(q+b)^2},$$

b being any constant whatever and the limits of the integral the same as before. From the equation

$$\sqrt{\pi} = e^{-b^2}\int_{-\infty}^{+\infty} dq e^{-(q^2+2bq)}$$

we conclude, by making $b^2 = kt$

$$e^{kt} = \frac{1}{\sqrt{\pi}}\int_{-\infty}^{+\infty} dq e^{-q^2} e^{-2q\sqrt{kt}},$$

hence the preceding value of u or $e^{-x} e^{kt}$ is equivalent to

$$\frac{1}{\sqrt{\pi}}\int_{-\infty}^{+\infty} dq e^{-q^2} e^{-(x+2q\sqrt{kt})};$$

we might also suppose u equal to the function

$$ae^{-nx} e^{kn^2t},$$

a and n being any two constants; and we should find in the same way that this function is equivalent to

$$\frac{a}{\sqrt{\pi}}\int_{-\infty}^{+\infty} dq e^{-q^2} e^{-n(x+2q\sqrt{kt})}.$$

We can therefore in general take as the value of u the sum of an infinite number of such values, and we shall have

$$u = \int_{-\infty}^{+\infty} dq e^{-q^2} \left(a_1 e^{-n_1(x+2q\sqrt{kt})} + a_2 e^{-n_2(x+2q\sqrt{kt})} + a_3 e^{-n_3(x+2q\sqrt{kt})} + \&c.\right)$$

The constants a_1, a_2, a_3, &c., and n_1, n_2, n_3, &c. being undetermined, the series represents any function whatever of $x + 2q\sqrt{kt}$; we have therefore $u = \int dq e^{-q^2} \phi (x + 2q\sqrt{kt})$. The integral should be taken from $q = -\infty$ to $q = +\infty$, and the value of u will necessarily satisfy the equation $\frac{du}{dt} = k\frac{d^2u}{dx^2}$. This integral which contains one arbitrary function was not known when we had undertaken our researches on the theory of heat, which were transmitted to the Institute of France in the month of December, 1807: it has been given by M. Laplace[1], in a work which forms part of volume VIII of the Mémoires de l'École Polytechnique; we apply it simply to the determination of the linear movement of heat. From it we conclude

$$v = e^{-ht}\int_{-\infty}^{+\infty} dq e^{-q^2}\phi\,(x + 2q\sqrt{kt}) + e^{-x\sqrt{\frac{HL}{KS}}},$$

when $t = 0$ the value of u is $F(x) - e^{-x\sqrt{\frac{HL}{KS}}}$ or $f(x)$;
hence

$$f(x) = \int_{-\infty}^{+\infty} dq e^{-q^2}\phi(x) \text{ and } \phi(x) = \frac{1}{\sqrt{\pi}} f(x).$$

Thus the arbitrary function which enters into the integral, is determined by means of the given function $f(x)$, and we have the following equation, which contains the solution of the problem,

$$v = + e^{-x\sqrt{\frac{HL}{KS}}} + \frac{e^{-ht}}{\sqrt{\pi}}\int_{-\infty}^{+\infty} dq e^{-q^2} f(x + 2q\sqrt{kt}),$$

it is easy to represent this solution by a construction.

365. Let us apply the previous solution to the case in which all points of the line AB having the initial temperature 0, the end A is heated so as to be maintained continually at the temperature 1. It follows from this that $F(x)$ has a nul value when x differs from 0. Thus $f(x)$ is equal to $-e^{-x\sqrt{\frac{HL}{KS}}}$ whenever x differs from 0, and to 0 when x is nothing. On the other hand it is necessary that on making x negative, the value of $f(x)$ should change sign, so that we have the condition $f(-x) = -f(x)$. We thus know the nature of the discontinuous function $f(x)$; it becomes $-e^{-x\sqrt{\frac{HL}{KS}}}$ when x exceeds 0, and $+e^{x\sqrt{\frac{HL}{KS}}}$ when x is less than 0. We must now write instead of x the quantity $x + 2q\sqrt{kt}$. To find u or $\int_{-\infty}^{+\infty} dq e^{-q^2}\frac{1}{\sqrt{\pi}} f(x + 2q\sqrt{kt})$, we must first take the integral from

$$x + 2q\sqrt{kt} = 0 \text{ to } x + 2q\sqrt{kt} = \infty,$$

[1] *Journal de l'École Polytechnique*, Tome VIII. pp. 235—244, Paris, 1809. Laplace shews also that the complete integral of the equation contains only one arbitrary function, but in this respect he had been anticipated by Poisson. [A. F.]

and then from

$$x + 2q\sqrt{kt} = -\infty \text{ to } x + 2q\sqrt{kt} = 0.$$

For the first part, we have

$$-\frac{1}{\sqrt{\pi}}\int dq e^{-q^2} e^{-(x+2q\sqrt{kt})\sqrt{\frac{HL}{KS}}},$$

and replacing k by its value $\frac{K}{CD}$ we have

$$-\int\frac{dq}{\sqrt{\pi}} e^{-q^2} e^{-\left(x+2q\sqrt{\frac{Kt}{CD}}\right)\sqrt{\frac{HL}{KS}}},$$

or
$$-\frac{1}{\sqrt{\pi}} e^{-x\sqrt{\frac{HL}{KS}}}\int dq e^{-q^2} e^{-2q\sqrt{\frac{HLt}{CDS}}},$$

or
$$-\frac{e^{-x\sqrt{\frac{HL}{KS}}}}{\sqrt{\pi}} e^{\frac{HL}{CDS}t}\int dq e^{-\left(q+\sqrt{\frac{HLt}{CDS}}\right)^2}.$$

Denoting the quantity $q + \sqrt{\frac{HLt}{CDS}}$ by r the preceding expression becomes

$$-\frac{e^{-x\sqrt{\frac{HL}{KS}}}}{\sqrt{\pi}} e^{\frac{HLt}{CDS}}\int dr e^{-r^2},$$

this integral $\int dr e^{-r^2}$ must be taken by hypothesis from

$$x + 2q\sqrt{\frac{Kt}{CD}} = 0 \text{ to } x + 2q\sqrt{\frac{Kt}{CD}} = \infty,$$

or from $$q = -\frac{x}{2\sqrt{\frac{Kt}{CD}}} \text{ to } q = \infty,$$

or from $$r = \sqrt{\frac{HLt}{CDS}} - \frac{x}{2\sqrt{\frac{Kt}{CD}}} \text{ to } r = \infty.$$

The second part of the integral is

$$\frac{1}{\sqrt{\pi}}\int dq e^{-q} e^{\left(x+2q\sqrt{\frac{Kt}{CD}}\right)\sqrt{\frac{HL}{KS}}},$$

or $$\frac{1}{\sqrt{\pi}} e^{x\sqrt{\frac{HL}{KS}}} \int dq e^{-q^2} e^{2q\sqrt{\frac{HLt}{CDS}}},$$

or $$\frac{1}{\sqrt{\pi}} e^{x\sqrt{\frac{HL}{KS}}} e^{\frac{HLt}{CDS}} \int dr e^{-r^2},$$

denoting by r the quantity $q-\sqrt{\frac{HLt}{CDS}}$. The integral $\int dr e^{-r}$ must be taken by hypothesis from

$$x+2q\sqrt{\frac{Kt}{CD}}=-\infty \text{ to } x+2q\sqrt{\frac{Kt}{CD}}=0,$$

or from $$q=-\infty \text{ to } q=-\frac{x}{2\sqrt{\frac{Kt}{CD}}}, \text{ that is}$$

from $$r=-\infty \text{ to } r=-\sqrt{\frac{HLt}{CDS}}-\frac{x}{2\sqrt{\frac{Kt}{CD}}}.$$

The two last limits may, from the nature of the function e^{-r^2}, be replaced by these:

$$r=\sqrt{\frac{HLt}{CDS}}+\frac{x}{2\sqrt{\frac{Kt}{CD}}}, \text{ and } r=\infty.$$

It follows from this that the value of u is expressed thus:

$$u=e^{x\sqrt{\frac{HL}{KS}}} e^{\frac{HLt}{CDS}} \int dr e^{-r^2} - e^{-x\sqrt{\frac{HL}{KS}}} e^{\frac{HLt}{CDS}} \int dr e^{-r^2},$$

the first integral must be taken from

$$r=\sqrt{\frac{HLt}{CDS}}+\frac{x}{2\sqrt{\frac{Kt}{CD}}} \text{ to } r=\infty,$$

and the second from

$$r=\sqrt{\frac{HLt}{CDS}}-\frac{x}{2\sqrt{\frac{Kt}{CD}}} \text{ to } r=\infty.$$

Let us represent now the integral $\frac{1}{\sqrt{\pi}}\int dr e^{-r^2}$ from $r=R$ to $r=\infty$ by $\psi(R)$, and we shall have

$$u = e^{\frac{HLt}{CDS}} e^{x\sqrt{\frac{HL}{KS}}} \psi\left(\sqrt{\frac{HLt}{CDS}} + \frac{x}{2\sqrt{\frac{Kt}{CD}}}\right)$$

$$- e^{\frac{HLt}{CDS}} e^{-x\sqrt{\frac{HL}{KS}}} \psi\left(\sqrt{\frac{HLt}{CDS}} - \frac{x}{2\sqrt{\frac{Kt}{CD}}}\right),$$

hence u', which is equivalent to $u e^{-\frac{HLt}{CDS}}$, is expressed by

$$e^{x\sqrt{\frac{HL}{KS}}} \psi\left(\sqrt{\frac{HLt}{CDS}} + \frac{x}{2\sqrt{\frac{Kt}{CD}}}\right)$$

$$- e^{-x\sqrt{\frac{HL}{KS}}} \psi\left(\sqrt{\frac{HLt}{CDS}} - \frac{x}{2\sqrt{\frac{Kt}{CD}}}\right),$$

and

$$v = e^{-x\sqrt{\frac{HL}{KS}}} - e^{-x\sqrt{\frac{HL}{KS}}} \psi\left(\sqrt{\frac{HLt}{CDS}} - \frac{x}{2\sqrt{\frac{Kt}{CD}}}\right)$$

$$+ e^{x\sqrt{\frac{HL}{KS}}} \psi\left(\sqrt{\frac{HLt}{CDS}} + \frac{x}{2\sqrt{\frac{Kt}{CD}}}\right).$$

The function denoted by $\psi(R)$ has been known for some time, and we can easily calculate, either by means of convergent series, or by continued fractions, the values which this function receives, when we substitute for R given quantities; thus the numerical application of the solution is subject to no difficulty[1].

[1] The following references are given by Riemann:

Kramp. *Analyse des réfractions astronomiques et terrestres.* Leipsic and Paris, An. VII. 4to. Table I. at the end contains the values of the integral $\int_k^\infty e^{-\beta^2} d\beta$ from $k=0{\cdot}00$ to $k=3{\cdot}00$.

Legendre. *Traité des fonctions elliptiques et des intégrales Eulériennes.* Tome II.

366. If H be made nothing, we have

$$v = 1 - \left\{\psi\left(-\frac{x}{2\sqrt{\frac{Kt}{CD}}}\right) - \psi\left(\frac{x}{2\sqrt{\frac{Kt}{CD}}}\right)\right\}.$$

This equation represents the propagation of heat in an infinite bar, all points of which were first at temperature 0, except those at the extremity which is maintained at the constant temperature 1. We suppose that heat cannot escape through the external surface of the bar; or, which is the same thing, that the thickness of the bar is infinitely great. This value of v indicates therefore the law according to which heat is propagated in a solid, terminated by an infinite plane, supposing that this infinitely thick wall has first at all parts a constant initial temperature 0, and that the surface is submitted to a constant temperature 1. It will not be quite useless to point out several results of this solution.

Denoting by $\phi(R)$ the integral $\frac{1}{\sqrt{\pi}}\int dr\, e^{-r^2}$ taken from $r=0$ to $r=R$, we have, when R is a positive quantity,

$$\psi(R) = \tfrac{1}{2} - \phi(R) \text{ and } \psi(-R) = \tfrac{1}{2} + \phi(R),$$

hence

$$\psi(-R) - \psi(R) = 2\phi(R) \text{ and } v = 1 - 2\phi\left(\frac{x}{2\sqrt{\frac{Kt}{CD}}}\right);$$

developing the integral $\phi(R)$ we have

$$\phi(R) = \frac{1}{\sqrt{\pi}}\left(R - \frac{1}{1}\frac{1}{3}R^3 + \frac{1}{\lfloor 2}\frac{1}{5}R^5 - \frac{1}{\lfloor 3}\frac{1}{7}R^7 + \&c.\right);$$

Paris, 1826. 4to. pp. 520—1. Table of the values of the integral $\int dx\left(\log\frac{1}{x}\right)^{-\frac{1}{2}}$. The first part for values of $\left(\log\frac{1}{x}\right)$ from 0·00 to 0·50; the second part for values of x from 0·80 to 1·00.

Encke. *Astronomisches Jahrbuch für* 1834. Berlin, 1832, 8vo. Table I. at the end gives the values of $\frac{2}{\sqrt{\pi}}\int_0^t e^{-t^2}dt$ from $t=0{\cdot}00$ to $t=2{\cdot}00$. [A. F.]

hence

$$\frac{1}{2}v\sqrt{\pi} = \frac{1}{2}\sqrt{\pi} - \frac{x}{2\sqrt{\frac{Kt}{CD}}} + \frac{1}{1}\frac{1}{3}\left(\frac{x}{2\sqrt{\frac{Kt}{CD}}}\right)^3 - \frac{1}{\underline{|2}}\frac{1}{5}\left(\frac{x}{2\sqrt{\frac{Kt}{CD}}}\right)^5 + \&c.$$

1st, if we suppose x nothing, we find $v=1$; 2nd, if x not being nothing, we suppose $t=0$, the sum of the terms which contain x represents the integral $\int dr e^{-r^2}$ taken from $r=0$ to $r=\infty$, and consequently is equal to $\frac{1}{2}\sqrt{\pi}$; therefore v is nothing; 3rd, different points of the solid situated at different depths x_1, x_2, x_3, &c. arrive at the same temperature after different times t_1, t_2, t_3, &c. which are proportional to the squares of the lengths x_1, x_2, x_3, &c.; 4th, in order to compare the quantities of heat which during an infinitely small instant cross a section S situated in the interior of the solid at a distance x from the heated plane, we must take the value of the quantity $-KS\frac{dv}{dx}$ and we have

$$-KS\frac{dv}{dx} = \frac{2KS}{2\sqrt{\frac{Kt}{CD}}}\sqrt{\pi}\left\{1 - \frac{1}{1}\left(\frac{x}{2\sqrt{\frac{Kt}{CD}}}\right)^2 + \frac{1}{\underline{|2}}\left(\frac{x}{2\sqrt{\frac{Kt}{CD}}}\right)^4 - \&c.\right\}$$

$$= S\frac{\sqrt{CDK}}{\sqrt{\pi t}}e^{-\frac{x}{2\sqrt{\frac{Kt}{CD}}}};$$

thus the expression of the quantity $\frac{dv}{dx}$ is entirely disengaged from the integral sign. The preceding value at the surface of the heated solid becomes $S\frac{\sqrt{CDK}}{\sqrt{\pi t}}$, which shews how the flow of heat at the surface varies with the quantities C, D, K, t; to find how much heat the source communicates to the solid during the lapse of the time t, we must take the integral

$$\int S \frac{\sqrt{CDK}}{\sqrt{\pi}} \frac{dt}{\sqrt{t}} \text{ or } \frac{2S\sqrt{CDK}}{\sqrt{\pi}} \sqrt{t}:$$

thus the heat acquired increases proportionally to the square root of the time elapsed.

367. By a similar analysis we may treat the problem of the diffusion of heat, which also depends on the integration of the equation $\frac{dv}{dt} = k\frac{d^2v}{dx^2} - hv$. Representing by $f(x)$ the initial temperature of a point in the line situated at a distance x from the origin, we proceed to determine what ought to be the temperature of the same point after a time t. Making $v = e^{-ht}z$, we have $\frac{dz}{dt} = k\frac{d^2z}{dt^2}$, and consequently $z = \int_{-\infty}^{+\infty} dq\, e^{-q^2}\phi(x + 2q\sqrt{kt})$. When $t = 0$, we must have

$$v = f(x) = \int_{-\infty}^{+\infty} dq\, e^{-q^2}\phi(x) \quad \text{or} \quad \phi(x) = \frac{1}{\sqrt{\pi}} f(x);$$

hence

$$v = \frac{e^{-kt}}{\sqrt{\pi}} \int_{-\infty}^{+\infty} dq\, e^{-q^2} f(x + 2q\sqrt{kt}).$$

To apply this general expression to the case in which a part of the line from $x = -\alpha$ to $x = \alpha$ is uniformly heated, all the rest of the solid being at the temperature 0, we must consider that the factor $f(x + 2q\sqrt{kt})$ which multiplies e^{-q^2} has, according to hypothesis, a constant value 1, when the quantity which is under the sign of the function is included between $-\alpha$ and α, and that all the other values of this factor are nothing. Hence the integral $\int dq\, e^{-q^2}$ ought to be taken from $x + 2q\sqrt{kt} = -\alpha$ to $x + 2q\sqrt{kt} = \alpha$, or from $q = \frac{-x-\alpha}{2\sqrt{kt}}$ to $q = \frac{-x+\alpha}{2\sqrt{kt}}$. Denoting as above by $\frac{1}{\sqrt{\pi}}\psi(R)$ the integral $\int dr\, e^{-r^2}$ taken from $r = R$ to $r = \infty$, we have

$$v = e^{-kt}\left\{\psi\left(\frac{-x-\alpha}{2\sqrt{kt}}\right) - \psi\left(\frac{-x+\alpha}{2\sqrt{kt}}\right)\right\}.$$

368. We shall next apply the general equation

$$v = \frac{e^{-ht}}{\sqrt{\pi}} \int_{-\infty}^{+\infty} dq\, e^{-q^2} f(x + 2q\sqrt{kt})$$

to the case in which the infinite bar, heated by a source of constant intensity 1, has arrived at fixed temperatures and is then cooling freely in a medium maintained at the temperature 0. For this purpose it is sufficient to remark that the initial function denoted by $f(x)$ is equivalent to $e^{-x\sqrt{\frac{h}{k}}}$ so long as the variable x which is under the sign of the function is positive, and that the same function is equivalent to $e^{x\sqrt{\frac{h}{k}}}$ when the variable which is affected by the symbol f is less than 0. Hence

$$v = \frac{e^{-ht}}{\sqrt{\pi}} \left(\int dq\, e^{-q^2} e^{-x\sqrt{\frac{h}{k}}} e^{-2q\sqrt{ht}} + \int dq\, e^{-q^2} e^{x\sqrt{\frac{h}{k}}} e^{2q\sqrt{ht}} \right),$$

the first integral must be taken from

$$x + 2q\sqrt{kt} = 0 \text{ to } x + 2q\sqrt{kt} = \infty,$$

and the second from

$$x + 2q\sqrt{kt} = -\infty \text{ to } x + 2q\sqrt{kt} = 0.$$

The first part of the value of v is

$$\frac{e^{-ht}}{\sqrt{\pi}} e^{-x\sqrt{\frac{h}{k}}} \int dq\, e^{-q^2} e^{-2q\sqrt{ht}},$$

or
$$\frac{e^{-x\sqrt{\frac{h}{k}}}}{\sqrt{\pi}} \int dq e^{-(q+\sqrt{ht})^2},$$

or
$$\frac{e^{-x\sqrt{\frac{h}{k}}}}{\sqrt{\pi}} \int dr e^{-r^2};$$

making $r = q + \sqrt{ht}$. The integral should be taken from

$$q = \frac{-x}{2\sqrt{kt}} \text{ to } q = \infty,$$

or from
$$r = \sqrt{ht} - \frac{x}{2\sqrt{kt}} \text{ to } r = \infty.$$

The second part of the value of v is

$$\frac{e^{-ht}}{\sqrt{\pi}} e^{x\sqrt{\frac{h}{k}}} \int dq\, e^{-q^2} e^{2q\sqrt{ht}} \quad \text{or} \quad e^{\sqrt{\frac{h}{k}}} \int dr\, e^{-r^2};$$

making $r = q - \sqrt{ht}$. The integral should be taken from

$$r = -\infty \text{ to } r = -\sqrt{ht} - \frac{x}{2\sqrt{kt}},$$

or from $$r = \sqrt{ht} + \frac{x}{2\sqrt{kt}} \text{ to } r = \infty,$$

whence we conclude the following expression:

$$v = e^{-x\sqrt{\frac{h}{k}}} \psi\left(\sqrt{ht} - \frac{x}{2\sqrt{kt}}\right) + e^{x\sqrt{\frac{h}{k}}} \psi\left(\sqrt{ht} + \frac{x}{2\sqrt{kt}}\right).$$

369. We have obtained (Art. 367) the equation

$$v = e^{-ht}\left\{\psi\left(\frac{-x+\alpha}{2\sqrt{kt}}\right) - \psi\left(\frac{-x+\alpha}{2\sqrt{kt}}\right)\right\},$$

to express the law of diffusion of heat in a bar of small thickness, heated uniformly at its middle point between the given limits

$$x = -\alpha, \quad x = +\alpha.$$

We had previously solved the same problem by following a different method, and we had arrived, on supposing $\alpha = 1$, at the equation

$$v = \frac{2}{\pi} e^{-ht} \int_{-\infty}^{+\infty} \frac{dq}{q} \cos qx \sin q\, e^{-q^2kt}, \text{ (Art. 348).}$$

To compare these two results we shall suppose in each $x = 0$; denoting again by $\psi(R)$ the integral $\int dr\, e^{-r^2}$ taken from $r = 0$ to $r = R$, we have

$$v = e^{-ht}\left\{\psi\left(\frac{-\alpha}{2\sqrt{kt}}\right) - \psi\left(\frac{\alpha}{2\sqrt{kt}}\right)\right\},$$

or $$v = \frac{2e^{-ht}}{\sqrt{\pi}}\left\{\frac{\alpha}{2\sqrt{kt}} - \frac{1}{1}\frac{1}{3}\left(\frac{\alpha}{2\sqrt{kt}}\right)^3 + \frac{1}{\underline{|2}}\frac{1}{5}\left(\frac{\alpha}{2\sqrt{kt}}\right)^5 - \&c.\right\};$$

on the other hand we ought to have

$$v = \frac{2}{\pi} e^{-ht} \int \frac{dq}{q} \sin q \, e^{-q^2kt},$$

or

$$v = \frac{2}{\pi} e^{-ht} \int dq e^{-q^2kt} \left(1 - \frac{q^2}{\lfloor 3} + \frac{q^4}{\lfloor 5} - \&c.\right).$$

Now the integral $\int due^{-u^2} u^{2m}$ taken from $u=0$ to $u=\infty$ has a known value, m being any positive integer. We have in general

$$\int_0^\infty due^{-u^2} u^{2m} = \frac{1.3.5.7\ldots(2m-1)}{2.2.2.2\ldots\quad 2} \frac{1}{2} \sqrt{\pi}.[1]$$

The preceding equation gives then, on making $q^2kt = u^2$,

$$v = \frac{2e^{-ht}}{\pi\sqrt{kt}} \int due^{-u^2} \left(1 - \frac{u^2}{\lfloor 3} \frac{1}{kt} + \frac{u^4}{\lfloor 5} \frac{1}{k^2t^2} - \&c.\right),$$

or

$$v = \frac{2e^{-ht}}{\sqrt{\pi}} \left[\frac{1}{2\sqrt{kt}} - \frac{1}{1}\frac{1}{3}\left(\frac{1}{2\sqrt{kt}}\right)^3 + \frac{1}{\lfloor 2}\frac{1}{5}\left(\frac{1}{2\sqrt{kt}}\right)^5 - \&c.\right].$$

This equation is the same as the preceding when we suppose $\alpha=1$. We see by this that integrals which we have obtained by different processes, lead to the same convergent series, and we arrive thus at two identical results, whatever be the value of x.

We might, in this problem as in the preceding, compare the quantities of heat which, in a given instant, cross different sections of the heated prism, and the general expression of these quantities contains no sign of integration; but passing by these remarks, we shall terminate this section by the comparison of the different forms which we have given to the integral of the equation which represents the diffusion of heat in an infinite line.

370. To satisfy the equation $\frac{du}{dt} = k\frac{d^2u}{dx^2}$, we may assume $u = e^{-x} e^{kt}$, or in general $u = e^{-nx} e^{n^2kt}$, whence we deduce easily (Art. 364) the integral

$$u = \int_{-\infty}^{+\infty} dq e^{-q^2} \phi\,(x + 2q\sqrt{kt}).$$

[1] Cf. Riemann, § 18.

From the known equation

$$\sqrt{\pi} = \int_{-\infty}^{+\infty} dq e^{-q^2},$$

we conclude

$\sqrt{\pi} = \int_{-\infty}^{+\infty} dq\, e^{-(q+a)^2}$, a being any constant; we have therefore

$$e^{a^2} = \frac{1}{\sqrt{\pi}} \int dq\, e^{-q^2} e^{-2aq}, \quad \text{or}$$

$$e^{a^2} = \frac{1}{\sqrt{\pi}} \int dq\, e^{-q^2} \left(1 - 2aq + \frac{2^2 a^2 q^2}{\underline{|2}} - \frac{2^3 a^3 q^3}{\underline{|3}} + \text{etc.}\right).$$

This equation holds whatever be the value of a. We may develope the first member; and by comparison of the terms we shall obtain the already known values of the integral $\int dq\, e^{-q^2} q^n$. This value is nothing when n is odd, and we find when n is an even number $2m$,

$$\int_{-\infty}^{+\infty} dq\, e^{-q^2} q^{2m} = \frac{1.3.5.7 \ldots (2m-1)}{2.2.2.2 \ldots \quad 2} \sqrt{\pi}.$$

371. We have employed previously as the integral of the equation $\frac{du}{dt} = k \frac{d^2u}{dx^2}$ the expression

$$u = a_1 e^{-n_1^2 kt} \cos n_1 x + a_2 e^{-n_2^2 kt} \cos n_2 x + a_3 e^{-n_3^2 kt} \cos n_3 x + \&c.;$$

or this,

$$u = a_1 e^{-n_1^2 kt} \sin n_1 x + a_2 e^{-n_2^2 kt} \sin n_2 x + a_3 e^{-n_3^2 kt} \sin n_3 x + \&c.$$

a_1, a_2, a_3, &c., and n_1, n_2, n_3, &c., being two series of arbitrary constants. It is easy to see that each of these expressions is equivalent to the integral

$$\int dq\, e^{-q^2} \sin n\, (x + 2q\sqrt{kt}), \quad \text{or} \quad \int dq\, e^{-q^2} \cos n\, (x + 2q\sqrt{kt}).$$

In fact, to determine the value of the integral

$$\int_{-\infty}^{+\infty} dq\, e^{-q^2} \sin (x + 2q\sqrt{kt});$$

we shall give it the following form

$$\int dq\, e^{-q^2} \sin x \cos 2q\sqrt{kt} + \int dq\, e^{-q^2} \cos x \sin 2q\sqrt{kt}\,;$$

or else,

$$\int_{-\infty}^{+\infty} dq\, e^{-q^2} \sin x \left(\frac{e^{2q\sqrt{-kt}}}{2} + \frac{e^{-2q\sqrt{-kt}}}{2}\right)$$

$$+ \int_{-\infty}^{\infty} dq\, e^{-q^2} \cos x \left(\frac{e^{2q\sqrt{-kt}}}{2\sqrt{-1}} - \frac{e^{-2q\sqrt{-kt}}}{2\sqrt{-1}}\right),$$

which is equivalent to

$$e^{-kt} \sin x \left(\tfrac{1}{2}\int dq\, e^{-(q-\sqrt{-kt})^2} + \tfrac{1}{2}\int dq\, e^{-(q+\sqrt{-kt})^2}\right)$$

$$+ e^{-kt} \cos x \left(\frac{1}{2\sqrt{-1}}\int dq\, e^{-(q-\sqrt{-kt})^2} - \frac{1}{2\sqrt{-1}}\int dq\, e^{-(q+\sqrt{-kt})^2}\right),$$

the integral $\int dq\, e^{-(q\pm\sqrt{-kt})^2}$ taken from $q = -\infty$ to $q = \infty$ is $\sqrt{\pi}$, we have therefore for the value of the integral $\int dq\, e^{-q^2} \sin(x + 2q\sqrt{kt})$, the quantity $\sqrt{\pi}\, e^{-kt} \sin x$, and in general

$$\sqrt{\pi}\, e^{-n^2kt} \sin nx = \int dq\, e^{-q^2} \sin n\,(x + 2q\sqrt{kt}),$$

we could determine in the same manner the integral

$$\int_{-\infty}^{+\infty} dq\, e^{-q^2} \cos n\,(x + 2q\sqrt{kt}),$$

the value of which is $\sqrt{\pi}\, e^{-n^2kt} \cos nx$.

We see by this that the integral

$$e^{-n_1^2kt}(a_1 \sin n_1x + b_1 \cos n_1x) + e^{-n_2^2kt}(a_2 \sin n_2x + b_2 \cos n_2x)$$
$$+ e^{-n_3^2kt}(a_3 \sin n_3x + b_3 \cos n_3x) + \&c.$$

is equivalent to

$$\frac{1}{\sqrt{\pi}}\int_{-\infty}^{+\infty} dq\, e^{-q^2} \begin{Bmatrix} a_1 \sin n_1(x + 2q\sqrt{kt}) + a_2 \sin n_2(x + 2q\sqrt{kt}) + \&c. \\ b_1 \cos n_1(x + 2q\sqrt{kt}) + b_2 \cos n_2(x + 2q\sqrt{kt}) + \&c. \end{Bmatrix}.$$

The value of the series represents, as we have seen previously, any function whatever of $x + 2q\sqrt{kt}$; hence the general integral can be expressed thus

$$v = \int dq\, e^{-q^2} \phi\,(x + 2q\sqrt{kt}).$$

The integral of the equation $\frac{du}{dt} = k\frac{d^2u}{dx^2}$ may besides be presented under diverse other forms[1]. All these expressions are necessarily identical.

SECTION II.

Of the free movement of heat in an infinite solid.

372. The integral of the equation $\frac{dv}{dt} = \frac{K}{CD}\frac{d^2v}{dx^2}$ (*a*) furnishes immediately that of the equation with four variables

$$\frac{dv}{dt} = \frac{K}{CD}\left(\frac{d^2v}{dx^2} + \frac{d^2v}{dy^2} + \frac{d^2v}{dz^2}\right) \ldots\ldots\ldots\ldots\ldots\ldots (A),$$

as we have already remarked in treating the question of the propagation of heat in a solid cube. For which reason it is sufficient in general to consider the effect of the diffusion in the linear case. When the dimensions of bodies are not infinite, the distribution of heat is continually disturbed by the passage from the solid medium to the elastic medium; or, to employ the expressions proper to analysis, the function which determines the temperature must not only satisfy the partial differential equation and the initial state, but is further subjected to conditions which depend on the form of the surface. In this case the integral has a form more difficult to ascertain, and we must examine the problem with very much more care in order to pass from the case of one linear co-ordinate to that of three orthogonal co-ordinates: but when the solid mass is not interrupted, no accidental condition opposes itself to the free diffusion of heat. Its movement is the same in all directions.

[1] See an article by Sir W. Thomson, "On the Linear Motion of Heat," Part I, *Camb. Math. Journal*, Vol. III. pp. 170—174. [A. F.]

The variable temperature v of a point of an infinite line is expressed by the equation

$$v = \frac{1}{\sqrt{\pi}} \int_{-\infty}^{+\infty} dq\, e^{-q^2} f(x + 2q\sqrt{t}) \ldots\ldots\ldots\ldots\ldots (i).$$

x denotes the distance between a fixed point 0, and the point m, whose temperature is equal to v after the lapse of a time t. We suppose that the heat cannot be dissipated through the external surface of the infinite bar, and that the initial state of the bar is expressed by the equation $v = f(x)$. The differential equation, which the value of v must satisfy, is

$$\frac{dv}{dt} = \frac{K}{CD} \frac{d^2v}{dx^2} \ldots\ldots\ldots\ldots\ldots\ldots\ldots\ldots\ldots (a).$$

But to simplify the investigation, we write

$$\frac{dv}{dt} = \frac{d^2v}{dx^2} \ldots\ldots\ldots\ldots\ldots\ldots\ldots\ldots\ldots (b);$$

which assumes that we employ instead of t another unknown equal to $\frac{Kt}{CD}$.

If in $f(x)$, a function of x and constants, we substitute $x + 2n\sqrt{t}$ for x, and if, after having multiplied by $\frac{dn}{\sqrt{\pi}} e^{-n^2}$, we integrate with respect to n between infinite limits, the expression

$$\frac{1}{\sqrt{\pi}} \int_{-\infty}^{+\infty} dn\, e^{-n^2} f(x + 2n\sqrt{t})$$

satisfies, as we have proved above, the differential equation (b); that is to say the expression has the property of giving the same value for the second fluxion with respect to x, and for the first fluxion with respect to t. From this it is evident that a function of three variables $f(x, y, z)$ will enjoy a like property, if we substitute for x, y, z the quantities

$$x + 2n\sqrt{t},\ y + 2p\sqrt{t},\ z + 2q\sqrt{t},$$

provided we integrate after having multiplied by

$$\frac{dn}{\sqrt{\pi}} e^{-n^2}, \frac{dp}{\sqrt{\pi}} e^{-p^2}, \frac{dq}{\sqrt{\pi}} e^{-q^2}.$$

In fact, the function which we thus form,

$$\pi^{-\frac{3}{2}}\int dn \int dp \int dq\, e^{-(n^2+p^2+q^2)} f(x+2n\sqrt{t},\ y+2p\sqrt{t},\ z+2q\sqrt{t}),$$

gives three terms for the fluxion with respect to t, and these three terms are those which would be found by taking the second fluxion with respect to each of the three variables x, y, z.

Hence the equation

$$v=\pi^{-\frac{3}{2}}\int dn \int dp \int dq\, e^{-(n^2+p^2+q^2)} f(x+2n\sqrt{t},\ y+2p\sqrt{t},\ z+2q\sqrt{t}) \quad \ldots\ldots\ldots\ldots(I),$$

gives a value of v which satisfies the partial differential equation

$$\frac{dv}{dt}=\frac{d^2v}{dx^2}+\frac{d^2v}{dy^2}+\frac{d^2v}{dz^2} \quad \ldots\ldots\ldots\ldots (B).$$

373. Suppose now that a formless solid mass (that is to say one which fills infinite space) contains a quantity of heat whose actual distribution is known. Let $v=F(x, y, z)$ be the equation which expresses this initial and arbitrary state, so that the molecule whose co-ordinates are x, y, z has an initial temperature equal to the value of the given function $F(x, y, z)$. We can imagine that the initial heat is contained in a certain part of the mass whose first state is given by means of the equation $v=F(x, y, z)$, and that all other points have a nul initial temperature.

It is required to ascertain what the system of temperatures will be after a given time. The variable temperature v must consequently be expressed by a function $\phi(x, y, z, t)$ which ought to satisfy the general equation (A) and the condition $\phi(x, y, z, 0) = F(x, y, z)$. Now the value of this function is given by the integral

$$v=\pi^{-\frac{3}{2}}\int dn \int dp \int dq\, e^{-(n^2+p^2+q^2)} F(x+2n\sqrt{t},\ y+2p\sqrt{t},\ z+2q\sqrt{t}).$$

In fact, this function v satisfies the equation (A), and if in it we make $t=0$, we find

$$\pi^{-\frac{3}{2}}\int dn \int dp \int dq\, e^{-(n^2+p^2+q^2)} F(x, y, z),$$

or, effecting the integrations, $F(x, y, z)$.

374. Since the function v or $\phi(x, y, z, t)$ represents the initial state when in it we make $t=0$, and since it satisfies the differential equation of the propagation of heat, it represents also that state of the solid which exists at the commencement of the second instant, and making the second state vary, we conclude that the same function represents the third state of the solid, and all the subsequent states. Thus the value of v, which we have just determined, containing an entirely arbitrary function of three variables x, y, z, gives the solution of the problem; and we cannot suppose that there is a more general expression, although otherwise the same integral may be put under very different forms.

Instead of employing the equation

$$v=\frac{1}{\sqrt{\pi}}\int dq\, e^{-q^2} f(x+2q\sqrt{t}),$$

we might give another form to the integral of the equation $\frac{dv}{dt}=\frac{d^2v}{dx^2}$; and it would always be easy to deduce from it the integral which belongs to the case of three dimensions. The result which we should obtain would necessarily be the same as the preceding.

To give an example of this investigation we shall make use of the particular value which has aided us in forming the exponential integral.

Taking then the equation $\frac{dv}{dt}=\frac{d^2v}{dx^2}$ ... (b), let us give to v the very simple value $e^{-n^2t}\cos nx$, which evidently satisfies the differential equation (b). In fact, we derive from it $\frac{dv}{dt}=-n^2v$ and $\frac{d^2v}{dx^2}=-n^2v$. Hence also, the integral

$$\int_{-\infty}^{+\infty} dn\, e^{-n^2t}\cos nx$$

belongs to the equation (b); for this value of v is formed of the sum of an infinity of particular values. Now, the integral

$$\int_{-\infty}^{+\infty} dn\, e^{-n^2t}\cos nx$$

is known, and is known to be equivalent to $\frac{e^{-\frac{x^2}{4t}}}{\sqrt{\pi}\sqrt{t}}$ (see the following article). Hence this last function of x and t agrees also with the differential equation (b). It is besides very easy to verify directly that the particular value $\frac{e^{-\frac{x^2}{4t}}}{\sqrt{t}}$ satisfies the equation in question.

The same result will occur if we replace the variable x by $x-\alpha$, α being any constant. We may then employ as a particular value the function $\frac{A\,e^{-\frac{(x-\alpha)^2}{4t}}}{\sqrt{t}}$, in which we assign to α any value whatever. Consequently the sum $\int d\alpha f(\alpha)\frac{A\,e^{-\frac{(x-\alpha)^2}{4t}}}{\sqrt{t}}$ also satisfies the differential equation (b); for this sum is composed of an infinity of particular values of the same form, multiplied by arbitrary constants. Hence we can take as a value of v in the equation $\frac{dv}{dt}=\frac{d^2v}{dx^2}$ the following,

$$v=\int_{-\infty}^{+\infty} d\alpha f(\alpha)\frac{A\,e^{-\frac{(x-\alpha)^2}{t}}}{\sqrt{t}},$$

A being a constant coefficient. If in the last integral we suppose $\frac{(x-\alpha)^2}{4t}=q^2$, making also $A=\frac{1}{2\sqrt{\pi}}$, we shall have

$$v=\int_{-\infty}^{+\infty} d\alpha f(\alpha)\frac{e^{-\frac{(x-\alpha)^2}{4t}}}{2\sqrt{\pi}\sqrt{t}}\dots\dots\dots\dots\dots\dots\dots (i),$$

or

$$v=\frac{1}{\sqrt{\pi}}\int_{-\infty}^{+\infty} dq\, e^{-q^2} f(x+2q\sqrt{t})\dots\dots\dots\dots (i).$$

We see by this how the employment of the particular values

$$e^{-n^2t}\cos nx \quad\text{or}\quad \frac{e^{-\frac{x^2}{4t}}}{\sqrt{t}}$$

leads to the integral under a finite form.

375. The relation in which these two particular values are to each other is discovered when we evaluate the integral[1]

$$\int_{-\infty}^{+\infty} dn\, e^{-n^2 t} \cos nx.$$

To effect the integration, we might develope the factor $\cos nx$ and integrate with respect to n. We thus obtain a series which represents a known development; but the result may be derived more easily from the following analysis. The integral $\int dn\, e^{-n^2 t} \cos nx$ is transformed to $\int dp\, e^{-p^2} \cos 2pu$, by assuming $n^2 t = p^2$ and $nx = 2pu$. We thus have

$$\int_{-\infty}^{+\infty} dn\, e^{-n^2 t} \cos nx = \frac{1}{\sqrt{t}} \int_{-\infty}^{+\infty} dp\, e^{-p^2} \cos 2pu.$$

We shall now write

$$\int dp\, e^{-p^2} \cos 2pu = \tfrac{1}{2} \int dp\, e^{-p^2 + 2pu\sqrt{-1}} + \tfrac{1}{2} \int dp\, e^{-p^2 - 2pu\sqrt{-1}}$$

$$= \tfrac{1}{2} e^{-u^2} \int dp\, e^{-p^2 + 2pu\sqrt{-1} + u^2} + \tfrac{1}{2} e^{-u^2} \int dp\, e^{-p^2 - 2pu\sqrt{-1} + u^2}$$

$$= \tfrac{1}{2} e^{-u^2} \int dp\, e^{-(p - u\sqrt{-1})^2} + \tfrac{1}{2} e^{-u^2} \int dp\, e^{-(p + u\sqrt{-1})^2}.$$

Now each of the integrals which enter into these two terms is equal to $\sqrt{\pi}$. We have in fact in general

$$\sqrt{\pi} = \int_{-\infty}^{+\infty} dq\, e^{-q^2},$$

and consequently

$$\sqrt{\pi} = \int_{-\infty}^{+\infty} dq\, e^{-(q+b)^2},$$

whatever be the constant b. We find then on making

$$b = \mp u\sqrt{-1}, \quad \int dq\, e^{-q^2} \cos 2qu = e^{-u^2} \sqrt{\pi},$$

hence

$$\int_{-\infty}^{+\infty} dn\, e^{-n^2 t} \cos nx = \frac{e^{-u^2} \sqrt{\pi}}{\sqrt{t}},$$

[1] The value is obtained by a different method in Todhunter's *Integral Calculus*, § 375. [A. F.]

and putting for u its value $\frac{x}{2\sqrt{t}}$, we have

$$\int_{-\infty}^{+\infty} dn\, e^{-n^2t} \cos nx = \frac{e^{-\frac{x^2}{4t}}}{\sqrt{t}} \sqrt{\pi}.$$

Moreover the particular value $\frac{e^{-\frac{x^2}{4t}}}{\sqrt{t}}$ is simple enough to present itself directly without its being necessary to deduce it from the value $e^{-n^2t} \cos nx$. However it may be, it is certain that the function $\frac{e^{-\frac{x^2}{4t}}}{\sqrt{t}}$ satisfies the differential equation $\frac{dv}{dt} = \frac{d^2v}{dx^2}$; it is the same consequently with the function $\frac{e^{-\frac{(x-\alpha)^2}{4t}}}{\sqrt{t}}$, whatever the quantity α may be.

376. To pass to the case of three dimensions, it is sufficient to multiply the function of x and t, $\frac{e^{-\frac{(x-\alpha)^2}{4t}}}{\sqrt{t}}$, by two other similar functions, one of y and t, the other of z and t; the product will evidently satisfy the equation

$$\frac{dv}{dt} = \frac{d^2v}{dx^2} + \frac{d^2v}{dy^2} + \frac{d^2v}{dz^2}.$$

We shall take then for v the value thus expressed:

$$v = t^{-\frac{3}{2}} e^{-\frac{(x-\alpha)^2+(y-\beta)^2+(z-\gamma)^2}{4t}}.$$

If now we multiply the second member by $d\alpha$, $d\beta$, $d\gamma$, and by any function whatever $f(\alpha, \beta, \gamma)$ of the quantities α, β, γ, we find, on indicating the integration, a value of v formed of the sum of an infinity of particular values multiplied by arbitrary constants.

It follows from this that the function v may be thus expressed:

$$v = \int_{-\infty}^{+\infty} d\alpha \int_{-\infty}^{+\infty} d\beta \int_{-\infty}^{+\infty} d\gamma\, f(\alpha, \beta, \gamma)\, t^{-\frac{3}{2}} e^{-\frac{(\alpha-x)^2+(\beta-y)^2+(\gamma-z)^2}{4t}} \quad \ldots.(j).$$

This equation contains the general integral of the proposed equation (A): the process which has led us to this integral ought to be remarked since it is applicable to a great variety of cases; it is useful chiefly when the integral must satisfy conditions relative to the surface. If we examine it attentively we perceive that the transformations which it requires are all indicated by the physical nature of the problem. We can also, in equation (j), change the variables. By taking

$$\frac{(\alpha-x)^2}{4t}=n^2,\quad \frac{(\beta-y)^2}{4t}=p^2,\quad \frac{(\gamma-z)^2}{4t}=q^2,$$

we have, on multiplying the second member by a constant coefficient A,

$$v=2^3A\int dn\int dp\int dq\, e^{-(n^2+p^2+q^2)}f(x+2n\sqrt{t},\ y+2p\sqrt{t},\ z+2q\sqrt{t}).$$

Taking the three integrals between the limits $-\infty$ and $+\infty$, and making $t=0$ in order to ascertain the initial state, we find $v=2^3A\pi^{\frac{3}{2}}f(x,y,z)$. Thus, if we represent the known initial temperatures by $F(x,y,z)$, and give to the constant A the value $2^{-3}\pi^{-\frac{3}{2}}$, we arrive at the integral

$$v=\pi^{-\frac{3}{2}}\int_{-\infty}^{+\infty}dn\int_{-\infty}^{+\infty}dp\int_{-\infty}^{+\infty}dq\, e^{-n^2}e^{-p^2}e^{-q^2}F(x+2n\sqrt{t},\ y+2p\sqrt{t},\ z+2q\sqrt{t}),$$

which is the same as that of Article 372.

The integral of equation (A) may be put under several other forms, from which that is to be chosen which suits best the problem which it is proposed to solve.

It must be observed in general, in these researches, that two functions $\phi(x,y,z,t)$ are the same when they each satisfy the differential equation (A), and when they are equal for a definite value of the time. It follows from this principle that integrals, which are reduced, when in them we make $t=0$, to the same arbitrary function $F(x,y,z)$, all have the same degree of generality; they are necessarily identical.

The second member of the differential equation (a) was multiplied by $\frac{K}{CD}$, and in equation (b) we supposed this coefficient equal to unity. To restore this quantity, it is sufficient to write

$\frac{Kt}{CD}$ instead of t, in the integral (i) or in the integral (j). We shall now indicate some of the results which follow from these equations.

377. The function which serves as the exponent of the number e* can only represent an absolute number, which follows from the general principles of analysis, as we have proved explicitly in Chapter II., section IX. If in this exponent we replace the unknown t by $\frac{Kt}{CD}$, we see that the dimensions of K, C, D and t, with reference to unit of length, being -1, 0, -3, and 0, the dimension of the denominator $\frac{Kt}{CD}$ is 2 the same as that of each term of the numerator, so that the whole dimension of the exponent is 0. Let us consider the case in which the value of t increases more and more; and to simplify this examination let us employ first the equation

$$v = \int d\alpha f(\alpha) \frac{e^{-\frac{(\alpha - x)^2}{4t}}}{2\sqrt{\pi t}} \ldots\ldots\ldots\ldots\ldots\ldots (i),$$

which represents the diffusion of heat in an infinite line. Suppose the initial heat to be contained in a given portion of the line, from $x = -h$ to $x = +g$, and that we assign to x a definite value X, which fixes the position of a certain point m of that line. If the time t increase without limit, the terms $\frac{-\alpha^2}{4t}$ and $\frac{+2\alpha X}{4t}$ which enter into the exponent will become smaller and smaller absolute numbers, so that in the product $e^{-\frac{x^2}{4t}} e^{-\frac{2\alpha x}{4t}} e^{-\frac{\alpha^2}{4t}}$ we can omit the two last factors which sensibly coincide with unity. We thus find

$$v = \frac{e^{-\frac{x^2}{4t}}}{2\sqrt{\pi}\sqrt{t}} \int_{-h}^{+g} d\alpha f(\alpha) \ldots\ldots\ldots\ldots\ldots\ldots (y).$$

This is the expression of the variable state of the line after a very long time; it applies to all parts of the line which are less distant from the origin than the point m. The definite integral

* In such quantities as $e^{-\frac{x^2}{4t}}$. [A. F.]

$\int_{-h}^{+g} d\alpha f(\alpha)$ denotes the whole quantity of heat B contained in the solid, and we see that the primitive distribution has no influence on the temperatures after a very long time. They depend only on the sum B, and not on the law according to which the heat has been distributed.

378. If we suppose a single element ω situated at the origin to have received the initial temperature f, and that all the others had initially the temperature 0, the product ωf will be equal to the integral $\int_{-h}^{+g} d\alpha f(\alpha)$ or B. The constant f is exceedingly great since we suppose the line ω very small.

The equation $v = \dfrac{e^{-\frac{x^2}{4t}}}{2\sqrt{\pi}\sqrt{t}}\,\omega f$ represents the movement which would take place, if a single element situated at the origin had been heated. In fact, if we give to x any value a, not infinitely small, the function $\dfrac{e^{-\frac{x^2}{4t}}}{\sqrt{t}}$ will be nothing when we suppose $t = 0$. The same would not be the case if the value of x were nothing. In this case the function $\dfrac{e^{-\frac{x^2}{4t}}}{\sqrt{t}}$ receives on the contrary an infinite value when $t = 0$. We can ascertain distinctly the nature of this function, if we apply the general principles of the theory of curved surfaces to the surface whose equation is

$$z = \frac{e^{-\frac{x^2}{4y}}}{\sqrt{y}}.$$

The equation $v = \dfrac{e^{-\frac{x^2}{4t}}}{2\sqrt{\pi}\sqrt{t}}\,\omega f$ expresses then the variable temperature at any point of the prism, when we suppose the whole initial heat collected into a single element situated at the origin. This hypothesis, although special, belongs to a general problem, since after a sufficiently long time, the variable state of the solid is always the same as if the initial heat had been collected at the origin. The law according to which the heat was distributed, has much influence on the variable temperatures of the prism; but this effect becomes weaker and weaker, and ends with being quite insensible.

379. It is necessary to remark that the reduced equation (y) does not apply to that part of the line which lies beyond the point m whose distance has been denoted by X.

In fact, however great the value of the time may be, we might choose a value of x such that the term $e^{\frac{2ax}{4t}}$ would differ sensibly from unity, so that this factor could not then be suppressed. We must therefore imagine that we have marked on either side of the origin 0 two points, m and m', situated at a certain distance X or $-X$, and that we increase more and more the value of the time, observing the successive states of the part of the line which is included between m and m'. These variable states converge more and more towards that which is expressed by the equation

$$v = \frac{e^{-\frac{x^2}{4t}}}{2\sqrt{\pi}\sqrt{t}} \int_{-h}^{+g} d\alpha\, f(\alpha) \ldots\ldots\ldots\ldots\ldots\ldots (y).$$

Whatever be the value assigned to X, we shall always be able to find a value of the time so great that the state of the line $m'om$ does not differ sensibly from that which the preceding equation (y) expresses.

If we require that the same equation should apply to other parts more distant from the origin, it will be necessary to suppose a value of the time greater than the preceding.

The equation (y) which expresses in all cases the final state of any line, shews that after an exceedingly long time, the different points acquire temperatures almost equal, and that the temperatures of the same point end by varying in inverse ratio of the square root of the times elapsed since the commencement of the diffusion. The decrements of the temperature of any point whatever always become proportional to the increments of the time.

380. If we made use of the integral

$$v = \int \frac{d\alpha\, f(\alpha)\, e^{-\frac{(\alpha-x)^2}{4kt}}}{2\sqrt{\pi k t}} \ldots\ldots\ldots\ldots\ldots\ldots\ldots (l)$$

to ascertain the variable state of the points of the line situated at a great distance from the heated portion, and in order to express the ultimate condition suppressed also the factor $e^{-\frac{a^2-2ax}{4kt}}$, the results which we should obtain would not be exact. In fact, supposing that the heated portion extends only from $a=0$ to $a=g$ and that the limit g is very small with respect to the distance x of the point whose temperature we wish to determine; the quantity $-\frac{(a-x)^2}{4kt}$ which forms the exponent reduces in fact to $-\frac{x^2}{4kt}$; that is to say the ratio of the two quantities $\frac{(a-x)^2}{4kt}$ and $\frac{x^2}{4kt}$ approaches more nearly to unity as the value of x becomes greater with respect to that of a: but it does not follow that we can replace one of these quantities by the other in the exponent of e. In general the omission of the subordinate terms cannot thus take place in exponential or trigonometrical expressions. The quantities arranged under the symbols of sine or cosine, or under the exponential symbol e, are always absolute numbers, and we can omit only the parts of those numbers whose value is extremely small; their relative values are here of no importance. To decide if we may reduce the expression

$$\int_0^g d\alpha f(\alpha)\, e^{-\frac{(\alpha-x)^2}{4kt}} \text{ to } e^{-\frac{x^2}{4kt}} \int_0^g d\alpha f(\alpha),$$

we must not examine whether the ratio of x to a is very great, but whether the terms $\frac{2\alpha x}{4kt}$, $\frac{-a^2}{4kt}$ are very small numbers. This condition always exists when t the time elapsed is extremely great; but it does not depend on the ratio $\frac{x}{a}$.

381. Suppose now that we wish to ascertain how much time ought to elapse in order that the temperatures of the part of the solid included between $x=0$ and $x=X$, may be represented very nearly by the reduced equation

$$v = \frac{e^{-\frac{x^2}{4kt}}}{2\sqrt{\pi kt}} \int_{-h}^{+g} d\alpha f(\alpha),$$

and that 0 and g may be the limits of the portion originally heated.

The exact solution is given by the equation

$$v=\int_0^g \frac{d\alpha f(\alpha)\, e^{-\frac{(\alpha-x)^2}{4kt}}}{2\sqrt{\pi kt}} \quad \ldots\ldots\ldots\ldots\ldots\ldots\ldots (i),$$

and the approximate solution is given by the equation

$$v=\frac{e^{-\frac{x^2}{4kt}}}{2\sqrt{\pi kt}}\int_0^g d\alpha f(\alpha) \quad \ldots\ldots\ldots\ldots\ldots\ldots\ldots (y),$$

k denoting the value $\frac{K}{CD}$ of the conducibility. In order that the equation (y) may be substituted for the preceding equation (i), it is in general requisite that the factor $e^{\frac{2\alpha x-\alpha^2}{4kt}}$, which is that which we omit, should differ very little from unity; for if it were 1 or $\frac{1}{2}$ we might apprehend an error equal to the value calculated or to the half of that value. Let then $e^{\frac{2\alpha x-\alpha^2}{4kt}}=1+\omega$, ω being a small fraction, as $\frac{1}{100}$ or $\frac{1}{1000}$; from this we derive the condition

$$\frac{2\alpha x-\alpha^2}{4kt}=\omega, \quad \text{or } t=\frac{1}{\omega}\left(\frac{2\alpha x-\alpha^2}{4k}\right),$$

and if the greatest value g which the variable α can receive is very small with respect to x, we have $t=\frac{1}{\omega}\frac{gx}{2k}$.

We see by this result that the more distant from the origin the points are whose temperatures we wish to determine by means of the reduced equation, the more necessary it is for the value of the time elapsed to be great. Thus the heat tends more and more to be distributed according to a law independent of the primitive heating. After a certain time, the diffusion is sensibly effected, that is to say the state of the solid depends on nothing more than the quantity of the initial heat, and not on the distribution which was made of it. The temperatures of points sufficiently near to the origin are soon represented without error by the reduced equation (y); but it is not the same with points very distant from the source. We can then make use of that equation only when the time elapsed is extremely long. Numerical applications make this remark more perceptible.

382. Suppose that the substance of which the prism is formed is iron, and that the portion of the solid which has been heated is a decimetre in length, so that $g = 0{\cdot}1$. If we wish to ascertain what will be, after a given time, the temperature of a point m whose distance from the origin is a metre, and if we employ for this investigation the approximate integral (y), we shall commit an error greater as the value of the time is smaller. This error will be less than the hundredth part of the quantity sought, if the time elapsed exceeds three days and a half.

In this case the distance included between the origin 0 and the point m, whose temperature we are determining, is only ten times greater than the portion heated. If this ratio is one hundred instead of being ten, the reduced integral (y) will give the temperature nearly to less than one hundredth part, when the value of the time elapsed exceeds one month. In order that the approximation may be admissible, it is necessary in general, 1st that the quantity $\frac{2ax - a^2}{4kt}$ should be equal to but a very small fraction as $\frac{1}{100}$ or $\frac{1}{1000}$ or less; 2nd, that the error which must follow should have an absolute value very much less than the small quantities which we observe with the most sensitive thermometers.

When the points which we consider are very distant from the portion of the solid which was originally heated, the temperatures which it is required to determine are extremely small; thus the error which we should commit in employing the reduced equation would have a very small absolute value; but it does not follow that we should be authorized to make use of that equation. For if the error committed, although very small, exceeds or is equal to the quantity sought; or even if it is the half or the fourth, or an appreciable part, the approximation ought to be rejected. It is evident that in this case the approximate equation (y) would not express the state of the solid, and that we could not avail ourselves of it to determine the ratios of the simultaneous temperatures of two or more points.

383. It follows from this examination that we ought not to conclude from the integral $v = \frac{1}{2\sqrt{\pi k t}} \int_0^g d\alpha f(\alpha)\, e^{-\frac{(\alpha - x)^2}{4kt}}$ that the law of the primitive distribution has no influence on the temperature of points very distant from the origin. The resultant effect of this distribution soon ceases to have influence on the points near to the heated portion; that is to say their temperature depends on nothing more than the quantity of the initial heat, and not on the distribution which was made of it: but greatness of distance does not concur to efface the impress of the distribution, it preserves it on the contrary during a very long time and retards the diffusion of heat. Thus the equation

$$v = \frac{1}{\sqrt{\pi}} \frac{e^{-\frac{x^2}{4kt}}}{\sqrt{4kt}} \int_0^g d\alpha f(\alpha)$$

only after an immense time represents the temperatures of points extremely remote from the heated part. If we applied it without this condition, we should find results double or triple of the true results, or even incomparably greater or smaller; and this would not only occur for very small values of the time, but for great values, such as an hour, a day, a year. Lastly this expression would be so much the less exact, all other things being equal, as the points were more distant from the part originally heated.

384. When the diffusion of heat is effected in all directions, the state of the solid is represented as we have seen by the integral

$$v = \iiint \frac{d\alpha\, d\beta\, d\gamma}{2^3 \sqrt{\pi^3 k^3 t^3}} e^{-\frac{(\alpha - x)^2 + (\beta - y)^2 + (\gamma - z)^2}{4kt}} f(\alpha, \beta, \gamma) \ldots\ldots (j).$$

If the initial heat is contained in a definite portion of the solid mass, we know the limits which comprise this heated part, and the quantities α, β, γ, which vary under the integral sign, cannot receive values which exceed those limits. Suppose then that we mark on the three axes six points whose distances are $+X$, $+Y$, $+Z$, and $-X$, $-Y$, $-Z$, and that we consider the successive states of the solid included within the six planes which cross the axes at these distances; we see that the exponent of e under the sign of integration, reduces to $-\left(\frac{x^2+y^2+z^2}{4kt}\right)$, when the value of the time increases without limit. In fact, the terms such as $\frac{2\alpha x}{4kt}$ and $\frac{\alpha^2}{4kt}$ receive in this case very small absolute values, since the numerators are included between fixed limits, and the denominators increase to infinity. Thus the factors which we omit differ extremely little from unity. Hence the variable state of the solid, after a great value of the time, is expressed by

$$v = \frac{e^{-\frac{x^2+y^2+z^2}{4kt}}}{2^3\sqrt{\pi^3k^3t^3}}\int d\alpha \int d\beta \int d\gamma\, f(\alpha, \beta, \gamma).$$

The factor $\int d\alpha \int d\beta \int d\gamma\, f(\alpha, \beta, \gamma)$ represents the whole quantity of heat B which the solid contains. Thus the system of temperatures depends not upon the initial distribution of heat, but only on its quantity. We might suppose that all the initial heat was contained in a single prismatic element situated at the origin, whose extremely small orthogonal dimensions were $\omega_1, \omega_2, \omega_3$. The initial temperature of this element would be denoted by an exceedingly great number f, and all the other molecules of the solid would have a nul initial temperature. The product $\omega_1\omega_2\omega_3 f$ is equal in this case to the integral

$$\int d\alpha \int d\beta \int d\gamma\, f(\alpha, \beta, \gamma).$$

Whatever be the initial heating, the state of the solid which corresponds to a very great value of the time, is the same as if all the heat had been collected into a single element situated at the origin.

385. Suppose now that we consider only the points of the solid whose distance from the origin is very great with respect to the dimensions of the heated part; we might first imagine that this condition is sufficient to reduce the exponent of e in the general integral. The exponent is in fact

$$-\frac{(\alpha-x)^2+(\beta-y)^2+(\gamma-z)^2}{4kt};$$

and the variables α, β, γ are, by hypothesis, included between finite limits, so that their values are always extremely small with respect to the greater co-ordinate of a point very remote from the origin. It follows from this that the exponent of e is composed of two parts $M+\mu$, one of which is very small with respect to the other. But from the fact that the ratio $\frac{\mu}{M}$ is a very small fraction, we cannot conclude that the exponential $e^{M+\mu}$ becomes equal to e^M, or differs only from it by a quantity very small with respect to its actual value. We must not consider the relative values of M and μ, but only the absolute value of μ. In order that we may be able to reduce the exact integral (j) to the equation

$$v = B\frac{e^{-\frac{(x^2+y^2+z^2)}{4kt}}}{2\sqrt[3]{\pi^3k^3t^3}},$$

it is necessary that the quantity

$$\frac{2\alpha x + 2\beta y + 2\gamma z - \alpha^2 - \beta^2 - \gamma^2}{4kt},$$

whose dimension is 0, should always be a very small number. If we suppose that the distance from the origin to the point m, whose temperature we wish to determine, is very great with respect to the extent of the part which was at first heated, we should examine whether the preceding quantity is always a very small fraction ω. This condition must be satisfied to enable us to employ the approximate integral

$$v = B2^{-3}(\pi kt)^{-\frac{3}{2}}e^{-\frac{x^2}{4kt}}:$$

but this equation does not represent the variable state of that part of the mass which is very remote from the source of heat. It gives on the contrary a result so much the less exact, all other things being equal, as the points whose temperature we are determining are more distant from the source.

The initial heat contained in a definite portion of the solid mass penetrates successively the neighbouring parts, and spreads itself in all directions; only an exceedingly small quantity of it arrives at points whose distance from the origin is very great.

When we express analytically the temperature of these points, the object of the investigation is not to determine numerically these temperatures, which are not measurable, but to ascertain their ratios. Now these quantities depend certainly on the law according to which the initial heat has been distributed, and the effect of this initial distribution lasts so much the longer as the parts of the prism are more distant from the source. But if the terms which form part of the exponent, such as $\frac{2\alpha x}{4kt}$ and $\frac{\alpha^2}{4kt}$, have absolute values decreasing without limit, we may employ the approximate integrals.

This condition occurs in problems where it is proposed to determine the highest temperatures of points very distant from the origin. We can demonstrate in fact that in this case the values of the times increase in a greater ratio than the distances, and are proportional to the squares of these distances, when the points we are considering are very remote from the origin. It is only after having established this proposition that we can effect the reduction under the exponent. Problems of this kind are the object of the following section.

SECTION III.

Of the highest temperatures in an infinite solid.

386. We shall consider in the first place the linear movement in an infinite bar, a portion of which has been uniformly heated, and we shall investigate the value of the time which must elapse in order that a given point of the line may attain its highest temperature.

Let us denote by $2g$ the extent of the part heated, the middle of which corresponds with the origin 0 of the distances x. All the points whose distance from the axis of y is less than g and greater than $-g$, have by hypothesis a common initial temperature f, and all other sections have the initial temperature 0. We suppose that no loss of heat occurs at the external surface of the prism, or, which is the same thing, we assign to the section perpendicular to the axis infinite dimensions. It is required to ascertain what will be the time t which corresponds to the maximum of temperature at a given point whose distance is x.

We have seen, in the preceding Articles, that the variable temperature at any point is expressed by the equation

$$v = \frac{1}{2\sqrt{\pi k t}} \int d\alpha\, f(\alpha)\, e^{-\frac{(\alpha - x)^2}{4kt}}.$$

The coefficient k represents $\frac{K}{CD}$, K being the specific conducibility, C the capacity for heat, and D the density.

To simplify the investigation, make $k=1$, and in the result write kt or $\frac{Kt}{CD}$ instead of t. The expression for v becomes

$$v = \frac{1}{2\sqrt{\pi}} \frac{f}{\sqrt{t}} \int_{-g}^{+g} d\alpha\, e^{-\frac{(\alpha - x)^2}{4t}}.$$

This is the integral of the equation $\frac{dv}{dt} = \frac{d^2v}{dx^2}$. The function $\frac{dv}{dx}$ measures the velocity with which the heat flows along the axis of the prism. Now this value of $\frac{dv}{dx}$ is given in the actual problem without any integral sign. We have in fact

$$\frac{dv}{dx} = \frac{f}{2\sqrt{\pi t}} \int_{-g}^{+g} 2d\alpha \frac{\alpha - x}{4t} e^{-\frac{(\alpha - x)^2}{4t}},$$

or, effecting the integration,

$$\frac{dv}{dx} = \frac{f}{2\sqrt{\pi t}} \left\{ e^{-\frac{(x+g)^2}{4t}} - e^{-\frac{(x-g)^2}{4t}} \right\}.$$

387. The function $\frac{d^2v}{dx^2}$ may also be expressed without the sign of integration: now it is equal to a fluxion of the first order $\frac{dv}{dt}$; hence on equating to zero this value of $\frac{dv}{dt}$, which measures the instantaneous increase of the temperature at any point, we have the relation sought between x and t. We thus find

$$\frac{d^2v}{dx^2} = \frac{f}{2\sqrt{\pi t}} \left(\frac{-2(x+g)}{4t} e^{-\frac{(x+g)^2}{4t}} + \frac{2(x-g)}{4t} e^{-\frac{(x-g)^2}{4t}} \right) = \frac{dv}{dt},$$

which gives

$$(x+g)\,e^{-\frac{(x+g)^2}{4t}} = (x-g)\,e^{-\frac{(x-g)^2}{4t}};$$

whence we conclude

$$t = \frac{gx}{\log\left(\frac{x+g}{x-g}\right)}.$$

We have supposed $\frac{K}{CD}=1$. To restore the coefficient we must write $\frac{Kt}{CD}$ instead of t, and we have

$$t = \frac{gCD}{K}\frac{x}{\log\left(\frac{x+g}{x-g}\right)}.$$

The highest temperatures follow each other according to the law expressed by this equation. If we suppose it to represent the varying motion of a body which describes a straight line, x being the space passed over, and t the time elapsed, the velocity of the moving body will be that of the maximum of temperature.

When the quantity g is infinitely small, that is to say when the initial heat is collected into a single element situated at the origin, the value of t is reduced to $\frac{0}{0}$, and by differentiation or development in series we find $\frac{Kt}{CD}=\frac{x^2}{2}$.

We have left out of consideration the quantity of heat which escapes at the surface of the prism; we now proceed to take account of that loss, and we shall suppose the initial heat to be contained in a single element of the infinite prismatic bar.

388. In the preceding problem we have determined the variable state of an infinite prism a definite portion of which was affected throughout with an initial temperature f. We suppose that the initial heat was distributed through a finite space from $x=0$ to $x=b$.

We now suppose that the same quantity of heat bf is contained in an infinitely small element, from $x=0$ to $x=\omega$. The temperature of the heated layer will therefore be $\frac{fb}{\omega}$, and from this follows what was said before, that the variable state of the solid is expressed by the equation

$$v = \frac{fb}{\sqrt{\pi}} \frac{e^{\frac{-x^2}{4kt}}}{2\sqrt{kt}} \cdot e^{-ht} \text{ ....} (a);$$

this result holds when the coefficient $\frac{K}{CD}$ which enters into the differential equation $\frac{dv}{dt} = \frac{K}{CD}\frac{d^2v}{dx^2} - hv$, is denoted by k. As to the coefficient h, it is equal to $\frac{Hl}{CDS}$; S denoting the area of the section of the prism, l the contour of that section, and H the conducibility of the external surface.

Substituting these values in the equation (a) we have

$$v = \frac{bf}{\sqrt{\pi}} \frac{e^{-x^2\frac{CD}{4kt}}}{2\sqrt{\frac{Kt}{CD}}} e^{-\frac{Hl}{CDS}t} \text{ ....} (A);$$

f represents the mean initial temperature, that is to say, that which a single point would have if the initial heat were distributed equally between the points of a portion of the bar whose length is l, or more simply, unit of measure. It is required to determine the value t of the time elapsed, which corresponds to a maximum of temperature at a given point.

To solve this problem, it is sufficient to derive from equation (a) the value of $\frac{dv}{dt}$, and equate it to zero; we have

$$\frac{dv}{dt} = -hv + \frac{x^2}{4kt^2}v - \frac{1}{2}\frac{v}{t}, \text{ and } \frac{1}{t^2} - \frac{2k}{x^2}\frac{1}{t} = \frac{4hk}{x^2} \text{.........} (b),$$

hence the value θ, of the time which must elapse in order that the point situated at the distance x may attain its highest temperature, is expressed by the equation

$$\theta k = \frac{1}{\frac{1}{x^2} + \sqrt{\frac{1}{x^4} + \frac{4h}{kx^2}}} \text{ ....} (c).$$

To ascertain the highest temperature V, we remark that the exponent of e^{-1} in equation (a) is $ht+\frac{x^2}{4kt}$. Now equation (b) gives $ht=\frac{x^2}{4kt}-\frac{1}{2}$; hence $ht+\frac{x^2}{4kt}=\frac{x^2}{2kt}-\frac{1}{2}$, and putting for $\frac{1}{t}$ its known value, we have $ht+\frac{x^2}{4kt}=\sqrt{\frac{1}{4}+\frac{h}{k}x^2}$; substituting this exponent of e^{-1} in equation (a), we have

$$V=\frac{bf}{2\sqrt{\pi}}\frac{e^{-\sqrt{\frac{1}{4}+\frac{h}{k}x^2}}}{\sqrt{k\theta}};$$

and replacing $\sqrt{\theta k}$ by its known value, we find, as the expression of the maximum V,

$$V=\frac{bf}{2\sqrt{\pi}}e^{-\sqrt{\frac{h}{k}x^2+\frac{1}{4}}}\sqrt{\frac{1}{x^2}+\sqrt{\frac{4h}{k}\frac{1}{x^2}+\frac{1}{x^4}}}\quad\ldots\ldots\ldots\ldots(d).$$

The equations (c) and (d) contain the solution of the problem; let us replace h and k by their values $\frac{Hl}{CDS}$ and $\frac{K}{CD}$; let us also write $\frac{1}{2}g$ instead of $\frac{S}{l}$, representing by g the semi-thickness of the prism whose base is a square. We have to determine V and θ, the equations

$$V=\frac{bf}{2\sqrt{\pi}}\frac{e^{-\sqrt{\frac{2H}{Kg}x^2+\frac{1}{4}}}}{x}\sqrt{1+2\sqrt{\frac{2H}{Kg}x^2+\frac{1}{4}}}\quad\ldots\ldots\ldots(D),$$

$$\frac{K}{CD}\theta=\frac{x^2}{1+2\sqrt{\frac{2H}{Kg}x^2+\frac{1}{4}}}\quad\ldots\ldots\ldots\ldots\ldots\ldots(C).$$

These equations are applicable to the movement of heat in a thin bar, whose length is very great. We suppose the middle of this prism to have been affected by a certain quantity of heat bf which is propagated to the ends, and scattered through the convex surface. V denotes the maximum of temperature for the point whose distance from the primitive source is x; θ is the time which has elapsed since the beginning of the diffusion up to the instant at which the highest temperature V occurs. The coefficients C, H, K, D denote the same specific properties as in the preceding problems, and g is the half-side of the square formed by a section of the prism.

389. In order to make these results more intelligible by a numerical application, we may suppose that the substance of which the prism is formed is iron, and that the side $2g$ of the square is the twenty-fifth part of a metre.

We measured formerly, by our experiments, the values of H and K; those of C and D were already known. Taking the metre as the unit of length, and the sexagesimal minute as the unit of time, and employing the approximate values of H, K, C, D, we shall determine the values of V and θ corresponding to a given distance. For the examination of the results which we have in view, it is not necessary to know these coefficients with great precision.

We see at first that if the distance x is about a metre and a half or two metres, the term $\frac{2H}{Kg}x^2$, which enters under the radical, has a large value with reference to the second term $\frac{1}{4}$. The ratio of these terms increases as the distance increases.

Thus the law of the highest temperatures becomes more and more simple, according as the heat removes from the origin. To determine the regular law which is established through the whole extent of the bar, we must suppose the distance x to be very great, and we find

$$V = \frac{bf}{\sqrt{2\pi}} \frac{e^{-x\sqrt{\frac{2H}{Kg}}}}{\sqrt{x}} \left(\frac{2H}{Kg}\right)^{\frac{1}{4}} \quad \ldots\ldots\ldots\ldots\ldots\ldots\ldots(\delta),$$

$$\frac{K}{CD}\theta = \frac{x}{2\sqrt{\frac{2H}{Kg}}} \quad \text{or } \theta = \frac{CD\sqrt{g}}{2^{\frac{3}{2}}\sqrt{HK}}\, x \ldots\ldots\ldots(\gamma).$$

390. We see by the second equation that the time which corresponds to the maximum of temperature increases proportionally with the distance. Thus the velocity of the wave (if however we may apply this expression to the movement in question) is constant, or rather it more and more tends to become so, and preserves this property in its movement to infinity from the origin of heat.

We may remark also in the first equation that the quantity $fe^{-x\sqrt{\frac{2H}{Kg}}}$ expresses the permanent temperatures which the different points of the bar would take, if we affected the origin with a fixed temperature f, as may be seen in Chapter I., Article 76.

In order to represent to ourselves the value of V, we must therefore imagine that all the initial heat which the source contains is equally distributed through a portion of the bar whose length is b, or the unit of measure. The temperature f, which would result for each point of this portion, is in a manner the mean temperature. If we supposed the layer situated at the origin to be retained at a constant temperature f during an infinite time, all the layers would acquire fixed temperatures whose general expression is $fe^{-x\sqrt{\frac{2H}{Kg}}}$, denoting by x the distance of the layer. These fixed temperatures represented by the ordinates of a logarithmic curve are extremely small, when the distance is considerable; they decrease, as is known, very rapidly, according as we remove from the origin.

Now the equation (δ) shews that these fixed temperatures, which are the highest each point can acquire, much exceed the highest temperatures which follow each other during the diffusion of heat. To determine the latter maximum, we must calculate the value of the fixed maximum, multiply it by the constant number $\left(\frac{2H}{Kg}\right)^{\frac{1}{4}} \frac{1}{\sqrt{2\pi}}$, and divide by the square root of the distance x.

Thus the highest temperatures follow each other through the whole extent of the line, as the ordinates of a logarithmic curve divided by the square roots of the abscissæ, and the movement of the wave is uniform. According to this general law the heat collected at a single point is propagated in direction of the length of the solid.

391. If we regarded the conducibility of the external surface of the prism as nothing, or if the conducibility K or the thickness $2g$ were supposed infinite, we should obtain very different results.

We could then omit the term $\frac{2H}{Kg}x^2$, and we should have[1]

$$V = \frac{f}{\sqrt{2}\sqrt{e}\sqrt{\pi}}\frac{1}{x} \text{ and } \frac{K\theta}{CD} = \tfrac{1}{2}x^2.$$

In this case the value of the maximum is inversely proportional to the distance. Thus the movement of the wave would not be uniform. It must be remarked that this hypothesis is purely theoretical, and if the conducibility H is not nothing, but only an extremely small quantity, the velocity of the wave is not variable in the parts of the prism which are very distant from the origin. In fact, whatever be the value of H, if this value is given, as also those of K and g, and if we suppose that the distance x increases without limit, the term $\frac{2H}{Kg}x^2$ will always become much greater than $\frac{1}{4}$. The distances may at first be small enough for the term $\frac{2H}{Kg}x^2$ to be omitted under the radical. The times are then proportional to the squares of the distances; but as the heat flows in direction of the infinite length, the law of propagation alters, and the times become proportional to the distances. The initial law, that is to say, that which relates to points extremely near to the source, differs very much from the final law which is established in the very distant parts, and up to infinity: but, in the intermediate portions, the highest temperatures follow each other according to a mixed law expressed by the two preceding equations (D) and (C).

392. It remains for us to determine the highest temperatures for the case in which heat is propagated to infinity in every direction within the material solid. This investigation, in accordance with the principles which we have established, presents no difficulty.

When a definite portion of an infinite solid has been heated, and all other parts of the mass have the same initial temperature 0, heat is propagated in all directions, and after a certain time the state of the solid is the same as if the heat had been originally collected in a single point at the origin of co-ordinates. The time

[1] See equations (D) and (C), article 388, making $b=1$. [A. F.]

which must elapse before this last effect is set up is exceedingly great when the points of the mass are very distant from the origin. Each of these points which had at first the temperature 0 is imperceptibly heated; its temperature then acquires the greatest value which it can receive; and it ends by diminishing more and more, until there remains no sensible heat in the mass. The variable state is in general represented by the equation

$$v = \int da \int db \int dc \, \frac{e^{-\frac{(a-x)^2+(b-y)^2+(c-z)^2}{4t}}}{2^3 \pi^{\frac{3}{2}} t^{\frac{3}{2}}} f(a, b, c) \ldots\ldots\ldots (E).$$

The integrals must be taken between the limits

$$a = -a_1,\ a = a_2,\ b = -b_1,\ b = b_2,\ c = -c_1,\ c = c_2.$$

The limits $-a_1$, $+a_2$, $-b_1$, $+b_2$, $-c_1$, $+c_2$ are given; they include the whole portion of the solid which was originally heated. The function $f(a, b, c)$ is also given. It expresses the initial temperature of a point whose co-ordinates are a, b, c. The definite integrations make the variables a, b, c disappear, and there remains for v a function of x, y, z, t and constants. To determine the time θ which corresponds to a maximum of v, at a given point m, we must derive from the preceding equation the value of $\frac{dv}{dt}$; we thus form an equation which contains θ and the co-ordinates of the point m. From this we can then deduce the value of θ. If then we substitute this value of θ instead of t in equation (E), we find the value of the highest temperature V expressed in x, y, z and constants.

Instead of equation (E) let us write

$$v = \int da \int db \int dc \, P f(a, b, c),$$

denoting by P the multiplier of $f(a, b, c)$, we have

$$\frac{dv}{dt} = -\frac{3}{2}\frac{v}{t} + \int da \int db \int dc \frac{(a-x)^2 + (b-y)^2 + (c-z)^2}{4t^2} P f(a,b,c) \ldots (e).$$

393. We must now apply the last expression to points of the solid which are very distant from the origin. Any point whatever of the portion which contains the initial heat, having for coordinates the variables a, b, c, and the co-ordinates of the point m

whose temperature we wish to determine being x, y, z, the square of the distance between these two points is $(a-x)^2+(b-y)^2+(c-z)^2$; and this quantity enters as a factor into the second term of $\frac{dv}{dt}$.

Now the point m being very distant from the origin, it is evident that the distance Δ from any point whatever of the heated portion coincides with the distance D of the same point from the origin; that is to say, as the point m removes farther and farther from the primitive source, which contains the origin of co-ordinates, the final ratio of the distances D and Δ becomes 1.

It follows from this that in equation (e) which gives the value of $\frac{dv}{dt}$ the factor $(a-x)^2+(b-y)^2+(c-z)^2$ may be replaced by $x^2+y^2+z^2$ or r^2, denoting by r the distance of the point m from the origin. We have then

$$\frac{dv}{dt} = -\frac{3}{2}\frac{v}{t} + \frac{r^2}{4t^2}\int da \int db \int dc\, P f(a,b,c),$$

or

$$\frac{dv}{dt} = v\left(\frac{r^2}{4t^2} - \frac{3}{2t}\right).$$

If we put for v its value, and replace t by $\frac{Kt}{CD}$, in order to re-establish the coefficient $\frac{K}{CD}$ which we had supposed equal to 1, we have

$$\frac{dv}{dt} = \left\{\frac{r^2}{4\left(\frac{Kt}{CD}\right)^2} - \frac{3}{2\frac{Kt}{CD}}\right\} \int da \int db \int dc\, \frac{e^{-\frac{(a-x)^2+(b-y)^2+(c-z)^2}{4\frac{Kt}{CD}}}}{2^3\pi^{\frac{3}{2}}\left(\frac{K}{CD}\right)^{\frac{3}{2}}} f(a,b,c) \ldots (\alpha).$$

394. This result belongs only to the points of the solid whose distance from the origin is very great with respect to the greatest dimension of the source. It must always be carefully noticed that it does not follow from this condition that we can omit the variables a, b, c under the exponential symbol. They ought only to be omitted outside this symbol. In fact, the term which enters under the signs of integration, and which multiplies $f(a,b,c)$, is the

product of several factors, such as

$$e^{\frac{-a^2}{4\frac{Kt}{CD}}},\quad e^{\frac{2ax}{4\frac{Kt}{CD}}},\quad e^{\frac{-x^2}{4\frac{Kt}{CD}}}.$$

Now it is not sufficient for the ratio $\frac{x}{a}$ to be always a very great number in order that we may suppress the two first factors. If, for example, we suppose a equal to a decimetre, and x equal to ten metres, and if the substance in which the heat is propagated is iron, we see that after nine or ten hours have elapsed, the factor $e^{\frac{2ax}{4\frac{Kt}{CD}}}$ is still greater than 2; hence by suppressing it we should reduce the result sought to half its value. Thus the value of $\frac{dv}{dt}$, as it belongs to points very distant from the origin, and for any time whatever, ought to be expressed by equation (α). But it is not the same if we consider only extremely large values of the time, which increase in proportion to the squares of the distances: in accordance with this condition we must omit, even under the exponential symbol, the terms which contain a, b, or c. Now this condition holds when we wish to determine the highest temperature which a distant point can acquire, as we proceed to prove.

395. The value of $\frac{dv}{dt}$ must in fact be nothing in the case in question; we have therefore

$$\frac{r^2}{4\left(\frac{Kt}{CD}\right)^2} - \frac{3}{2\frac{Kt}{CD}} = 0, \quad \text{or} \quad \frac{K}{CD}t = \frac{1}{6}r^2.$$

Thus the time which must elapse in order that a very distant point may acquire its highest temperature is proportional to the square of the distance of this point from the origin.

If in the expression for v we replace the denominator $\frac{4Kt}{CD}$ by its value $\frac{2}{3}r^2$, the exponent of e^{-1} which is

$$\frac{(a-x)^2 + (b-y)^2 + (c-z)^2}{\frac{2}{3}r^2}$$

may be reduced to $\frac{3}{2}$, since the factors which we omit coincide with unity. Consequently we find

$$v = \frac{3^{\frac{3}{2}}}{(2\pi e)^{\frac{3}{2}}} \frac{1}{r^3} \int da \int db \int dc\, f(a, b, c).$$

The integral $\int da \int db \int dc\, f(a, b, c)$ represents the quantity of the initial heat: the volume of the sphere whose radius is r is $\frac{4}{3}\pi r^3$, so that denoting by f the temperature which each molecule of this sphere would receive, if we distributed amongst its parts all the initial heat, we shall have $v = \sqrt{\frac{6}{\pi e^3}} f$.

The results which we have developed in this chapter indicate the law according to which the heat contained in a definite portion of an infinite solid progressively penetrates all the other parts whose initial temperature was nothing. This problem is solved more simply than that of the preceding Chapters, since by attributing to the solid infinite dimensions, we make the conditions relative to the surface disappear, and the chief difficulty consists in the employment of those conditions. The general results of the movement of heat in a boundless solid mass are very remarkable, since the movement is not disturbed by the obstacle of surfaces. It is accomplished freely by means of the natural properties of heat. This investigation is, properly speaking, that of the irradiation of heat within the material solid.

SECTION IV.

Comparison of the integrals.

396. The integral of the equation of the propagation of heat presents itself under different forms, which it is necessary to compare. It is easy, as we have seen in the second section of this Chapter, Articles 372 and 376, to refer the case of three dimensions to that of the linear movement; it is sufficient therefore to integrate the equation

$$\frac{dv}{dt} = \frac{K}{CD} \frac{d^2v}{dx^2},$$

or the equation

$$\frac{dv}{dt} = \frac{d^2v}{dx^2} \quad \text{.................................} (a).$$

To deduce from this differential equation the laws of the propagation of heat in a body of definite form, in a ring for example, it was necessary to know the integral, and to obtain it under a certain form suitable to the problem, a condition which could be fulfilled by no other form. This integral was given for the first time in our Memoir sent to the Institute of France on the 21st of December, 1807 (page 124, Art. 84): it consists in the following equation, which expresses the variable system of temperatures of a solid ring:

$$v = \frac{1}{2\pi R} \Sigma \int d\alpha\, F(\alpha)\, e^{-\frac{i^2 t}{R^2}} \cos \frac{i(x-\alpha)}{R} \quad \text{............} (\alpha).$$

R is the radius of the mean circumference of the ring; the integral with respect to α must be taken from $\alpha = 0$ to $\alpha = 2\pi R$, or, which gives the same result, from $\alpha = -\pi R$ to $\alpha = \pi R$; i is any integer, and the sum Σ must be taken from $i = -\infty$ to $i = +\infty$; v denotes the temperature which would be observed after the lapse of a time t, at each point of a section separated by the arc x from that which is at the origin. We represent by $v = F(x)$ the initial temperature at any point of the ring. We must give to i the successive values

$$0,\ +1,\ +2,\ +3,\ \&\text{c.},\ \text{and}\ -1,\ -2,\ -3,\ \&\text{c.},$$

and instead of $\cos \frac{i(x-\alpha)}{R}$ write

$$\cos \frac{ix}{R} \cos \frac{i\alpha}{R} + \sin \frac{ix}{R} \sin \frac{i\alpha}{R}.$$

We thus obtain all the terms of the value of v. Such is the form under which the integral of equation (a) must be placed, in order to express the variable movement of heat in a ring (Chap. IV., Art. 241). We consider the case in which the form and extent of the generating section of the ring are such, that the points of the same section sustain temperatures sensibly equal. We suppose also that no loss of heat occurs at the surface of the ring.

397. The equation (α) being applicable to all values of R, we can suppose in it R infinite; in which case it gives the solution of the following problem. The initial state of a solid prism of small thickness and of infinite length, being known and expressed by $v = F(x)$, to determine all the subsequent states. Consider the radius R to contain numerically n times the unit radius of the trigonometrical tables. Denoting by q a variable which successively becomes dq, $2dq$, $3dq$, ... idq, &c., the infinite number n may be expressed by $\frac{1}{dq}$, and the variable number i by $\frac{q}{dq}$. Making these substitutions we find

$$v = \frac{1}{2\pi} \Sigma\, dq \int d\alpha\, F(\alpha)\, e^{-q^2 t} \cos q\,(x - \alpha).$$

The terms which enter under the sign Σ are differential quantities, so that the sign becomes that of a definite integral; and we have

$$v = \frac{1}{2\pi} \int_{-\infty}^{+\infty} d\alpha\, F(\alpha) \int_{-\infty}^{+\infty} dq\, e^{-q^2 t} \cos (qx - q\alpha) \ldots\ldots\ldots (\beta).$$

This equation is a second form of the integral of the equation (α); it expresses the linear movement of heat in a prism of infinite length (Chap. VII., Art. 354). It is an evident consequence of the first integral (α).

398. We can in equation (β) effect the definite integration with respect to q; for we have, according to a known lemma, which we have already proved (Art. 375),

$$\int_{-\infty}^{+\infty} dz\, e^{-z^2} \cos 2hz = e^{-h^2} \sqrt{\pi}.$$

Making then $z^2 = q^2 t$, we find

$$\int_{-\infty}^{+\infty} dq\, e^{-q^2 t} \cos (qx - q\alpha) = \frac{\sqrt{\pi}}{\sqrt{t}}\, e^{-\left(\frac{\alpha - x}{2\sqrt{t}}\right)^2}.$$

Hence the integral (β) of the preceding Article becomes

$$v = \int_{-\infty}^{+\infty} \frac{d\alpha\, F(\alpha)}{2\sqrt{\pi}\sqrt{t}}\, e^{-\left(\frac{\alpha - x}{2\sqrt{t}}\right)^2} \ldots\ldots\ldots\ldots\ldots\ldots (\gamma).$$

If we employ instead of α another unknown quantity β, making $\frac{\alpha - x}{2\sqrt{t}} = \beta$, we find

$$v = \frac{1}{\sqrt{\pi}} \int d\beta \, e^{-\beta^2} F(x + 2\beta\sqrt{t}) \ldots\ldots\ldots\ldots\ldots (\delta).$$

This form (δ) of the integral[1] of equation (a) was given in Volume VIII. of the *Mémoires de l'Ecole Polytechnique*, by M. Laplace, who arrived at this result by considering the infinite series which represents the integral.

Each of the equations (β), (γ), (δ) expresses the linear diffusion of heat in a prism of infinite length. It is evident that these are three forms of the same integral, and that not one can be considered more general than the others. Each of them is contained in the integral (α) from which it is derived, by giving to R an infinite value.

399. It is easy to develope the value of v deduced from equation (a) in series arranged according to the increasing powers of one or other variable. These developments are self-evident, and we might dispense with referring to them; but they give rise to remarks useful in the investigation of integrals. Denoting by ϕ', ϕ'', ϕ''', &c., the functions $\frac{d}{dx}\phi(x)$, $\frac{d^2}{dx^2}\phi(x)$, $\frac{d^3}{dx^3}\phi(x)$, &c., we have

$$\frac{dv}{dt} = v'', \text{ and } v = c + \int dt \, v'';$$

[1] A direct proof of the equivalence of the forms

$$\frac{1}{\sqrt{\pi}} \int_{-\infty}^{+\infty} d\beta \, e^{-\beta^2} \phi(x + 2\beta\sqrt{t}) \text{ and } e^{t\frac{d^2}{dx^2}} \phi(x), \quad \text{(see Art. 401)},$$

has been given by Mr Glaisher in the *Messenger of Mathematics*, June 1876, p. 30. Expanding $\phi(x + 2\beta\sqrt{t})$ by Taylor's Theorem, integrate each term separately: terms involving uneven powers of $\sqrt{t}$ vanish, and we have the second form; which is therefore equivalent to

$$\frac{1}{\pi} \int_{-\infty}^{\infty} d\alpha \int_0^{\infty} dq \, e^{-q^2 t} \cos q(\alpha - x) \, \phi(\alpha),$$

from which the first form may be derived as above. We have thus a slightly generalized form of Fourier's Theorem, p. 351. [A. F.]

here the constant represents any function of x. Putting for v'' its value $c'' + \int dt\, v^{\text{iv}}$, and continuing always similar substitutions, we find

$$v = c + \int dt\, v''$$

$$= c + \int dt \left(c'' + \int dt\, v^{\text{iv}}\right)$$

$$= c + \int dt \left[c'' + \int dt \left(c^{\text{iv}} + \int dt\, v^{\text{vi}}\right)\right],$$

or $$v = c + tc'' + \frac{t^2}{\lfloor 2} c^{\text{iv}} + \frac{t^3}{\lfloor 3} c^{\text{vi}} + \frac{t^4}{\lfloor 4} c^{\text{viii}} + \&\text{c.} \ldots\ldots\ldots\ldots (T).$$

In this series, c denotes an arbitrary function of x. If we wish to arrange the development of the value of v, according to ascending powers of x, we employ

$$\frac{d^2v}{dx^2} = \frac{dv}{dt},$$

and, denoting by $\phi_{,}$, $\phi_{,,}$, $\phi_{,,,}$, &c. the functions

$$\frac{d}{dt}\phi, \quad \frac{d^2}{dt^2}\phi, \quad \frac{d^3}{dt^3}\phi, \quad \&\text{c.},$$

we have first $v = a + bx + \int dx \int dx\, v_{,}$; a and b here represent any two functions of t. We can then put for v its value

$$a_{,} + b_{,}x + \int dx \int dx\, v_{,,};$$

and for $v_{,,}$ its value $a_{,,} + b_{,,}x + \int dx \int dx\, v_{,,,}$, and so on. By continued substitutions

$$v = a + bx + \int dx \int dx\, v_{,}$$

$$= a + bx + \int dx \int dx \left(a_{,} + b_{,}x + \int dx \int dx\, v_{,,}\right)$$

$$= a + bx + \int dx \int dx \left[a_{,} + b_{,}x + \int dx \int dx \left(a_{,,} + b_{,,}x + \int dx \int dx\, v_{,,,}\right)\right]$$

or $$v = a + \frac{x^2}{\lfloor 2} a_{,} + \frac{x^4}{\lfloor 4} a_{,,} + \frac{x^6}{\lfloor 6} a_{,,,} + \&c.$$

$$+ xb + \frac{x^3}{\lfloor 3} b_{,} + \frac{x^5}{\lfloor 5} b_{,,} + \&c. \ldots\ldots\ldots\ldots (X).$$

In this series, a and b denote two arbitrary functions of t.

If in the series given by equation (X) we put, instead of a and b, two functions $\phi(t)$ and $\psi(t)$, and develope them according to ascending powers of t, we find only a single arbitrary function of x, instead of two functions a and b. We owe this remark to M. Poisson, who has given it in Volume VI. of the *Mémoires de l'Ecole Polytechnique*, page 110.

Reciprocally, if in the series expressed by equation (T) we develope the function c according to powers of x, arranging the result with respect to the same powers of x, the coefficients of these powers are formed of two entirely arbitrary functions of t; which can be easily verified on making the investigation.

400. The value of v, developed according to powers of t, ought in fact to contain only one arbitrary function of x; for the differential equation (a) shews clearly that, if we knew, as a function of x, the value of v which corresponds to $t=0$, the other values of the function v which correspond to subsequent values of t, would be determined by this value.

It is no less evident that the function v, when developed according to ascending powers of x, ought to contain two completely arbitrary functions of the variable t. In fact the differential equation $\frac{d^2v}{dx^2} = \frac{dv}{dt}$ shews that, if we knew as a function of t the value of v which corresponds to a definite value of x, we could not conclude from it the values of v which correspond to all the other values of x. It would be necessary in addition, to give as a function of t the value of v which corresponds to a second value of x, for example, to that which is infinitely near to the first. All the other states of the function v, that is to say those which correspond to all the other values of x, would then be determined. The differential equation (a) belongs to a curved surface, the vertical ordinate of any point being v, and the two horizontal co-ordinates

x and and t. It follows evidently from this equation (a) that the form of the surface is determined, when we give the form of the vertical section in the plane which passes through the axis of x: and this follows also from the physical nature of the problem; for it is evident that, the initial state of the prism being given, all the subsequent states are determined. But we could not construct the surface, if it were only subject to passing through a curve traced on the first vertical plane of t and v. It would be necessary to know further the curve traced on a second vertical plane parallel to the first, to which it may be supposed extremely near. The same remarks apply to all partial differential equations, and we see that the order of the equation does not determine in all cases the number of the arbitrary functions.

401. The series (T) of Article 399, which is derived from the equation

$$\frac{dv}{dt} = \frac{d^2v}{dx^2} \quad \text{..........................} (a),$$

may be put under the form $v = e^{tD^2}\phi(x)$. Developing the exponential according to powers of D, and writing $\frac{d^i}{dx^i}$ instead of D^i, considering i as the order of the differentiation, we have

$$v = \phi(x) + t\frac{d^2}{dx^2}\phi(x) + \frac{t^2}{\lfloor 2}\frac{d^4}{dx^4}\phi(x) + \frac{t^3}{\lfloor 3}\frac{d^6}{dx^6}\phi(x) + \&c.$$

Following the same notation, the first part of the series (X) (Art. 399), which contains only even powers of x, may be expressed under the form $\cos(x\sqrt{-D})\,\phi(t)$. Develope according to powers of x, and write $\frac{d^i}{dt^i}$ instead of D^i, considering i as the order of the differentiation. The second part of the series (X) can be derived from the first by integrating with respect to x, and changing the function $\phi(t)$ into another arbitrary function $\psi(t)$. We have therefore

$$v = \cos(x\sqrt{-D})\,\phi(t) + W,$$

and

$$W = \left\{\int_0^x dx \cos(x\sqrt{-D})\right\}\psi(t).$$

This known abridged notation is derived from the analogy which exists between integrals and powers. As to the use made of it here, the object is to express series, and to verify them without any development. It is sufficient to differentiate under the signs which the notation employs. For example, from the equation $v = e^{tD^2}\phi(x)$, we deduce, by differentiation with respect to t only,

$$\frac{dv}{dt} = D^2 e^{tD^2}\phi(x) = D^2 v = \frac{d^2}{dx^2}v;$$

which shews directly that the series satisfies the differential equation (a). Similarly, if we consider the first part of the series (X), writing

$$v = \cos(x\sqrt{-D})\,\phi(t),$$

we have, differentiating twice with respect to x only,

$$\frac{d^2v}{dx^2} = D\cos(x\sqrt{-D})\,\phi(t) = Dv = \frac{dv}{dt}.$$

Hence this value of v satisfies the differential equation (a).

We should find in the same manner that the differential equation

$$\frac{d^2v}{dx^2} + \frac{d^2v}{dy^2} = 0 \text{..........................} (b),$$

gives as the expression for v in a series developed according to increasing powers of y,

$$v = \cos(yD)\,\phi(x).$$

We must develope with respect to y, and write $\frac{d}{dx}$ instead of D: from this value of v we deduce in fact,

$$\frac{d^2v}{dy^2} = -D^2\cos(yD)\,\phi(x) = -D^2v = -\frac{d^2}{dx^2}v.$$

The value $\sin(yD)\,\psi(x)$ satisfies also the differential equation; hence the general value of v is

$$v = \cos(yD)\,\phi(x) + W, \quad \text{where } W = \sin(yD)\,\psi(x).$$

402. If the proposed differential equation is

$$\frac{d^2v}{dt^2} = \frac{d^2v}{dx^2} + \frac{d^2v}{dy^2} \text{.......................} (c),$$

and if we wish to express v in a series arranged according to powers of t, we may denote by $D\phi$ the function

$$\frac{d^2}{dx^2}\phi + \frac{d^2}{dy^2}\phi;$$

and the equation being $\frac{d^2v}{dt^2} = Dv$, we have

$$v = \cos(t\sqrt{-D})\,\phi(x, y).$$

From this we infer that

$$\frac{d^2v}{dt^2} = Dv = \frac{d^2v}{dx^2} + \frac{d^2v}{dy^2}.$$

We must develope the preceding value of v according to powers of t, write $\left(\frac{d^2}{dx^2} + \frac{d^2}{dy^2}\right)^i$, instead of D^i, and then regard i as the order of differentiation.

The following value $\int dt \cos(t\sqrt{-D})\,\psi(x, y)$ satisfies the same condition; thus the most general value of v is

$$v = \cos(t\sqrt{-D})\,\phi(x, y) + W;$$

and $$W = \int dt \cos(t\sqrt{-D})\,\psi(x, y);$$

v is a function $f(x, y, t)$ of three variables. If we make $t = 0$, we have $f = (x, y, 0) = \phi(x, y)$; and denoting $\frac{d}{dt}f(x, y, t)$ by $f'(x, y, t)$, we have $f'(x, y, 0) = \psi(x, y)$.

If the proposed equation is

$$\frac{d^2v}{dt^2} + \frac{d^4v}{dx^4} = 0 \text{.......................} (d),$$

the value of v in a series arranged according to powers of t will be $v = \cos(tD^2)\,\phi(x, y)$, denoting $\frac{d^2}{dx^2}$ by D; for we deduce from this value

$$\frac{d^2v}{dt^2} = -D^4v = -\frac{d^4}{dx^4}v.$$

The general value of v, which can contain only two arbitrary functions of x and y, is therefore

$$v = \cos(tD^2)\,\phi(x, y) + W,$$

and $$W = \int_0^t dt \cos(tD^2)\,\psi(x, y).$$

Denoting v by $f(x, y, t)$, and $\frac{dv}{dt}$ by $f'(x, y, t)$, we have to determine the two arbitrary functions,

$$\phi(x, y) = f(x, y, 0), \quad \text{and } \psi(x, y) = f'(x, y, 0).$$

403. If the proposed differential equation is

$$\frac{d^2v}{dt^2} + \frac{d^4v}{dx^4} + 2\frac{d^4v}{dx^2\,dy^2} + \frac{d^4v}{dy^4} = 0 \quad \ldots\ldots\ldots\ldots\ldots(e),$$

we may denote by $D\phi$ the function $\frac{d^2\phi}{dx^2} + \frac{d^2\phi}{dy^2}$, so that $DD\phi$ or $D^2\phi$ can be formed by raising the binomial $\left(\frac{d^2}{dx^2} + \frac{d^2}{dy^2}\right)$ to the second degree, and regarding the exponents as orders of differentiation. Equation (e) then becomes $\frac{d^2v}{dt^2} + D^2v = 0$; and the value of v, arranged according to powers of t, is $\cos(tD)\,\phi(x, y)$; for from this we derive

$$\frac{d^2v}{dt^2} = -D^2v, \quad \text{or } \frac{d^2v}{dt^2} + \frac{d^4v}{dx^4} + 2\frac{d^4v}{dx^2\,dy^2} + \frac{d^4v}{dy^4} = 0.$$

The most general value of v being able to contain only two arbitrary functions of x and y, which is an evident consequence of the form of the equation, may be expressed thus:

$$v = \cos(tD)\,\phi(x, y) + \int dt \cos(tD)\,\psi(x, y).$$

The functions ϕ and ψ are determined as follows, denoting the function v by $f(x, y, t)$, and $\frac{d}{dt}f(x, y, t)$ by $f_1(x, y, t)$,

$$\phi(x, y) = f(x, y, 0), \qquad \psi(x, y) = f_1(x, y, 0).$$

Lastly, let the proposed differential equation be

$$\frac{dv}{dt} = a\frac{d^2v}{dx^2} + b\frac{d^4v}{dx^4} + c\frac{d^6v}{dx^6} + d\frac{d^8v}{dx^8} + \&c. \ldots\ldots\ldots\ldots\ldots (f),$$

the coefficients a, b, c, d are known numbers, and the order of the equation is indefinite.

The most general value of v can only contain one arbitrary function of x; for it is evident, from the very form of the equation, that if we knew, as a function of x, the value of v which corresponds to $t = 0$, all the other values of v, which correspond to successive values of t, would be determined. To express v, we should have therefore the equation $v = e^{tD}\phi(x)$.

We denote by $D\phi$ the expression

$$a\frac{d^2\phi}{dx^2} + b\frac{d^4\phi}{dx^4} + c\frac{d^6\phi}{dx^6} + \&c.;$$

that is to say, in order to form the value of v, we must develop according to powers of t, the quantity

$$e^{t(a\alpha^2 + b\alpha^4 + c\alpha^6 + d\alpha^8 + \&c.)},$$

and then write $\frac{d}{dx}$ instead of α, considering the powers of α as orders of differentiation. In fact, this value of v being differentiated with respect to t only, we have

$$\frac{dv}{dt} = \frac{de^{tD}}{dt}\phi(x) = Dv = a\frac{d^2v}{dx^2} + b\frac{d^4v}{dx^4} + c\frac{d^6v}{dx^6} + \&c.$$

It would be useless to multiply applications of the same process. For very simple equations we can dispense with abridged expressions; but in general they supply the place of very complex investigations. We have chosen, as examples, the preceding equations, because they all relate to physical phenomena whose analytical expression is analogous to that of the movement of heat. The two first, (a) and (b), belong to the theory of heat; and the three following (*c*), (*d*), (*e*), to dynamical problems; the last (*f*) expresses what the movement of heat would be in solid bodies, if the instantaneous transmission were not limited to an extremely small distance. We have an example of this kind of problem in the movement of luminous heat which penetrates diaphanous media.

404. We can obtain by different means the integrals of these equations: we shall indicate in the first place that which results from the use of the theorem enunciated in Art. 361, which we now proceed to recal.

If we consider the expression

$$\int_{-\infty}^{+\infty} d\alpha\, \phi(\alpha) \int_{-\infty}^{+\infty} d\phi \cos(px - p\alpha), \ldots\ldots\ldots\ldots\ldots\ldots (a)$$

we see that it represents a function of x; for the two definite integrations with respect to α and p make these variables disappear, and a function of x remains. The nature of the function will evidently depend on that which we shall have chosen for $\phi(\alpha)$. We may ask what the function $\phi(\alpha)$, ought to be, in order that after two definite integrations we may obtain a given function $f(x)$. In general the investigation of the integrals suitable for the expression of different physical phenomena, is reducible to problems similar to the preceding. The object of these problems is to determine the arbitrary functions under the signs of the definite integration, so that the result of this integration may be a given function. It is easy to see, for example, that the general integral of the equation

$$\frac{dv}{dt} = a\frac{d^2v}{dx^2} + b\frac{d^4v}{dx^4} + c\frac{d^6v}{dx^6} + d\frac{d^8v}{dx^8} + \&c. \ldots\ldots\ldots\ldots\ldots (f)$$

would be known if, in the preceding expression (*a*), we could determine $\phi(\alpha)$, so that the result of the equation might be a given function $f(x)$. In fact, we form directly a particular value of v, expressed thus,

$$v = e^{-mt}\cos px,$$

and we find this condition,

$$m = ap^2 + bp^4 + cp^6 + \&c.$$

We might then also take

$$v = e^{-mt}\cos(px - p\alpha),$$

giving to the constant α any value. We have similarly

$$v = \int d\alpha\, \phi(\alpha)\, e^{-t(ap^2+bp^4+cp^6+\&c)}\cos(px - p\alpha).$$

It is evident that this value of v satisfies the differential equation (f); it is merely the sum of particular values.

Further, supposing $t = 0$, we ought to find for v an arbitrary function of x. Denoting this function by $f(x)$, we have

$$f(x) = \int d\alpha\, \phi(\alpha) \int dp \cos(px - p\alpha).$$

Now it follows from the form of the equation (f), that the most general value of v can contain only one arbitrary function of x. In fact, this equation shews clearly that if we know as a function of x the value of v for a given value of the time t, all the other values of v which correspond to other values of the time, are necessarily determined. It follows rigorously that if we know, as a function of t and x, a value of v which satisfies the differential equation; and if further, on making $t = 0$, this function of x and t becomes an entirely arbitrary function of x, the function of x and t in question is the general integral of equation (f). The whole problem is therefore reduced to determining, in the equation above, the function $\phi(\alpha)$, so that the result of two integrations may be a given function $f(x)$. It is only necessary, in order that the solution may be general, that we should be able to take for $f(x)$ an entirely arbitrary and even discontinuous function. It is merely required therefore to know the relation which must always exist between the given function $f(x)$ and the unknown function $\phi(\alpha)$. Now this very simple relation is expressed by the theorem of which we are speaking. It consists in the fact that when the integrals are taken between infinite limits, the function $\phi(\alpha)$ is $\frac{1}{2\pi} f(\alpha)$; that is to say, that we have the equation

$$f(x) = \frac{1}{2\pi}\int_{-\infty}^{+\infty} d\alpha\, f(\alpha) \int_{-\infty}^{+\infty} dp \cos(px - p\alpha) \ldots\ldots\ldots\ldots (B).$$

From this we conclude as the general integral of the proposed equation (f),

$$v=\frac{1}{2\pi}\int_{-\infty}^{+\infty}d\alpha f(\alpha)\int_{-\infty}^{+\infty}dp\, e^{-t(ap^2+bp^4+cp^6+\&c)}\cos(px-p\alpha)\ \ldots(c).$$

405. If we propose the equation

$$\frac{d^2v}{dt^2}+\frac{d^4v}{dx^4}=0\ldots\ldots\ldots\ldots\ldots\ldots(d),$$

which expresses the transverse vibratory movement of an elastic plate[1], we must consider that, from the form of this equation, the most general value of v can contain only two arbitrary functions of x: for, denoting this value of v by $f(x, t)$, and the function $\frac{d}{dt}f(x,t)$ by $f'(x,t)$, it is evident that if we knew $f(x, 0)$ and $f'(x, 0)$, that is to say, the values of v and $\frac{dv}{dt}$ at the first instant, all the other values of v would be determined.

This follows also from the very nature of the phenomenon. In fact, consider a rectilinear elastic lamina in its state of rest: x is the distance of any point of this plate from the origin of coordinates; the form of the lamina is very slightly changed, by drawing it from its position of equilibrium, in which it coincided with the axis of x on the horizontal plane; it is then abandoned to its own forces excited by the change of form. The displacement is supposed to be arbitrary, but very small, and such that the initial form given to the lamina is that of a curve drawn on a vertical plane which passes through the axis of x. The system will successively change its form, and will continue to move in the vertical plane on one side or other of the line of equilibrium. The most general condition of this motion is expressed by the equation

$$\frac{d^2v}{dt^2}+\frac{d^4v}{dx^4}=0\ \ldots\ldots\ldots\ldots\ldots\ldots(d).$$

Any point m, situated in the position of equilibrium at a distance x from the origin 0, and on the horizontal plane, has, at

[1] An investigation of the general equation for the lateral vibration of a thin elastic rod, of which (d) is a particular case corresponding to no permanent internal tension, the angular motions of a section of the rod being also neglected, will be found in Donkin's *Acoustics*, Chap. IX. §§ 169—177. [A. F.]

the end of the time t, been removed from its place through the perpendicular height v. This variable flight v is a function of x and t. The initial value of v is arbitrary; it is expressed by any function $\phi(x)$. Now, the equation (d) deduced from the fundamental principles of dynamics shews that the second fluxion of v, taken with respect to t, or $\frac{d^2v}{dt^2}$, and the fluxion of the fourth order taken with respect to x, or $\frac{d^4v}{dx^4}$ are two functions of x and t, which differ only in sign. We do not enter here into the special question relative to the discontinuity of these functions; we have in view only the analytical expression of the integral.

We may suppose also, that after having arbitrarily displaced the different points of the lamina, we impress upon them very small initial velocities, in the vertical plane in which the vibrations ought to be accomplished. The initial velocity given to any point m has an arbitrary value. It is expressed by any function $\psi(x)$ of the distance x.

It is evident that if we have given the initial form of the system or $\phi(x)$ and the initial impulses or $\psi(x)$, all the subsequent states of the system are determinate. Thus the function v or $f(x, t)$, which represents, after any time t, the corresponding form of the lamina, contains two arbitrary functions $\phi(x)$ and $\psi(x)$.

To determine the function sought $f(x, t)$, consider that in the equation

$$\frac{d^2v}{dt^2} + \frac{d^4v}{dx^4} = 0 \quad\text{.............................}(d)$$

we can give to v the very simple value

$$u = \cos q^2t \cos qx,$$

or else

$$u = \cos q^2t \cos(qx - q\alpha);$$

denoting by q and α any quantities which contain neither x nor t. We therefore also have

$$u = \int d\alpha\, F(\alpha) \int dq \cos q^2t \cos(qx - q\alpha),$$

$F(\alpha)$ being any function, whatever the limits of the integrations may be. This value of v is merely a sum of particular values.

Supposing now that $t=0$, the value of v must necessarily be that which we have denoted by $f(x, 0)$ or $\phi(x)$. We have therefore

$$\phi(x)=\int d\alpha\, F(\alpha) \int dq \cos (qx-q\alpha).$$

The function $F(\alpha)$ must be determined so that, when the two integrations have been effected, the result shall be the arbitrary function $\phi(x)$. Now the theorem expressed by equation (B) shews that when the limits of both integrals are $-\infty$ and $+\infty$, we have

$$F(\alpha)=\frac{1}{2\pi} \phi(\alpha).$$

Hence the value of u is given by the following equation:

$$u=\frac{1}{2\pi} \int_{-\infty}^{+\infty} d\alpha\, \phi(\alpha) \int_{-\infty}^{+\infty} dq \cos q^2 t \cos (qx-q\alpha).$$

If this value of u were integrated with respect to t, the ϕ in it being changed to ψ, it is evident that the integral (denoted by W) would again satisfy the proposed differential equation (d), and we should have

$$W=\frac{1}{2\pi} \int d\alpha\, \psi(\alpha) \int dq \frac{1}{q^2} \sin q^2 t \cos (qx-q\alpha).$$

This value W becomes nothing when $t=0$; and if we take the expression

$$\frac{dW}{dt}=\frac{1}{2\pi} \int_{-\infty}^{+\infty} d\alpha\, \psi(\alpha) \int_{-\infty}^{+\infty} dq \cos q^2 t \cos (qx-q\alpha),$$

we see that on making $t=0$ in it, it becomes equal to $\psi(x)$. The same is not the case with the expression $\frac{du}{dt}$; it becomes nothing when $t=0$, and u becomes equal to $\phi(x)$ when $t=0$.

It follows from this that the integral of equation (d) is

$$v=\frac{1}{2\pi} \int_{-\infty}^{+\infty} d\alpha\, \phi(\alpha) \int_{-\infty}^{+\infty} dq \cos q^2 t \cos (qx-q\alpha)+W=u+W,$$

and

$$W=\frac{1}{2\pi} \int_{-\infty}^{+\infty} d\alpha\, \psi(\alpha) \int_{-\infty}^{+\infty} dq \frac{1}{q^2} \sin q^2 t \cos (qx-q\alpha).$$

In fact, this value of v satisfies the differential equation (d); also when we make $t=0$, it becomes equal to the entirely arbitrary function $\phi(x)$; and when we make $t=0$ in the expression $\frac{dv}{dt}$, it reduces to a second arbitrary function $\psi(x)$. Hence the value of v is the complete integral of the proposed equation, and there cannot be a more general integral.

406. The value of v may be reduced to a simpler form by effecting the integration with respect to q. This reduction, and that of other expressions of the same kind, depends on the two results expressed by equations (1) and (2), which will be proved in the following Article.

$$\int_{-\infty}^{+\infty} dq \cos q^2 t \cos qz = \frac{\sqrt{\pi}}{\sqrt{t}} \sin\left(\frac{\pi}{4} + \frac{z^2}{4t}\right) \ldots\ldots\ldots\ldots\ldots (1).$$

$$\int_{-\infty}^{+\infty} dq \sin q^2 t \cos qz = \frac{\sqrt{\pi}}{\sqrt{t}} \sin\left(\frac{\pi}{4} - \frac{z^2}{4t}\right) \ldots\ldots\ldots\ldots\ldots (2).$$

From this we conclude

$$u = \frac{1}{2\sqrt{\pi t}} \int d\alpha\, \phi(\alpha) \sin\left\{\frac{\pi}{4} + \frac{(x-\alpha)^2}{4t}\right\} \ldots\ldots\ldots\ldots (\delta).$$

Denoting $\frac{\alpha - x}{2\sqrt{t}}$ by another unknown μ, we have

$$\alpha = x + 2\mu\sqrt{t}, \quad d\alpha = 2d\mu\sqrt{t}.$$

Putting in place of $\sin\left(\frac{\pi}{4} + \mu^2\right)$ its value

$$\frac{1}{\sqrt{2}} \sin \mu^2 + \frac{1}{\sqrt{2}} \cos \mu^2,$$

we have

$$u = \frac{1}{\sqrt{2\pi}} \int_{-\infty}^{+\infty} d\mu\, (\sin \mu^2 + \cos \mu^2)\, \phi(\alpha + 2\mu\sqrt{t}) \ldots\ldots\ldots (\delta').$$

We have proved in a special memoir that (δ) or (δ'), the integrals of equation (d), represent clearly and completely the motion of the different parts of an infinite elastic lamina. They contain the distinct expression of the phenomenon, and readily explain all its laws. It is from this point of view chiefly that we have proposed them to the attention of geometers. They shew how oscillations are propagated and set up through the whole extent of the lamina, and how the effect of the initial displacement, which is arbitrary and fortuitous, alters more and more as it recedes from the origin, soon becoming insensible, and leaving only the existence of the action of forces proper to the system, the forces namely of elasticity.

407. The results expressed by equations (1) and (2) depend upon the definite integrals

$$\int dx \cos x^2, \quad \text{and} \int dx \sin x^2;$$

let
$$g = \int_{-\infty}^{+\infty} dx \cos x^2, \text{ and } h = \int_{-\infty}^{+\infty} dx \sin x^2;$$

and regard g and h as known numbers. It is evident that in the two preceding equations we may put $y + b$ instead of x, denoting by b any constant whatever, and the limits of the integral will be the same. Thus we have

$$g = \int_{-\infty}^{+\infty} dy \cos (y^2 + 2by + b^2), \quad h = \int_{-\infty}^{+\infty} dy \sin (y^2 + 2by + b^2),$$

$$g = \int dy \begin{Bmatrix} \cos y^2 \cos 2by \cos b^2 - \cos y^2 \sin 2by \sin b^2 \\ - \sin y^2 \sin 2by \cos b^2 - \sin y^2 \cos 2by \sin b^2 \end{Bmatrix}.$$

Now it is easy to see that all the integrals which contain the factor $\sin 2by$ are nothing, if the limits are $-\infty$ and $+\infty$; for $\sin 2by$ changes sign at the same time as y. We have therefore

$$g = \cos b^2 \int dy \cos y^2 \cos 2by - \sin b^2 \int dy \sin y^2 \cos 2by \ldots\ldots\ldots (a).$$

The equation in h also gives

$$h = \int dy \begin{Bmatrix} \sin y^2 \cos 2by \cos b^2 + \cos y^2 \cos 2by \sin b^2 \\ + \cos y^2 \sin 2by \cos b^2 - \sin y^2 \sin 2by \sin b^2 \end{Bmatrix};$$

and, omitting also the terms which contain $\sin 2by$, we have

$$h = \cos b^2 \int dy \sin y^2 \cos 2by + \sin b^2 \int dy \cos y^2 \cos 2by \ldots\ldots\ldots (b).$$

The two equations (*a*) and (*b*) give therefore for g and h the two integrals

$$\int dy \sin y^2 \cos 2by \text{ and } \int dy \cos y^2 \cos 2by,$$

which we shall denote respectively by A and B. We may now make

$$y^2 = p^2 t, \text{ and } 2by = pz; \text{ or } \quad y = p\sqrt{t}, \quad b = \frac{z}{2\sqrt{t}}:$$

we have therefore

$$\sqrt{t}\int dp \cos p^2 t \cos pz = A, \quad \sqrt{t}\int dp \sin p^2 t \cos pz = B.$$

The values[1] of g and h are derived immediately from the known result

$$\sqrt{\pi} = \int_{-\infty}^{+\infty} dx\, e^{-x^2}.$$

The last equation is in fact an identity, and consequently does not cease to be so, when we substitute for x the quantity

$$y\left(\frac{1+\sqrt{-1}}{\sqrt{2}}\right).$$

The substitution gives

$$\sqrt{\pi} = \frac{1+\sqrt{-1}}{\sqrt{2}}\int dy\, e^{-y^2\sqrt{-1}} = \frac{1+\sqrt{-1}}{\sqrt{2}}\int dy\,(\cos y^2 - \sqrt{-1}\sin y^2).$$

Thus the real part of the second member of the last equation is $\sqrt{\pi}$ and the imaginary part nothing. Whence we conclude

$$\sqrt{\pi} = \frac{1}{\sqrt{2}}\left(\int dy \cos y^2 + \int dy \sin y^2\right),$$

[1] More readily from the known results given in § 360, viz.—

$$\int_0^\infty \frac{du \sin u}{\sqrt{u}} = \sqrt{\frac{\pi}{2}}. \quad \text{Let } u = z^2,\ \tfrac{1}{2}\frac{du}{\sqrt{u}} = dz, \text{ then}$$

$$\int_0^\infty dz \sin z^2 = \tfrac{1}{2}\sqrt{\frac{\pi}{2}}. \text{ and } \int_{-\infty}^{+\infty} dz \sin z^2 = 2\int_0^\infty dz \sin z^2 = \sqrt{\frac{\pi}{2}}.$$

So for the cosine from $\int_0^\infty \frac{du \cos u}{\sqrt{u}} = \sqrt{\frac{\pi}{2}}$. [R. L. E.]

and $$0 = \int dy \cos y^2 - \int dy \sin y^2,$$

or $$\int_{-\infty}^{+\infty} dy \cos y^2 = g = \sqrt{\frac{\pi}{2}}, \quad \int dy \sin y^2 = h = \sqrt{\frac{\pi}{2}}.$$

It remains only to determine, by means of the equations (*a*) and (*b*), the values of the two integrals

$$\int dy \cos y^2 \cos 2by \quad \text{and} \int dy \sin y^2 \sin 2by.$$

They can be expressed thus:

$$A = \int dy \cos y^2 \cos 2by = h \sin b^2 + g \cos b^2,$$

$$B = \int dy \sin y^2 \cos 2by = h \cos b^2 - g \sin b^2;$$

whence we conclude

$$\int dp \cos p^2 t \cos pz = \frac{\sqrt{\pi}}{\sqrt{t}} \frac{1}{\sqrt{2}} \left(\cos \frac{z^2}{4t} + \sin \frac{z^2}{4t}\right),$$

$$\int dp \sin p^2 t \cos pz = \frac{\sqrt{\pi}}{\sqrt{t}} \frac{1}{\sqrt{2}} \left(\cos \frac{z^2}{4t} - \sin \frac{z^2}{4t}\right);$$

writing $\sin \frac{\pi}{4}$, or $\cos \frac{\pi}{4}$ instead of $\sqrt{\frac{1}{2}}$, we have

$$\int dp \cos p^2 t \cos pz = \frac{\sqrt{\pi}}{\sqrt{t}} \sin \left(\frac{\pi}{4} + \frac{z^2}{4t}\right) \dots\dots\dots\dots (1)$$

and $$\int dp \sin p^2 t \cos pz = \frac{\sqrt{\pi}}{\sqrt{t}} \sin \left(\frac{\pi}{4} - \frac{z^2}{4t}\right) \dots\dots\dots\dots (2).$$

408. The proposition expressed by equation (*B*) Article 404, or by equation (*E*) Article 361, which has served to discover the integral (δ) and the preceding integrals, is evidently applicable to a very great number of variables. In fact, in the general equation

$$f(x) = \frac{1}{2\pi} \int_{-\infty}^{+\infty} d\alpha f(\alpha) \int_{-\infty}^{+\infty} dp \cos (px - p\alpha),$$

or $$f(x) = \frac{1}{2\pi} \int_{-\infty}^{+\infty} dp \int_{-\infty}^{+\infty} d\alpha \cos (px - p\alpha) f(\alpha),$$

we can regard $f(x)$ as a function of the two variables x and y. The function $f(\alpha)$ will then be a function of α and y. We shall now regard this function $f(\alpha, y)$ as a function of the variable y, and we then conclude from the same theorem (B), Article 404,

that $$f(\alpha, y) = \frac{1}{2\pi}\int_{-\infty}^{+\infty} f(\alpha, \beta)\int dq \cos(qy - q\beta).$$

We have therefore, for the purpose of expressing any function whatever of the two variables x and y, the following equation

$$f(x, y) = \left(\frac{1}{2\pi}\right)^2 \int_{-\infty}^{+\infty} d\alpha \int_{-\infty}^{+\infty} d\beta\, f(\alpha, \beta) \int_{-\infty}^{+\infty} dp \cos(px - p\alpha)$$

$$\int_{-\infty}^{+\infty} dq \cos(qy - q\beta) \ldots (BB).$$

We form in the same manner the equation which belongs to functions of three variables, namely,

$$f(x, y, z) = \left(\frac{1}{2\pi}\right)^3 \int d\alpha \int d\beta \int d\gamma\, f(\alpha, \beta, \gamma)$$

$$\int dp \cos(px - p\alpha) \int dq \cos(qy - q\beta) \int dr \cos(rz - r\gamma) \ldots\ldots (BBB),$$

each of the integrals being taken between the limits $-\infty$ and $+\infty$.

It is evident that the same proposition extends to functions which include any number whatever of variables. It remains to show how this proportion is applicable to the discovery of the integrals of equations which contain more than two variables.

409. For example, the differential equation being

$$\frac{d^2v}{dt^2} = \frac{d^2v}{dx^2} + \frac{d^2v}{dy^2} \ldots\ldots\ldots\ldots\ldots\ldots (c),$$

we wish to ascertain the value of v as a function of (x, y, t), such that; 1st, on supposing $t = 0$, v or $f(x, y, t)$ becomes an arbitrary function $\phi(x, y)$ of x and y; 2nd, on making $t = 0$ in the value of $\frac{dv}{dt}$, or $f'(x, y, t)$, we find a second entirely arbitrary function $\psi(x, y)$.

From the form of the differential equation (c) we can infer that the value of v which satisfies this equation and the two preceding conditions is necessarily the general integral. To discover this integral, we first give to v the particular value

$$v = \cos mt \cos px \cos qy.$$

The substitution of v gives the condition $m = \sqrt{p^2 + q^2}$.

It is no less evident that we may write

$$v = \cos p(x - \alpha) \cos q(y - \beta) \cos t\sqrt{p^2 + q^2},$$

or

$$v = \int d\alpha \int d\beta \, F(\alpha, \beta) \int dp \cos(px - p\alpha) \int dq \cos(qy - q\beta) \cos t\sqrt{p^2 + q^2},$$

whatever be the quantities p, q, α, β and $F(\alpha, \beta)$, which contain neither x, y, nor t. In fact this value of t is merely the sum of particular values.

If we suppose $t = 0$, v necessarily becomes $\phi(x, y)$. We have therefore

$$\phi(x, y) = \int d\alpha \int d\beta \, F(\alpha, \beta) \int dp \cos(px - p\alpha) \int dq \cos(qy - q\beta).$$

Thus the problem is reduced to determining $F(\alpha, \beta)$, so that the result of the indicated integrations may be $\phi(x, y)$. Now, on comparing the last equation with equation (*BB*), we find

$$\phi(x, y) = \left(\frac{1}{2\pi}\right)^2 \int_{-\infty}^{+\infty} d\alpha \int_{-\infty}^{+\infty} d\beta \, \phi(\alpha, \beta) \int_{-\infty}^{+\infty} dp \cos(px - p\alpha) \int_{-\infty}^{+\infty} dq \cos(qy - q\beta).$$

Hence the integral may be expressed thus:

$$v = \left(\frac{1}{2\pi}\right)^2 \int d\alpha \int d\beta \, \phi(\alpha, \beta) \int dp \cos(px - p\alpha) \int dq \cos(qy - q\beta) \cos t\sqrt{p^2 + q^2}.$$

We thus obtain a first part u of the integral; and, denoting by W the second part, which ought to contain the other arbitrary function $\psi(x, y)$, we have

$$v = u + W,$$

and we must take W to be the integral $\int u dt$, changing only ϕ into ψ. In fact, u becomes equal to $\phi(x, y)$, when t is made $=0$; and at the same time W becomes nothing, since the integration, with respect to t, changes the cosine into a sine.

Further, if we take the value of $\frac{dv}{dt}$, and make $t=0$, the first part, which then contains a sine, becomes nothing, and the second part becomes equal to $\psi(x, y)$. Thus the equation $v=u+W$ is the complete integral of the proposed equation.

We could form in the same manner the integral of the equation

$$\frac{d^2v}{dt^2}=\frac{d^2v}{dx^2}+\frac{d^2v}{dy^2}+\frac{d^2v}{dz^2}.$$

It would be sufficient to introduce a new factor

$$\frac{1}{2\pi}\cos(rz-r\gamma),$$

and to integrate with respect to r and γ.

410. Let the proposed equation be $\frac{d^2v}{dx^2}+\frac{d^2v}{dy^2}+\frac{d^2v}{dz^2}=0$; it is required to express v as a function $f(x, y, z)$, such that, 1st, $f(x, y, 0)$ may be an arbitrary function $\phi(x, y)$; 2nd, that on making $z=0$ in the function $\frac{d}{dz}f(x, y, z)$ we may find a second arbitrary function $\psi(x, y)$. It evidently follows, from the form of the differential equation, that the function thus determined will be the complete integral of the proposed equation.

To discover this equation we may remark first that the equation is satisfied by writing $v=\cos px\cos qy\, e^{mz}$, the exponents p and q being any numbers whatever, and the value of m being $\pm\sqrt{p^2+q^2}$.

We might then also write

$$v=\cos(px-p\alpha)\cos(qy-q\beta)\left(e^{z\sqrt{p^2+q^2}}+e^{-z\sqrt{p^2+q^2}}\right),$$

or

$$v=\int d\alpha \int d\beta\, F(\alpha,\beta) \int dp \int dq \cos(px-p\alpha)\cos(qy-q\beta)$$
$$(e^{z\sqrt{p^2+q^2}}+e^{-z\sqrt{p^2+q^2}}).$$

If z be made equal to 0, we have, to determine $F(\alpha,\beta)$, the following condition

$$\phi(x,y)=\int d\alpha \int d\beta\, F(\alpha,\beta)\int dp\int dq \cos(px-p\alpha)\cos(qy-q\beta);$$

and, on comparing with the equation (BB), we see that

$$F(\alpha,\beta)=\left(\frac{1}{2\pi}\right)^2 \phi(\alpha,\beta);$$

we have then, as the expression of the first part of the integral,

$$u=\left(\frac{1}{2\pi}\right)^2\int d\alpha\int d\beta\,\phi(\alpha,\beta)\int dp\cos(px-p\alpha)\int dq\cos(qy-q\beta)$$
$$(e^{z\sqrt{p^2+q^2}}+e^{-z\sqrt{p^2+q^2}}).$$

The value of u reduces to $\phi(x,y)$ when $z=0$, and the same substitution makes the value of $\frac{du}{dx}$ nothing.

We might also integrate the value of u with respect to z, and give to the integral the following form in which ψ is a new arbitrary function:

$$W=\left(\frac{1}{2\pi}\right)^2\int d\alpha\int d\beta\,\psi(\alpha,\beta)\int dp\cos(px-p\alpha)\int dq\cos(qy-q\beta)$$
$$\frac{e^{z\sqrt{p^2+q^2}}-e^{-z\sqrt{p^2+q^2}}}{\sqrt{p^2+q^2}}.$$

The value of W becomes nothing when $z=0$, and the same substitution makes the function $\frac{dW}{dz}$ equal to $\psi(x,y)$. Hence the general integral of the proposed equation is $v=u+W$.

411. Lastly, let the equation be

$$\frac{d^2v}{dt^2}+\frac{d^4v}{dx^4}+2\frac{d^4v}{dx^2\,dy^2}+\frac{d^4v}{dy^4}=0\ldots\ldots\ldots\ldots\ldots(e),$$

it is required to determine v as a function $f(x, y, t)$, which satisfies the proposed equation (e) and the two following conditions: namely, 1st, the substitution $t=0$ in $f(x, y, t)$ must give an arbitrary function $\phi(x, y)$; 2nd, the same substitution in $\frac{d}{dt}f(x, y, t)$ must give a second arbitrary function $\psi(x, y)$.

It evidently follows from the form of equation (e), and from the principles which we have explained above, that the function v, when determined so as to satisfy the preceding conditions, will be the complete integral of the proposed equation. To discover this function we write first,

$$v = \cos px \cos qy \cos mt,$$

whence we derive

$$\frac{d^2v}{dt^2} = -m^2v, \quad \frac{d^4v}{dx^4} = p^4v, \quad \frac{d^4v}{dx^2\,dy^2} = p^2q^2v, \quad \frac{d^4v}{dy^4} = q^4v.$$

We have then the condition $m = p^2 + q^2$. Thus we can write

$$v = \cos px \cos qy \cos t\,(p^2 + q^2),$$

or

$$v = \cos(px - p\alpha)\cos(qy - q\beta)\cos(p^2t + q^2t),$$

or

$$v = \int d\alpha \int d\beta\, F(\alpha, \beta) \int dp \int dq \cos(px - p\alpha)\cos(qy - q\beta)\cos(p^2t + q^2t).$$

When we make $t=0$, we must have $v = \phi(x, y)$; which serves to determine the function $F(\alpha, \beta)$. If we compare this with the general equation (BB), we find that, when the integrals are taken between infinite limits, the value of $F(\alpha, \beta)$ is $\left(\frac{1}{2\pi}\right)^2 \phi(\alpha, \beta)$. We have therefore, as the expression of the first part u of the integral,

$$u = \left(\frac{1}{2\pi}\right)^2 \int d\alpha \int d\beta\, \phi(\alpha, \beta) \int dp \int dq \cos(px - p\alpha)\cos(qy - q\beta)\cos(p^2t + q^2t).$$

Integrating the value of u with respect to t, the second arbitrary function being denoted by ψ, we shall find the other part W of the integral to be expressed thus:

$$W=\left(\frac{1}{2\pi}\right)^2\int d\alpha\int d\beta\,\psi(\alpha,\beta)\int dp\int dq\cos(px-p\alpha)\cos(qy-q\beta)\,\frac{\sin(p^2t+q^2t)}{p^2+q^2}.$$

If we make $t=0$ in u and in W, the first function becomes equal to $\phi(x,y)$, and the second nothing; and if we also make $t=0$ in $\frac{d}{dt}u$ and in $\frac{d}{dt}W$, the first function becomes nothing, and the second becomes equal to $\psi(x,y)$: hence $v=u+W$ is the general integral of the proposed equation.

412. We may give to the value of u a simpler form by effecting the two integrations with respect to p and q. For this purpose we use the two equations (1) and (2) which we have proved in Art. 407, and we obtain the following integral,

$$u=\frac{1}{2\pi}\int_{-\infty}^{+\infty}d\alpha\int_{-\infty}^{+\infty}d\beta\,\phi(\alpha,\beta)\,\frac{1}{4t}\sin\frac{(x-\alpha)^2+(y-\beta)^2}{4t}.$$

Denoting by u the first part of the integral, and by W the second, which ought to contain another arbitrary function, we have

$$W=\int_0^t dt\,u \text{ and } v=u+W.$$

If we denote by μ and ν two new unknowns, such that we have

$$\frac{\alpha-x}{2\sqrt{t}}=\mu,\quad \frac{\beta-y}{2\sqrt{t}}=\nu,$$

and if we substitute for α, β, $d\alpha$, $d\beta$ their values

$$x+2\mu\sqrt{t},\quad y+2\nu\sqrt{t},\quad 2d\mu\sqrt{t},\quad 2d\nu\sqrt{t},$$

we have this other form of the integral,

$$v=\frac{1}{\pi}\int_{-\infty}^{+\infty}d\mu\int_{-\infty}^{+\infty}d\nu\sin(\mu^2+\nu^2)\,\phi(x+2\mu\sqrt{t},\ y+2\nu\sqrt{t})+W.$$

We could not multiply further these applications of our formulæ without diverging from our chief subject. The preceding examples relate to physical phenomena, whose laws were unknown and difficult to discover; and we have chosen them because the integrals of these equations, which have hitherto been fruitlessly sought for, have a remarkable analogy with those which express the movement of heat.

413. We might also, in the investigation of the integrals, consider first series developed according to powers of one variable, and sum these series by means of the theorems expressed by the equations (*B*), (*BB*). The following example of this analysis, taken from the theory of heat itself, appeared to us to be worthy of notice.

We have seen, Art. 399, that the general value of u derived from the equation

$$\frac{dv}{dt}=\frac{d^2v}{dx^2}\text{.......}\ \text{....................}(a),$$

developed in series, according to increasing powers of the variable t, contains one arbitrary function only of x; and that when developed in series according to increasing powers of x, it contains two completely arbitrary functions of t.

The first series is expressed thus:

$$v=\phi(x)+t\frac{d^2}{dx^2}\phi(x)+\frac{t^2}{\lfloor 2}\frac{d^4}{dx^4}\phi(x)+\&\text{c.}\text{.........}(T).$$

The integral denoted by (β), Art. 397, or

$$v=\frac{1}{2\pi}\int d\alpha\,\phi(\alpha)\int dp\,e^{-p^2t}\cos(px-p\alpha),$$

represents the sum of this series, and contains the single arbitrary function $\phi(x)$.

The value of v, developed according to powers of x, contains two arbitrary functions $f(t)$ and $F(t)$, and is thus expressed:

$$v=f(t)+\frac{x^2}{\lfloor 2}\frac{d}{dt}f(t)+\frac{x^4}{\lfloor 4}\frac{d^2}{dt^2}f(t)+\&\text{c.},$$

$$+xF(t)+\frac{x^3}{\lfloor 3}\frac{d}{dt}F(t)+\frac{x^5}{\lfloor 5}\frac{d^2}{dt^2}F(t)+\&\text{c.}\text{........}(X).$$

There is therefore, independently of equation (β), another form of the integral which represents the sum of the last series, and which contains two arbitrary functions, $f(t)$ and $F(t)$.

It is required to discover this second integral of the proposed equation, which cannot be more general than the preceding, but which contains two arbitrary functions.

We can arrive at it by summing each of the two series which enter into equation (X). Now it is evident that if we knew, in the form of a function of x and t, the sum of the first series which contains $f(t)$, it would be necessary, after having multiplied it by dx, to take the integral with respect to x, and to change $f(t)$ into $F(t)$. We should thus find the second series. Further, it would be enough to ascertain the sum of the odd terms which enter into the first series: for, denoting this sum by μ, and the sum of all the other terms by ν, we have evidently

$$\nu = \int_0^x dx \int_0^x dx \frac{d\mu}{dt}.$$

It remains then to find the value of μ. Now the function $f(t)$ may be thus expressed, by means of the general equation (B),

$$f(t) = \frac{1}{2\pi} \int d\alpha\, f(\alpha) \int dp \cos(pt - p\alpha) \quad \ldots\ldots\ldots (B).$$

It is easy to deduce from this the values of the functions

$$\frac{d^2}{dt^2} f(t), \quad \frac{d^4}{dt^4} f(t), \quad \frac{d^6}{dt^6} f(t), \quad \&c.$$

It is evident that differentiation is equivalent to writing in the second member of equation (B), under the sign $\int dp$, the respective factors $-p^2, +p^4, -p^6$, &c.

We have then, on writing once the common factor $\cos(pt - p\alpha)$,

$$\mu = \frac{1}{2\pi} \int d\alpha\, f(\alpha) \int dp \cos(pt - p\alpha) \left(1 - \frac{p^2 x^4}{\lfloor 4} + \frac{p^4 x^8}{\lfloor 8} - \&c.\right).$$

Thus the problem consists in finding the sum of the series which enters into the second member, which presents no difficulty. In fact, if y be the value of this series, we conclude

$$\frac{d^4 y}{dx^4} = -p^2 + \frac{p^4 x^4}{\lfloor 4} - \frac{p^6 x^8}{\lfloor 8} + \&c., \quad \text{or} \quad \frac{d^4 y}{dx^4} = -p^2 y.$$

Integrating this linear equation, and determining the arbitrary constants, so that, when x is nothing, y may be 1, and

$$\frac{dy}{dx},\ \frac{d^2y}{dx^2},\ \frac{d^3y}{dx^3},$$

may be nothing, we find, as the sum of the series,

$$y = \frac{1}{2}\left(e^{x\sqrt{\frac{p}{2}}} + e^{-x\sqrt{\frac{p}{2}}}\right) \cos x \sqrt{\frac{p}{2}}.$$

It would be useless to refer to the details of this investigation; it is sufficient to state the result, which gives, as the integral sought,

$$v = \frac{2}{\pi} \int d\alpha\, f(\alpha) \int dq\, q \left\{ \cos 2q^2 (t-\alpha) \left(e^{qx} + e^{-qx}\right) \cos qx \right.$$

$$\left. - \sin 2q^2 (t-\alpha) \left(e^{qx} - e^{-qx}\right) \sin qx \right\} + W \ldots\ldots (\beta\beta).$$

The term W is the second part of the integral; it is formed by integrating the first part with respect to x, from $x = 0$ to $x = x$, and by changing f into F. Under this form the integral contains two completely arbitrary functions $f(t)$ and $F(t)$. If, in the value of v, we suppose x nothing, the term W becomes nothing by hypothesis, and the first part u of the integral becomes $f(t)$. If we make the same substitution $x = 0$ in the value of $\frac{dv}{dx}$ it is evident that the first part $\frac{du}{dx}$ will become nothing, and that the second, $\frac{dW}{dx}$, which differs only from the first by the function F being substituted for f, will be reduced to $F(t)$. Thus the integral expressed by equation $(\beta\beta)$ satisfies all the conditions, and represents the sum of the two series which form the second member of the equation (X).

This is the form of the integral which it is necessary to select in several problems of the theory of heat[1]; we see that it is very different from that which is expressed by equation (β), Art. 397.

[1] See the article by Sir W. Thomson, "On the Linear Motion of Heat," Part II. Art. 1. *Camb. Math. Journal*, Vol. III. pp. 206—8. [A. F.]

414. We may employ very different processes of investigation to express, by definite integrals, the sums of series which represent the integrals of differential equations. The form of these expressions depends also on the limits of the definite integrals. We will cite a single example of this investigation, recalling the result of Art. 311. If in the equation which terminates that Article we write $x + t\sin u$ under the sign of the function ϕ, we have

$$\frac{1}{\pi}\int_0^\pi du\, \phi(x + t\sin u) = \phi(x) + \frac{t^2}{2^2}\phi''(x) + \frac{t^4}{2^2.4^2}\phi^{\text{iv}}(x) + \frac{t^6}{2^2.4^2.6^2}\phi^{\text{vi}}(x) + \&c.$$

Denoting by v the sum of the series which forms the second member, we see that, to make one of the factors 2^2, 4^2, 6^2, &c. disappear in each term, we must differentiate once with respect to t, multiply the result by t, and differentiate a second time with respect to t. We conclude from this that v satisfies the partial differential equation

$$\frac{d^2v}{dx^2} = \frac{1}{t}\frac{d}{dt}\left(t\frac{dv}{dt}\right), \quad \text{or} \quad \frac{d^2v}{dx^2} = \frac{d^2v}{dt^2} + \frac{1}{t}\frac{dv}{dt}.$$

We have therefore, to express the integral of this equation,

$$v = \frac{1}{\pi}\int_0^\pi du\, \phi(x + t\sin u) + W.$$

The second part W of the integral contains a new arbitrary function.

The form of this second part W of the integral differs very much from that of the first, and may also be expressed by definite integrals. The results, which are obtained by means of definite integrals, vary according to the processes of investigation by which they are derived, and according to the limits of the integrals.

415. It is necessary to examine carefully the nature of the general propositions which serve to transform arbitrary functions: for the use of these theorems is very extensive, and we derive from them directly the solution of several important physical problems, which could be treated by no other method. The

following proofs, which we gave in our first researches, are very suitable to exhibit the truth of these propositions.

In the general equation

$$f(x) = \frac{1}{\pi}\int_{-\infty}^{+\infty} d\alpha\, f(\alpha) \int_0^{+\infty} dp \cos(p\alpha - px),$$

which is the same as equation (B), Art. 404, we may effect the integration with respect to p, and we find

$$f(x) = \frac{1}{\pi}\int_{-\infty}^{+\infty} d\alpha\, f(\alpha) \frac{\sin(p\alpha - px)}{\alpha - x}.$$

We ought then to give to p, in the last expression, an infinite value; and, this being done, the second member will express the value of $f(x)$. We shall perceive the truth of this result by means of the following construction. Examine first the definite integral $\int_0^\infty dx \frac{\sin x}{x}$, which we know to be equal to $\frac{1}{2}\pi$, Art. 356. If we construct above the axis of x the curve whose ordinate is $\sin x$, and that whose ordinate is $\frac{1}{x}$, and then multiply the ordinate of the first curve by the corresponding ordinate of the second, we may consider the product to be the ordinate of a third curve whose form it is very easy to ascertain.

Its first ordinate at the origin is 1, and the succeeding ordinates become alternately positive or negative; the curve cuts the axis at the points where $x = \pi,\ 2\pi,\ 3\pi$, &c., and it approaches nearer and nearer to this axis.

A second branch of the curve, exactly like the first, is situated to the left of the axis of y. The integral $\int_0^\infty dx \frac{\sin x}{x}$ is the area included between the curve and the axis of x, and reckoned from $x = 0$ up to a positive infinite value of x.

The definite integral $\int_0^\infty dx \frac{\sin px}{x}$, in which p is supposed to be any positive number, has the same value as the preceding. In fact, let $px = z$; the proposed integral will become $\int_0^\infty dz \frac{\sin z}{z}$, and, consequently, it is also equal to $\frac{1}{2}\pi$. This proposition is true, whatever positive number p may be. If we suppose, for example, $p=10$, the curve whose ordinate is $\frac{\sin 10x}{x}$ has sinuosities very much closer and shorter than the sinuosities whose ordinate is $\frac{\sin x}{x}$; but the whole area from $x=0$ up to $x=\infty$ is the same.

Suppose now that the number p becomes greater and greater, and that it increases without limit, that is to say, becomes infinite. The sinuosities of the curve whose ordinate is $\frac{\sin px}{x}$ are infinitely near. Their base is an infinitely small length equal to $\frac{\pi}{p}$. That being so, if we compare the positive area which rests on one of these intervals $\frac{\pi}{p}$ with the negative area which rests on the following interval, and if we denote by X the finite and sufficiently large abscissa which answers to the beginning of the first arc, we see that the abscissa x, which enters as a denominator into the expression $\frac{\sin px}{x}$ of the ordinate, has no sensible variation in the double interval $\frac{2\pi}{p}$, which serves as the base of the two areas. Consequently the integral is the same as if x were a constant quantity. It follows that the sum of the two areas which succeed each other is nothing.

The same is not the case when the value of x is infinitely small, since the interval $\frac{2\pi}{p}$ has in this case a finite *ratio* to the value of x. We know from this that the integral $\int_0^\infty dx\,\frac{\sin px}{x}$, in which we suppose p to be an infinite number, is wholly formed out of the sum of its first terms which correspond to extremely small values of x. When the abscissa has a finite value X, the area does not vary, since the parts which compose it destroy each other two by two alternately. We express this result by writing

$$\int_0^\infty dx\,\frac{\sin px}{x}=\int_0^\omega dx\,\frac{\sin px}{x}=\tfrac{1}{2}\pi.$$

The quantity ω, which denotes the limit of the second integral, has an infinitely small value; and the value of the integral is the same when the limit is ω and when it is ∞.

416. This assumed, take the equation

$$f(x) = \frac{1}{\pi}\int_{-\infty}^{+\infty} d\alpha f(\alpha) \frac{\sin p(\alpha - x)}{\alpha - x}, \quad (p = \infty).$$

Having laid down the axis of the abscissæ α, construct above that axis the curve ff, whose ordinate is $f(\alpha)$. The form of this curve is entirely arbitrary; it might have ordinates existing only in one or several parts of its course, all the other ordinates being nothing.

Construct also above the same axis of abscissæ a curved line ss whose ordinate is $\frac{\sin pz}{z}$, z denoting the abscissa and p a very great positive number. The centre of this curve, or the point which corresponds to the greatest ordinate p, may be placed at the origin O of the abscissæ α, or at the end of any abscissa whatever. We suppose this centre to be successively displaced, and to be transferred to all points of the axis of α, towards the right, departing from the point O. Consider what occurs in a certain position of the second curve, when the centre has arrived at the point x, which terminates an abscissa x of the first curve.

The value of x being regarded as constant, and α being the only variable, the ordinate of the second curve becomes

$$\frac{\sin p(\alpha - x)}{\alpha - x}.$$

If then we *link together* the two curves, for the purpose of forming a third, that is to say, if we multiply each ordinate of the second, and represent the product by an ordinate of a third curve drawn above the axis of α, this product is

$$f(\alpha)\frac{\sin p(\alpha - x)}{\alpha - x}.$$

The whole area of the third curve, or the area included between this curve and the axis of abscissæ, may then be expressed by

$$\int_{-\infty}^{+\infty} d\alpha f(\alpha) \frac{\sin p(\alpha - x)}{\alpha - x}.$$

Now the number p being infinitely great, the second curve has all its sinuosities infinitely near; we easily see that for all points which are at a finite distance from the point x, the definite integral, or the whole area of the third curve, is formed of equal parts alternately positive or negative, which destroy each other two by two. In fact, for one of these points situated at a certain distance from the point x, the value of $f(\alpha)$ varies infinitely little when we increase the distance by a quantity less than $\frac{2\pi}{p}$. The same is the case with the denominator $\alpha - x$, which measures that distance. The area which corresponds to the interval $\frac{2\pi}{p}$ is therefore the same as if the quantities $f(\alpha)$ and $\alpha - x$ were not variables. Consequently it is nothing when $\alpha - x$ is a finite magnitude. Hence the definite integral may be taken between limits as near as we please, and it gives, between those limits, the same result as between infinite limits. The whole problem is reduced then to taking the integral between points infinitely near, one to the left, the other to the right of that where $\alpha - x$ is nothing, that is to say from $\alpha = x - \omega$ to $\alpha = x + \omega$, denoting by ω a quantity infinitely small. In this interval the function $f(\alpha)$ does not vary, it is equal to $f(x)$, and may be placed outside the symbol of integration. Hence the value of the expression is the product of $f(x)$ by

$$\int d\alpha \frac{\sin p(\alpha - x)}{\alpha - x},$$

taken between the limits $\alpha - x = -\omega$, and $\alpha - x = \omega$.

Now this integral is equal to π, as we have seen in the preceding article; hence the definite integral is equal to $\pi f(x)$, whence we obtain the equation

$$f(x) = \frac{1}{2\pi}\int_{-\infty}^{+\infty} d\alpha f(\alpha) \frac{2 \sin p(\alpha - x)}{\alpha - x}, \quad (p = \infty)$$

$$= \frac{1}{2\pi}\int_{-\infty}^{+\infty} d\alpha f(\alpha) \int_{-\infty}^{+\infty} dp \cos(px - p\alpha) \ldots\ldots (B).$$

417. The preceding proof supposes that notion of infinite quantities which has always been admitted by geometers. It would be easy to offer the same proof under another form, examining the changes which result from the continual increase of the

factor p under the symbol $\sin p(\alpha - x)$. These considerations are too well known to make it necessary to recall them.

Above all, it must be remarked that the function $f(x)$, to which this proof applies, is entirely arbitrary, and not subject to a continuous law. We might therefore imagine that the enquiry is concerning a function such that the ordinate which represents it has no existing value except when the abscissa is included between two given limits a and b, all the other ordinates being supposed nothing; so that the curve has no form or trace except above the interval from $x = a$ to $x = b$, and coincides with the axis of α in all other parts of its course.

The same proof shews that we are not considering here infinite values of x, but definite actual values. We might also examine on the same principles the cases in which the function $f(x)$ becomes infinite, for singular values of x included between the given limits; but these have no relation to the chief object which we have in view, which is to introduce into the integrals arbitrary functions; it is impossible that any problem in nature should lead to the supposition that the function $f(x)$ becomes infinite, when we give to x a singular value included between given limits.

In general the function $f(x)$ represents a succession of values or ordinates each of which is arbitrary. An infinity of values being given to the abscissa x, there are an equal number of ordinates $f(x)$. All have actual numerical values, either positive or negative or nul.

We do not suppose these ordinates to be subject to a common law; they succeed each other in any manner whatever, and each of them is given as if it were a single quantity.

It may follow from the very nature of the problem, and from the analysis which is applicable to it, that the passage from one ordinate to the following is effected in a continuous manner. But special conditions are then concerned, and the general equation (B), considered by itself, is independent of these conditions. It is rigorously applicable to discontinuous functions.

Suppose now that the function $f(x)$ coincides with a certain analytical expression, such as $\sin x$, e^{-x^2}, or $\phi(x)$, when we give to x a value included between the two limits a and b, and that all the values of $f(x)$ are nothing when x is not included between a and b; the limits of integration with respect to α, in the preceding equation (B), become then $\alpha = a$, $\alpha = b$; since the result is the same as for the limits $\alpha = -\infty$, $\alpha = \infty$, every value of $\phi(\alpha)$ being nothing by hypothesis, when α is not included between a and b. We have then the equation

$$f(x) = \frac{1}{2\pi}\int_a^b d\alpha\,\phi(\alpha)\int_{-\infty}^{+\infty} dp\cos(px - p\alpha)\ldots\ldots\ldots\ldots(B').$$

The second member of this equation (B') is a function of the variable x; for the two integrations make the variables α and p disappear, and x only remains with the constants a and b. Now the function equivalent to the second member is such, that on substituting for x any value included between a and b, we find the same result as on substituting this value of x in $\phi(x)$; and we find a nul result if, in the second member, we substitute for x any value not included between a and b. If then, keeping all the other quantities which form the second member, we replaced the limits a and b by nearer limits a' and b', each of which is included between a and b, we should change the function of x which is equal to the second member, and the effect of the change would be such that the second member would become nothing whenever we gave to x a value not included between a' and b'; and, if the value of x were included between a' and b', we should have the same result as on substituting this value of x in $\phi(x)$.

We can therefore vary at will the limits of the integral in the second member of equation (B'). This equation exists always for values of x included between any limits a and b, which we may have chosen; and, if we assign any other value to x, the second member becomes nothing. Let us represent $\phi(x)$ by the variable ordinate of a curve of which x is the abscissa; the second member, whose value is $f(x)$, will represent the variable ordinate of a second curve whose form will depend on the limits a and b. If these limits are $-\infty$ and $+\infty$, the two curves, one of which has $\phi(x)$ for ordinate, and the other $f(x)$, coincide exactly through the whole extent of their course. But, if we give other values a and b to these limits, the two curves coincide exactly through every part of their course which corresponds to the interval from $x = a$ to $x = b$. To right and left of this interval, the second curve coincides precisely at every point with the axis of x. This result is very remarkable, and determines the true sense of the proposition expressed by equation (B).

418. The theorem expressed by equation (Π) Art. 234 must be considered under the same point of view. This equation serves to develope an arbitrary function $f(x)$ in a series of sines or cosines of multiple arcs. The function $f(x)$ denotes a function completely arbitrary, that is to say a succession of given values, subject or not to a common law, and answering to all the values of x included between 0 and any magnitude X.

The value of this function is expressed by the following equation,

$$f(x) = \frac{1}{2\pi} \Sigma \int_a^b d\alpha f(\alpha) \cos \frac{2i\pi}{X} (x - \alpha) \dots\dots\dots\dots (A).$$

The integral, with respect to α, must be taken between the limits $\alpha = a$, and $\alpha = b$; each of these limits a and b is any quantity whatever included between 0 and X. The sign Σ affects the integer number i, and indicates that we must give to i every integer value negative or positive, namely,

$$\dots -5, \ -4, \ -3, \ -2, \ -1, \ 0, \ +1, \ +2, \ +3, \ +4, \ +5, \dots$$

and must take the sum of the terms arranged under the sign Σ. After these integrations the second member becomes a function of the variable x only, and of the constants a and b. The general proposition consists in this: 1st, that the value of the second member, which would be found on substituting for x a quantity included between a and b, is equal to that which would be obtained on substituting the same quantity for x in the function $f(x)$; 2nd, every other value of x included between 0 and X, but not included between a and b, being substituted in the second member, gives a nul result.

Thus there is no function $f(x)$, or part of a function, which cannot be expressed by a trigonometric series.

The value of the second member is periodic, and the interval of the period is X, that is to say, the value of the second member does not change when $x + X$ is written instead of x. All its values in succession are renewed at intervals X.

The trigonometrical series equal to the second member is convergent; the meaning of this statement is, that if we give to the variable x any value whatever, the sum of the terms of the series approaches more and more, and infinitely near to, a definite limit. This limit is 0, if we have substituted for x a quantity included between 0 and X, but not included between a and b; but if the quantity substituted for x is included between a and b, the limit of the series has the same value as $f(x)$. The last function is subject to no condition, and the line whose ordinate it represents may have any form; for example, that of a contour formed of a series of straight lines and curved lines. We see by this that the limits a and b, the whole interval X, and the nature of the function being arbitrary, the proposition has a very extensive signification; and, as it not only expresses an analytical property, but leads also to the solution of several important problems in nature, it was necessary to consider it under different points of view, and to indicate its chief applications. We have given several proofs of this theorem in the course of this work. That which we shall refer to in one of the following Articles (Art. 424) has the advantage of being applicable also to non-periodic functions.

If we suppose the interval X infinite, the terms of the series become differential quantities; the sum indicated by the sign Σ becomes a definite integral, as was seen in Arts. 353 and 355, and equation (A) is transformed into equation (B). Thus the latter equation (B) is contained in the former, and belongs to the case in which the interval X is infinite: the limits a and b are then evidently entirely arbitrary constants.

419. The theorem expressed by equation (B) presents also divers analytical applications, which we could not unfold without quitting the object of this work; but we will enunciate the principle from which these applications are derived.

We see that, in the second member of the equation

$$f(x) = \frac{1}{2\pi}\int_a^b d\alpha\, f(\alpha) \int_{-\infty}^{+\infty} dp \cos(px - p\alpha) \ldots\ldots\ldots\ldots\ldots\ldots (B),$$

the function $f(x)$ is so transformed, that the symbol of the function f affects no longer the variable x, but an auxiliary variable α. The variable x is only affected by the symbol cosine. It follows from this, that in order to differentiate the function $f(x)$ with respect to x, as many times as we wish, it is sufficient to differentiate the second member with respect to x under the symbol cosine. We then have, denoting by i any integer number whatever,

$$\frac{d^{2i}}{dx^{2i}} f(x) = \pm \int d\alpha f(\alpha) \int dp\, p^{2i} \cos(px - p\alpha).$$

We take the upper sign when i is even, and the lower sign when i is odd. Following the same rule relative to the choice of sign

$$\frac{d^{2i+1}}{dx^{2i+1}} f(x) = \mp \frac{1}{2\pi} \int d\alpha f(\alpha) \int dp\, p^{2i+1} \sin(px - p\alpha).$$

We can also integrate the second member of equation (B) several times in succession, with respect to x; it is sufficient to write in front of the symbol sine or cosine a negative power of p.

The same remark applies to finite differences and to summations denoted by the sign Σ, and in general to analytical operations which may be effected upon trigonometrical quantities. The chief characteristic of the theorem in question, is to transfer the general sign of the function to an auxiliary variable, and to place the variable x under the trigonometrical sign. The function $f(x)$ acquires in a manner, by this transformation, all the properties of trigonometrical quantities; differentiations, integrations, and summations of series thus apply to functions in general in the same manner as to exponential trigonometrical functions. For which reason the use of this proposition gives directly the integrals of partial differential equations with constant coefficients. In fact, it is evident that we could satisfy these equations by particular exponential values; and since the theorems of which we are speaking give to the general and arbitrary functions the character of exponential quantities, they lead easily to the expression of the complete integrals.

The same transformation gives also, as we have seen in Art. 413, an easy means of summing infinite series, when these series contain successive differentials, or successive integrals of the same function; for the summation of the series is reduced, by what precedes, to that of a series of algebraic terms.

420. We may also employ the theorem in question for the purpose of substituting under the general form of the function a binomial formed of a real part and an imaginary part. This analytical problem occurs at the beginning of the calculus of partial differential equations; and we point it out here since it has a direct relation to our chief object.

If in the function $f(x)$ we write $\mu+\nu\sqrt{-1}$ instead of x, the result consists of two parts $\phi+\sqrt{-1}\,\psi$. The problem is to determine each of these functions ϕ and ψ in terms of μ and ν. We shall readily arrive at the result if we replace $f(x)$ by the expression

$$\frac{1}{2\pi}\int d\alpha\, f(\alpha)\int dp \cos(px-p\alpha),$$

for the problem is then reduced to the substitution of $\mu+\nu\sqrt{-1}$ instead of x under the symbol cosine, and to the calculation of the real term and the coefficient of $\sqrt{-1}$. We thus have

$$f(x)=f(\mu+\nu\sqrt{-1})=\frac{1}{2\pi}\int d\alpha\quad(\alpha)\int dp\cos\left[p(\mu-\alpha)+p\nu\sqrt{-1}\right]$$

$$=\frac{1}{4\pi}\int d\alpha\, f(\alpha)\int dp\left\{\cos(p\mu-p\alpha)\left(e^{p\nu}+e^{-p\nu}\right)\right.$$

$$\left.+\sqrt{-1}\sin(p\mu-p\alpha)\left(e^{p\nu}-e^{-p\nu}\right)\right\};$$

hence $$\phi=\frac{1}{4\pi}\int d\alpha\, f(\alpha)\int dp\cos(p\mu-p\alpha)\left(e^{p\nu}+e^{-p\nu}\right),$$

$$\psi=\frac{1}{4\pi}\int d\alpha\, f(\alpha)\int dp\sin(p\mu-p\alpha)\left(e^{p\nu}-e^{-p\nu}\right).$$

Thus all the functions $f(x)$ which can be imagined, even those which are not subject to any law of continuity, are reduced to the form $M+N\sqrt{-1}$, when we replace the variable x in them by the binomial $\mu+\nu\sqrt{-1}$.

421. To give an example of the use of the last two formulæ, let us consider the equation $\frac{d^2v}{dx^2}+\frac{d^2v}{dy^2}=0$, which relates to the uniform movement of heat in a rectangular plate. The general integral of this equation evidently contains two arbitrary functions. Suppose then that we know in terms of x the value of v when $y=0$, and that we also know, as another function of x, the value of $\frac{dv}{dy}$ when $y=0$, we can deduce the required integral from that of the equation

$$\frac{d^2v}{dt^2}=\frac{d^2v}{dx^2},$$

which has long been known; but we find imaginary quantities under the functional signs: the integral is

$$v=\phi\left(x+y\sqrt{-1}\right)+\phi\left(x-y\sqrt{-1}\right)+W.$$

The second part W of the integral is derived from the first by integrating with respect to y, and changing ϕ into ψ.

It remains then to transform the quantities $\phi(x+y\sqrt{-1})$ and $\phi(x-y\sqrt{-1})$, in order to separate the real parts from the imaginary parts. Following the process of the preceding Article we find for the first part u of the integral,

$$u=\frac{1}{2\pi}\int_{-\infty}^{+\infty}d\alpha\, f(\alpha)\int_{-\infty}^{+\infty}dp\cos\left(px-p\alpha\right)\left(e^{py}+e^{-py}\right),$$

and consequently

$$W=\frac{1}{2\pi}\int_{-\infty}^{+\infty}d\alpha\, F(\alpha)\int_{-\infty}^{+\infty}\frac{dp}{p}\cos\left(px-p\alpha\right)\left(e^{py}-e^{-py}\right).$$

The complete integral of the proposed equation expressed in real terms is therefore $v=u+W$; and we perceive in fact, 1st, that it satisfies the differential equation; 2nd, that on making $y=0$ in it, it gives $v=f(x)$; 3rd, that on making $y=0$ in the function $\frac{dv}{dy}$, the result is $F(x)$.

422. We may also remark that we can deduce from equation (B) a very simple expression of the differential coefficient of the ith order, $\frac{d^i}{dx^i}f(x)$, or of the integral $\int^i dx^i f(x)$.

The expression required is a certain function of x and of the index i. It is required to ascertain this function under a form such that the number i may not enter it as an index, but as a quantity, in order to include, in the same formula, every case in which we assign to i any positive or negative value. To obtain it we shall remark that the expression

$$\cos\left(r + i\frac{\pi}{2}\right),$$

or

$$\cos r \cos\frac{i\pi}{2} - \sin r \sin\frac{i\pi}{2},$$

becomes successively

$$-\sin r, \quad -\cos r, \quad +\sin r, \quad +\cos r, \quad -\sin r, \quad \&c.,$$

if the respective values of i are 1, 2, 3, 4, 5, &c. The same results recur in the same order, when we increase the value of i. In the second member of the equation

$$f(x) = \frac{1}{2\pi}\int d\alpha f(\alpha)\int dp \cos(px - p\alpha),$$

we must now write the factor p^i before the symbol cosine, and add under this symbol the term $+i\frac{\pi}{2}$. We shall thus have

$$\frac{d^i}{dx^i}f(x) = \frac{1}{2\pi}\int_{-\infty}^{+\infty} d\alpha f(\alpha)\int_{-\infty}^{+\infty} dp\, p^i \cos\left(px - p\alpha + i\frac{\pi}{2}\right).$$

The number i, which enters into the second member, may be any positive or negative integer. We shall not press these applications to general analysis; it is sufficient to have shewn the use of our theorems by different examples. The equations of the fourth order, (d), Art. 405, and (e), Art. 411, belong as we have said to dynamical problems. The integrals of these equations were not yet known when we gave them in a *Memoir on the Vibrations of*

Elastic Surfaces, read at a sitting of the Academy of Sciences[1], 6th June, 1816 (Art. VI. §§ 10 and 11, and Art. VII. §§ 13 and 14). They consist in the two formulæ δ and δ', Art. 406, and in the two integrals expressed, one by the first equation of Art. 412, the other by the last equation of the same Article. We then gave several other proofs of the same results. This memoir contained also the integral of equation (*c*), Art. 409, under the form referred to in that Article. With regard to the integral $(\beta\beta)$ of equation (*a*), Art. 413, it is here published for the first time.

423. The propositions expressed by equations (*A*) and (*B'*), Arts. 418 and 417, may be considered under a more general point of view. The construction indicated in Arts. 415 and 416 applies not only to the trigonometrical function $\dfrac{\sin(p\alpha - px)}{\alpha - x}$; but suits all other functions, and supposes only that when the number p becomes infinite, we find the value of the integral with respect to α, by taking this integral between extremely near limits. Now this condition belongs not only to trigonometrical functions, but is applicable to an infinity of other functions. We thus arrive at the expression of an arbitrary function $f(x)$ under different very remarkable forms; but we make no use of these transformations in the special investigations which occupy us.

With respect to the proposition expressed by equation (*A*), Art. 418, it is equally easy to make its truth evident by constructions, and this was the theorem for which we employed them at first. It will be sufficient to indicate the course of the proof.

[1] The date is inaccurate. The memoir was read on June 8th, 1818, as appears from an abstract of it given in the *Bulletin des Sciences par la Société Philomatique*, September 1818, pp. 129—136, entitled, *Note relative aux vibrations des surfaces élastiques et au mouvement des ondes, par* M. Fourier. The reading of the memoir further appears from the *Analyse des travaux de l'Académie des Sciences pendant l'année* 1818, p. xiv, and its not having been published except in abstract, from a remark of Poisson at pp. 150—1 of his memoir *Sur les équations aux différences partielles*, printed in the *Mémoires de l'Académie des Sciences*, Tome III. (year 1818), Paris, 1820. The title, *Mémoire sur les vibrations des surfaces élastiques, par* M. Fourier, is given in the *Analyse*, p. xiv. The object, "to integrate several partial differential equations and to deduce from the integrals the knowledge of the physical phenomena to which these equations refer," is stated in the *Bulletin*, p. 129. [A. F.]

In equation (A), namely,

$$f(x)=\frac{1}{2\pi}\int_{-X}^{+X}d\alpha f(\alpha)\sum_{-\infty}^{+\infty}\cos 2i\pi\frac{\alpha-x}{X};$$

we can replace the sum of the terms arranged under the sign Σ by its value, which is derived from known theorems. We have seen different examples of this calculation previously, Section III., Chap. III. It gives as the result if we suppose, in order to simplify the expression, $2\pi=X$, and denote $\alpha-x$ by r,

$$\sum_{-j}^{+j}\cos jr=\cos jr+\sin jr\frac{\sin r}{\operatorname{versin} r}.$$

We must then multiply the second member of this equation by $d\alpha f(\alpha)$, suppose the number j infinite, and integrate from $\alpha=-\pi$ to $\alpha=+\pi$. The curved line, whose abscissa is α and ordinate $\cos jr$, being conjoined with the line whose abscissa is α and ordinate $f(\alpha)$, that is to say, when the corresponding ordinates are multiplied together, it is evident that the area of the curve *produced*, taken between any limits, becomes nothing when the number j increases without limit. Thus the first term $\cos jr$ gives a nul result.

The same would be the case with the term $\sin jr$, if it were not multiplied by the factor $\frac{\sin r}{\operatorname{versin} r}$; but on comparing the three curves which have a common abscissa α, and as ordinates $\sin jr$, $\frac{\sin r}{\operatorname{versin} r}$, $f(\alpha)$, we see clearly that the integral

$$\int d\alpha f(\alpha)\sin jr\frac{\sin r}{\operatorname{versin} r}$$

has no actual values except for certain intervals infinitely small, namely, when the ordinate $\frac{\sin r}{\operatorname{versin} r}$ becomes infinite. This will take place if r or $\alpha-x$ is nothing; and in the interval in which α differs infinitely little from x, the value of $f(\alpha)$ coincides with $f(x)$. Hence the integral becomes

$$2f(x)\int_0^\infty dr\sin jr\frac{r}{\frac{1}{2}r^2},\quad\text{or}\quad 4f(x)\int_0^\infty\frac{dr}{r}\sin jr,$$

which is equal to $2\pi f(x)$, Arts. 415 and 356. Whence we conclude the previous equation (A).

When the variable x is exactly equal to $-\pi$ or $+\pi$, the construction shews what is the value of the second member of the equation (A), $[\frac{1}{2}f(-\pi)$ or $\frac{1}{2}f(\pi)]$.

If the limits of integrations are not $-\pi$ and $+\pi$, but other numbers a and b, each of which is included between $-\pi$ and $+\pi$, we see by the same figure what the values of x are, for which the second member of equation (A) is nothing.

If we imagine that between the limits of integration certain values of $f(\alpha)$ become infinite, the construction indicates in what sense the general proposition must be understood. But we do not here consider cases of this kind, since they do not belong to physical problems.

If instead of restricting the limits $-\pi$ and $+\pi$, we give greater extent to the integral, selecting more distant limits a' and b', we know from the same figure that the second member of equation (A) is formed of several terms and makes the result of integration finite, whatever the function $f(x)$ may be.

We find similar results if we write $2\pi\frac{\alpha - x}{X}$ instead of r, the limits of integration being $-X$ and $+X$.

It must now be considered that the results at which we have arrived would also hold for an infinity of different functions of $\sin jr$. It is sufficient for these functions to receive values alternately positive and negative, so that the area may become nothing, when j increases without limit. We may also vary the factor $\frac{\sin r}{\text{versin}\, r}$, as well as the limits of integration, and we may suppose the interval to become infinite. Expressions of this kind are very general, and susceptible of very different forms. We cannot delay over these developments, but it was necessary to exhibit the employment of geometrical constructions; for they solve without any doubt questions which may arise on the extreme values, and on singular values; they would not have served to discover these theorems, but they prove them and guide all their applications.

424. We have yet to regard the same propositions under another aspect. If we compare with each other the solutions relative to the varied movement of heat in a ring, a sphere, a rectangular prism, a cylinder, we see that we had to develope an arbitrary function $f(x)$ in a series of terms, such as

$$a_1\phi(\mu_1 x) + a_2\phi(\mu_2 x) + a_3\phi(\mu_3 x) + \&c.$$

The function ϕ, which in the second member of equation (A) is a cosine or a sine, is replaced here by a function which may be very different from a sine. The numbers μ_1, μ_2, μ_3, &c. instead of being integers, are given by a transcendental equation, all of whose roots infinite in number are real.

The problem consisted in finding the values of the coefficients a_1, a_2, $a_3 \ldots a_i$; they have been arrived at by means of definite integrations which make all the unknowns disappear, except one. We proceed to examine specially the nature of this process, and the exact consequences which flow from it.

In order to give to this examination a more definite object, we will take as example one of the most important problems, namely, that of the varied movement of heat in a solid sphere. We have seen, Art. 290, that, in order to satisfy the initial distribution of the heat, we must determine the coefficients a_1, a_2, $a_3 \ldots a_i$, in the equation

$$xF(x) = a_1 \sin(\mu_1 x) + a_2 \sin(\mu_2 x) + a_3 \sin(\mu_3 x) + \&c. \ldots\ldots (e).$$

The function $F(x)$ is entirely arbitrary; it denotes the value v of the given initial temperature of the spherical shell whose radius is x. The numbers μ_1, $\mu_2 \ldots \mu_i$ are the roots μ, of the transcendental equation

$$\frac{\mu X}{\tan \mu X} = 1 - hX \ldots\ldots\ldots\ldots\ldots\ldots\ldots (f).$$

X is the radius of the whole sphere; h is a known numerical coefficient having any positive value. We have rigorously proved in our earlier researches, that all the values of μ or the roots of the equation (f) are real[1]. This demonstration is derived from the

[1] The *Mémoires de l'Académie des Sciences*, Tome x, Paris 1831, pp. 119—146, contain *Remarques générales sur l'application des principes de l'analyse algébrique*

general theory of equations, and requires only that we should suppose known the form of the imaginary roots which every equation may have. We have not referred to it in this work, since its place is supplied by constructions which make the proposition more evident. Moreover, we have treated a similar problem analytically, in determining the varied movement of heat in a cylindrical body (Art. 308). This arranged, the problem consists in discovering numerical values for $a_1, a_2, a_3, \ldots a_i$, &c., such that the second member of equation (*e*) necessarily becomes equal to $x F(x)$, when we substitute in it for x any value included between 0 and the whole length X.

To find the coefficient a_i, we have multiplied equation (*e*) by $dx \sin \mu_i x$, and then integrated between the limits $x = 0$, $x = X$, and we have proved (Art. 291) that the integral

$$\int_0^X dx \sin \mu_i x \sin \mu_j x$$

has a null value whenever the indices i and j are not the same; that is to say when the numbers μ_i and μ_j are two different roots of the equation (*f*). It follows from this, that the definite integration making all the terms of the second member disappear, except that which contains a_i, we have to determine this coefficient, the equation

$$\int_0^X dx\,[x\,F(x) \sin \mu_i x] = a_i \int_0^X dx \sin \mu_i x \sin \mu_i x.$$

Substituting this value of the coefficient a_i in equation (*e*), we derive from it the identical equation (ϵ),

$$x\,F(x) = \Sigma \sin(\mu_i x) \frac{\int_0^X d\alpha\,.\,\alpha F(\alpha) \sin \mu_i \alpha}{\int_0^X d\beta \sin \mu_i \beta \sin \mu_i \beta} \quad \ldots\ldots\ldots\ldots\ldots (\epsilon).$$

aux équations transcendantes, by Fourier. The author shews that the imaginary roots of $\sec x = 0$ do not satisfy the equation $\tan x = 0$, since for them, $\tan x = \sqrt{-1}$. The equation $\tan x = 0$ is satisfied only by the roots of $\sin x = 0$, which are all real. It may be shewn also that the imaginary roots of $\sec x = 0$ do not satisfy the equation $x - m \tan x = 0$, where m is less than 1, but this equation is satisfied only by the roots of the equation $f(x) = x \cos x - m \sin x = 0$, which are all real. For if $f_{r+1}(x)$, $f_r(x)$, $f_{r-1}(x)$, are three successive differential coefficients of $f(x)$, the values of x which make $f_r(x) = 0$, make the signs of $f_{r+1}(x)$ and $f_{r-1}(x)$ different. Hence by Fourier's Theorem relative to the number of changes of sign of $f(x)$ and its successive derivatives, $f(x)$ can have no imaginary roots. [A. F.]

In the second member we must give to i all its values, that is to say we must successively substitute for μ_i, all the roots μ of the equation (f). The integral must be taken for α from $\alpha = 0$ to $\alpha = X$, which makes the unknown α disappear. The same is the case with β, which enters into the denominator in such a manner that the term $\sin \mu_i x$ is multiplied by a coefficient a_i whose value depends only on X and on the index i. The symbol Σ denotes that after having given to i its different values, we must write down the sum of all the terms.

The integration then offers a very simple means of determining the coefficients directly; but we must examine attentively the origin of this process, which gives rise to the following remarks.

1st. If in equation (e) we had omitted to write down part of the terms, for example, all those in which the index is an even number, we should still find, on multiplying the equation by $dx \sin \mu_i x$, and integrating from $x = 0$ to $x = X$, the same value of a_i, which has been already determined, and we should thus form an equation which would not be true; for it would contain only part of the terms of the general equation, namely, those whose index is odd.

2nd. The complete equation (e) which we obtain, after having determined the coefficients, and which does not differ from the equation referred to (Art. 291) in which we might make $t = 0$ and $v = f(x)$, is such that if we give to x any value included between 0 and X, the two members are necessarily equal; but we cannot conclude, as we have remarked, that this equality would hold, if choosing for the first member $x F(x)$ a function subject to a continuous law, such as $\sin x$ or $\cos x$, we were to give to x a value not included between 0 and X. In general the resulting equation (ϵ) ought to be applied to values of x, included between 0 and X. Now the process which determines the coefficient a_i does not explain why all the roots μ_i must enter into equation (e), nor why this equation refers solely to values of x, included between 0 and X.

To answer these questions clearly, it is sufficient to revert to the principles which serve as the foundation of our analysis.

We divide the interval X into an infinite number n of parts equal to dx, so that we have $ndx = X$, and writing $f(x)$ instead of $xF(x)$, we denote by $f_1, f_2, f_3 \ldots f_i \ldots f_n$, the values of $f(x)$, which correspond to the values dx, $2dx$, $3dx$, ... idx ... ndx, assigned to x; we make up the general equation (e) out of a number n of terms; so that n unknown coefficients enter into it, a_1, a_2, a_3, ... $a_i \ldots a_n$. This arranged, the equation (e) represents n equations of the first degree, which we should form by substituting successively for x, its n values dx, $2dx$, $3dx$,...ndx. This system of n equations contains f_1 in the first equation, f_2 in the second, f_3 in the third, f_n in the nth. To determine the first coefficient a_1, we multiply the first equation by σ_1, the second by σ_2, the third by σ_3, and so on, and add together the equations thus multiplied. The factors $\sigma_1, \sigma_2, \sigma_3, \ldots \sigma_n$ must be determined by the condition, that the sum of all the terms of the second members which contain a_2 must be nothing, and that the same shall be the case with the following coefficients $a_3, a_4, \ldots a_n$. All the equations being then added, the coefficient a_1 enters only into the result, and we have an equation for determining this coefficient. We then multiply all the equations anew by other factors $\rho_1, \rho_2, \rho_3, \ldots \rho_n$ respectively, and determine these factors so that on adding the n equations, all the coefficients may be eliminated, except a_2. We have then an equation to determine a_2. Similar operations are continued, and choosing always new factors, we successively determine all the unknown coefficients. Now it is evident that this process of elimination is exactly that which results from integration between the limits 0 and X. The series $\sigma_1, \sigma_2, \sigma_3, \ldots \sigma_n$ of the first factors is $dx \sin(\mu_1 dx)$, $dx \sin(\mu_1 2dx)$, $dx \sin(\mu_1 3dx) \ldots dx \sin(\mu_1 ndx)$. In general the series of factors which serves to eliminate all the coefficients except a_i, is $dx \sin(\mu_i dx)$, $dx \sin(\mu_i 2dx)$, $dx \sin(\mu_i 3dx) \ldots dx \sin(\mu_i ndx)$; it is represented by the general term $dx \sin(\mu_i x)$, in which we give successively to x all the values

$$dx, \quad 2dx, \quad 3dx, \ldots ndx.$$

We see by this that the process which serves to determine these coefficients, differs in no respect from the ordinary process of elimination in equations of the first degree. The number n of equations is equal to that of the unknown quantities $a_1, a_2, a_3 \ldots a_n$, and is the same as the number of given quantities $f_1, f_2, f_3 \ldots f_n$. The values found for the coefficients are those which must exist in order that the n equations may hold good together, that is to say in order that equation (ϵ) may be true when we give to x one of these n values included between 0 and X; and since the number n is infinite, it follows that the first member $f(x)$ necessarily coincides with the second, when the value of x substituted in each is included between 0 and X.

The foregoing proof applies not only to developments of the form

$$a_1 \sin(\mu_1 x) + a_2 \sin(\mu_2 x) + a_3 \sin(\mu_3 x) + \ldots + a_i \sin \mu_i x,$$

it applies to all the functions $\phi(\mu_i x)$ which might be substituted for $\sin(\mu_i x)$, maintaining the chief condition, namely, that the integral $\int_0^X dx\, \phi(\mu_i x)\, \phi(\mu_j x)$ has a nul value when i and j are different numbers.

If it be proposed to develope $f(x)$ under the form

$$f(x) = a + \frac{a_1 \cos x}{b_1 \sin x} + \frac{a_2 \cos 2x}{b_2 \sin 2x} + \ldots + \frac{a_i \cos ix}{b_i \cos ix} + \&c.,$$

the quantities $\mu_1, \mu_2, \mu_3 \ldots \mu_i$, &c. will be integers, and the condition

$$\int_0^X dx \cos\left(2\pi i \frac{x}{X}\right) \sin\left(2\pi j \frac{x}{X}\right) = 0,$$

always holding when the indices i and j are different numbers, we obtain, by determining the coefficients a_i, b_i, the general equation (II), page 206, which does not differ from equation (A) Art. 418.

425. If in the second member of equation (e) we omitted one or more terms which correspond to one or more roots μ_i of the equation (f), equation (ϵ) would not in general be true. To prove this, let us suppose a term containing μ_j and a_j not to be written in the second member of equation (e), we might multiply the n equations respectively by the factors

$$dx \sin(\mu_j dx),\ dx \sin(\mu_j 2dx),\ dx \sin(\mu_j 3dx) \ldots dx \sin(\mu_j n dx);$$

and adding them, the sum of all the terms of the second members would be nothing, so that not one of the unknown coefficients would remain. The result, formed of the sum of the first members, that is to say the sum of the values $f_1, f_2, f_3 \ldots f_n$, multiplied respectively by the factors

$$dx \sin(\mu_j dx),\ dx \sin(\mu_j 2dx),\ dx \sin(\mu_j 3dx) \ldots dx \sin(\mu_j ndx),$$

would be reduced to zero. This relation would then necessarily exist between the given quantities $f_1, f_2, f_3 \ldots f_n$; and they could not be considered entirely arbitrary, contrary to hypothesis. If these quantities $f_1, f_2, f_3 \ldots f_n$ have any values whatever, the relation in question cannot exist, and we cannot satisfy the proposed conditions by omitting one or more terms, such as $a_j \sin(\mu_j x)$ in equation (e).

Hence the function $f(x)$ remaining undetermined, that is to say, representing the system of an infinite number of arbitrary constants which correspond to the values of x included between 0 and X, it is necessary to introduce into the second member of equation (e) all the terms such as $a_j \sin(\mu_j x)$, which satisfy the condition

$$\int_0^X dx \sin \mu_i x \sin \mu_j x = 0,$$

the indices i and j being different; but if it happen that the function $f(x)$ is such that the n magnitudes $f_1, f_2, f_3 \ldots f_n$ are connected by a relation expressed by the equation

$$\int_0^X dx \sin \mu_j x f(x) = 0,$$

it is evident that the term $a_j \sin \mu_j x$ might be omitted in the equation (e).

Thus there are several classes of functions $f(x)$ whose development, represented by the second member of the equation (ϵ), does not contain certain terms corresponding to some of the roots μ. There are for example cases in which we omit all the terms whose index is even; and we have seen different examples of this in the course of this work. But this would not hold, if the function $f(x)$ had all the generality possible. In all these cases, we ought to suppose the second member of equation (e) to be complete, and the investigation shews what terms ought to be omitted, since their coefficients become nothing.

426. We see clearly by this examination that the function $f(x)$ represents, in our analysis, the system of a number n of separate quantities, corresponding to n values of x included between 0 and X, and that these n quantities have values *actual*, and consequently not *infinite*, chosen at will. All might be nothing, except one, whose value would be given.

It might happen that the series of the n values $f_1, f_2, f_3 \ldots f_n$ was expressed by a function subject to a continuous law, such as x or x^3, $\sin x$, or $\cos x$, or in general $\phi(x)$; the curve line OCO, whose ordinates represent the values corresponding to the abscissa x, and which is situated above the interval from $x = 0$ to $x = X$, coincides then in this interval with the curve whose ordinate is $\phi(x)$, and the coefficients $a_1, a_2, a_3 \ldots a_n$ of equation (e) determined by the preceding rule always satisfy the condition, that any value of x included between 0 and X, gives the same result when substituted in $\phi(x)$, and in the second member of equation (ϵ).

$F(x)$ represents the initial temperature of the spherical shell whose radius is x. We might suppose, for example, $F(x) = bx$, that is to say, that the initial heat increases proportionally to the distance, from the centre, where it is nothing, to the surface where it is bX. In this case $xF(x)$ or $f(x)$ is equal to bx^2; and applying to this function the rule which determines the coefficients, bx^2 would be developed in a series of terms, such as

$$a_1 \sin(\mu_1 x) + a_2 \sin(\mu_2 x) + a_3 \sin(\mu_3 x) + \ldots + a_n \sin(\mu_n x).$$

Now each term $\sin(\mu_i x)$, when developed according to powers of x, contains only powers of odd order, and the function bx^2 is a power of even order. It is very remarkable that this function bx^2, denoting a series of values given for the interval from 0 to X, can be developed in a series of terms, such as $a_i \sin(\mu_i x)$.

We have already proved the rigorous exactness of these results, which had not yet been presented in analysis, and we have shewn the true meaning of the propositions which express them. We have seen, for example, in Article 223, that the function $\cos x$ is developed in a series of sines of multiple arcs, so that in the equation which gives this development, the first member contains only even powers of the variable, and the second contains only odd powers. Reciprocally, the function $\sin x$, into which only odd powers enter, is resolved, Art. 225, into a series of cosines which contain only even powers.

In the actual problem relative to the sphere, the value of $xF(x)$ is developed by means of equation (ϵ). We must then, as we see in Art. 290, write in each term the exponential factor, which contains t, and we have to express the temperature v, which is a function of x and t, the equation

$$xv = \Sigma \sin(\mu_i x)\, e^{-K\mu_i^2 t} \frac{\int_0^X dx \sin(\mu_i \alpha)\, \alpha F(\alpha)}{\int_0^X d\beta \sin(\mu_i \beta) \sin(\mu_i \beta)} \quad \ldots\ldots\ldots (E).$$

The general solution which gives this equation (E) is wholly independent of the nature of the function $F(x)$ since this function represents here only an infinite multitude of arbitrary constants, which correspond to as many values of x included between 0 and X.

If we supposed the primitive heat to be contained in a part only of the solid sphere, for example, from $x=0$ to $x=\frac{1}{2}X$, and that the initial temperatures of the upper layers were nothing, it would be sufficient to take the integral

$$\int d\alpha \sin(\mu_i \alpha) f(\alpha),$$

between the limits $x=0$ and $x=\frac{1}{2}X$.

In general, the solution expressed by equation (E) suits all cases, and the form of the development does not vary according to the nature of the function.

Suppose now that having written $\sin x$ instead of $F(x)$ we have determined by integration the coefficients a_i, and that we have formed the equation

$$x \sin x = a_1 \sin \mu_1 x + a_2 \sin \mu_2 x + a_3 \sin \mu_3 x + \&c.$$

It is certain that on giving to x any value whatever included between 0 and X, the second member of this equation becomes equal to $x \sin x$; this is a necessary consequence of our process. But it nowise follows that on giving to x a value not included between 0 and X, the same equality would exist. We see the contrary very distinctly in the examples which we have cited, and, particular cases excepted, we may say that a function subject to a continuous law, which forms the first member of equations of this kind, does not coincide with the function expressed by the second member, except for values of x included between 0 and X.

Properly speaking, equation (ϵ) is an identity, which exists for all values which may be assigned to the variable x; each member of this equation representing a certain analytical function which coincides with a known function $f(x)$ if we give to the variable x values included between 0 and X. With respect to the existence of functions, which coincide for all values of the variable included between certain limits and differ for other values, it is proved by all that precedes, and considerations of this kind are a necessary element of the theory of partial differential equations.

Moreover, it is evident that equations (ϵ) and (E) apply not only to the solid sphere whose radius is X, but represent, one the initial state, the other the variable state of an infinitely extended solid, of which the spherical body forms part; and when in these equations we give to the variable x values greater than X, they refer to the parts of the infinite solid which envelops the sphere.

This remark applies also to all dynamical problems which are solved by means of partial differential equations.

427. To apply the solution given by equation (E) to the case in which a single spherical layer has been originally heated, all the other layers having nul initial temperature, it is sufficient to take the integral $\int d\alpha \sin(\mu_i\alpha)\, \alpha F(\alpha)$ between two very near limits, $\alpha = r$, and $\alpha = r + u$, r being the radius of the inner surface of the heated layer, and u the thickness of this layer.

We can also consider separately the resulting effect of the initial heating of another layer included between the limits $r + u$ and $r + 2u$; and if we add the variable temperature due to this second cause, to the temperature which we found when the first layer alone was heated, the sum of the two temperatures is that which would arise, if the two layers were heated at the same time. In order to take account of the two joint causes, it is sufficient to take the integral $\int dx \sin(\mu_i x)\, \alpha F(\alpha)$ between the limits $\alpha = r$ and $\alpha = r + 2u$. More generally, equation (E) being capable of being put under the form

$$v = \int_0^X dx \,.\, \alpha F(\alpha) \sin \mu_i \alpha \, \Sigma \frac{\sin \mu_i x \, e^{-K\mu_i^2 t}}{x \int_0^X d\beta \sin \mu_i \beta \sin \mu_i \beta},$$

we see that the whole effect of the heating of different layers is the sum of the partial effects, which would be determined separately, by supposing each of the layers to have been alone heated. The same consequence extends to all other problems of the theory of heat; it is derived from the very nature of equations, and the form of the integrals makes it evident. We see that the heat contained in each element of a solid body produces its distinct effect, as if that element had alone been heated, all the others having nul initial temperature. These separate states are in a manner superposed, and unite to form the general system of temperatures.

For this reason the form of the function which represents the initial state must be regarded as entirely arbitrary. The definite integral which enters into the expression of the variable temperature, having the same limits as the heated solid, shows expressly that we unite all the partial effects due to the initial heating of each element.

428. Here we shall terminate this section, which is devoted almost entirely to analysis. The integrals which we have obtained are not only general expressions which satisfy the differential equations; they represent in the most distinct manner the natural effect which is the object of the problem. This is the chief condition which we have always had in view, and without which the results of investigation would appear to us to be only useless transformations. When this condition is fulfilled, the integral is, properly speaking, *the equation of the phenomenon;* it expresses clearly the character and progress of it, in the same manner as the finite equation of a line or curved surface makes known all the properties of those forms. To exhibit the solutions, we do not consider one form only of the integral; we seek to obtain directly that which is suitable to the problem. Thus it is that the integral which expresses the

movement of heat in a sphere of given radius, is very different from that which expresses the movement in a cylindrical body, or even in a sphere whose radius is supposed infinite. Now each of these integrals has a definite form which cannot be replaced by another. It is necessary to make use of it, if we wish to ascertain the distribution of heat in the body in question. In general, we could not introduce any change in the form of our solutions, without making them lose their essential character, which is the representation of the phenomena.

The different integrals might be derived from each other, since they are co-extensive. But these transformations require long calculations, and almost always suppose that the form of the result is known in advance. We may consider in the first place, bodies whose dimensions are finite, and pass from this problem to that which relates to an unbounded solid. We can then substitute a definite integral for the sum denoted by the symbol Σ. Thus it is that equations (α) and (β), referred to at the beginning of this section, depend upon each other. The first becomes the second, when we suppose the radius R infinite. Reciprocally we may derive from the second equation (β) the solutions relating to bodies of limited dimensions.

In general, we have sought to obtain each result by the shortest way. The chief elements of the method we have followed are these:

1st. We consider at the same time the general condition given by the partial differential equation, and all the special conditions which determine the problem completely, and we proceed to form the analytical expression which satisfies all these conditions.

2nd. We first perceive that this expression contains an infinite number of terms, into which unknown constants enter, or that it is equal to an integral which includes one or more arbitrary functions. In the first instance, that is to say, when the general term is affected by the symbol Σ, we derive from the special conditions a definite transcendental equation, whose roots give the values of an infinite number of constants.

The second instance obtains when the general term becomes an infinitely small quantity; the sum of the series is then changed into a definite integral.

3rd. We can prove by the fundamental theorems of algebra, or even by the physical nature of the problem, that the transcendental equation has all its roots real, in number infinite.

4th. In elementary problems, the general term takes the form of a sine or cosine; the roots of the definite equation are either whole numbers, or real or irrational quantities, each of them included between two definite limits.

In more complex problems, the general term takes the form of a function given implicitly by means of a differential equation integrable or not. However it may be, the roots of the definite equation exist, they are real, infinite in number. This distinction of the parts of which the integral must be composed, is very important, since it shews clearly the form of the solution, and the necessary relation between the coefficients.

5th. It remains only to determine the constants which depend on the initial state; which is done by elimination of the unknowns from an infinite number of equations of the first degree. We multiply the equation which relates to the initial state by a differential factor, and integrate it between defined limits, which are most commonly those of the solid in which the movement is effected.

There are problems in which we have determined the coefficients by successive integrations, as may be seen in the memoir whose object is the temperature of dwellings. In this case we consider the exponential integrals, which belong to the initial state of the infinite solid: it is easy to obtain these integrals[1].

It follows from the integrations that all the terms of the second member disappear, except only that whose coefficient we wish to determine. In the value of this coefficient, the denominator becomes nul, and we always obtain a definite integral whose limits are those of the solid, and one of whose factors is the arbitrary function which belongs to the initial state. This form of the result is necessary, since the variable movement, which is the object of the problem, is compounded of all those which would have existed separately, if each point of the solid had alone been heated, and the temperature of every other point had been nothing.

[1] See section 11 of the sketch of this memoir, given by the author in the *Bulletin des Sciences par la Société Philomatique*, 1818, pp. 1—11. [A. F.]

When we examine carefully the process of integration which serves to determine the coefficients, we see that it contains a complete proof, and shews distinctly the nature of the results, so that it is in no way necessary to verify them by other investigations.

The most remarkable of the problems which we have hitherto propounded, and the most suitable for shewing the whole of our analysis, is that of the movement of heat in a cylindrical body. In other researches, the determination of the coefficients would require processes of investigation which we do not yet know. But it must be remarked, that, without determining the values of the coefficients, we can always acquire an exact knowledge of the problem, and of the natural course of the phenomenon which is its object; the chief consideration is that of *simple movements.*

6th. When the expression sought contains a definite integral, the unknown functions arranged under the symbol of integration are determined, either by the theorems which we have given for the expression of arbitrary functions in definite integrals, or by a more complex process, several examples of which will be found in the Second Part.

These theorems can be extended to any number of variables. They belong in some respects to an inverse method of definite integration; since they serve to determine under the symbols $\int$ and Σ unknown functions which must be such that the result of integration is a given function.

The same principles are applicable to different other problems of geometry, of general physics, or of analysis, whether the equations contain finite or infinitely small differences, or whether they contain both.

The solutions which are obtained by this method are complete, and consist of general integrals. No other integral can be more extensive. The objections which have been made to this subject are devoid of all foundation; it would be superfluous now to discuss them.

7th. We have said that each of these solutions gives *the equation proper to the phenomenon,* since it represents it distinctly throughout the whole extent of its course, and serves to determine with facility all its results numerically.

The functions which are obtained by these solutions are then composed of a multitude of terms, either finite or infinitely small: but the form of these expressions is in no degree arbitrary; it is determined by the physical character of the phenomenon. For this reason, when the value of the function is expressed by a series into which exponentials relative to the time enter, it is of necessity that this should be so, since the natural effect whose laws we seek, is really decomposed into distinct parts, corresponding to the different terms of the series. The parts express so many *simple movements* compatible with the special conditions; for each one of these movements, all the temperatures decrease, preserving their primitive ratios. In this composition we ought not to see a result of analysis due to the linear form of the differential equations, but an actual effect which becomes sensible in experiments. It appears also in dynamical problems in which we consider the causes which destroy motion; but it belongs necessarily to all problems of the theory of heat, and determines the nature of the method which we have followed for the solution of them.

8th. The mathematical theory of heat includes: first, the exact definition of all the elements of the analysis; next, the differential equations; lastly, the integrals appropriate to the fundamental problems. The equations can be arrived at in several ways; the same integrals can also be obtained, or other problems solved, by introducing certain changes in the course of the investigation. We consider that these researches do not constitute a method different from our own; but confirm and multiply its results.

9th. It has been objected, to the subject of our analysis, that the transcendental equations which determine the exponents having imaginary roots, it would be necessary to employ the terms which proceed from them, and which would indicate a periodic character in part of the phenomenon; but this objection has no foundation, since the equations in question have in fact all their roots real, and no part of the phenomenon can be periodic.

10th. It has been alleged that in order to solve with certainty problems of this kind, it is necessary to resort in all cases to a certain form of the integral which was denoted as general; and equation (γ) of Art. 398 was propounded under this designation; but this distinction has no foundation, and the use of a single integral would only have the effect, in most cases, of complicating the investigation unnecessarily. It is moreover evident that this integral (γ) is derivable from that which we gave in 1807 to determine the movement of heat in a ring of definite radius R; it is sufficient to give to R an infinite value.

11th. It has been supposed that the method which consists in expressing the integral by a succession of exponential terms, and in determining their coefficients by means of the initial state, does not solve the problem of a prism which loses heat unequally at its two ends; or that, at least, it would be very difficult to verify in this manner the solution derivable from the integral (γ) by long calculations. We shall perceive, by a new examination, that our method applies directly to this problem, and that a single integration even is sufficient[1].

12th. We have developed in series of sines of multiple arcs functions which appear to contain only even powers of the variable, $\cos x$ for example. We have expressed by convergent series or by definite integrals separate parts of different functions, or functions discontinuous between certain limits, for example that which measures the ordinate of a triangle. Our proofs leave no doubt of the exact truth of these equations.

13th. We find in the works of many geometers results and processes of calculation analogous to those which we have employed. These are particular cases of a general method, which had not yet been formed, and which it became necessary to establish in order to ascertain even in the most simple problems the mathematical laws of the distribution of heat. This theory required an analysis appropriate to it, one principal element of which is the analytical expression of *separate functions*, or of *parts of functions.*

By a *separate function*, or *part of a function*, we understand a function $f(x)$ which has values existing when the variable x is included between given limits, and whose value is always nothing, if the variable is not included between those limits. This function measures the ordinate of a line which includes a finite arc of

[1] See the Memoir referred to in note 1, p. 12. [A. F.]

arbitrary form, and coincides with the axis of abscissæ in all the rest of its course.

This motion is not opposed to the general principles of analysis; we might even find the first traces of it in the writings of Daniel Bernouilli, of Cauchy, of Lagrange and Euler. It had always been regarded as manifestly impossible to express in a series of sines of multiple arcs, or at least in a trigonometric convergent series, a function which has no existing values unless the values of the variable are included between certain limits, all the other values of the function being nul. But this point of analysis is fully cleared up, and it remains incontestable that separate functions, or parts of functions, are exactly expressed by trigonometric convergent series, or by definite integrals. We have insisted on this consequence from the origin of our researches up to the present time, since we are not concerned here with an abstract and isolated problem, but with a primary consideration intimately connected with the most useful and extensive considerations. Nothing has appeared to us more suitable than geometrical constructions to demonstrate the truth of these new results, and to render intelligible the forms which analysis employs for their expression.

14th. The principles which have served to establish for us the analytical theory of heat, apply directly to the investigation of the movement of waves in fluids, a part of which has been agitated. They aid also the investigation of the vibrations of elastic laminæ, of stretched flexible surfaces, of plane elastic surfaces of very great dimensions, and apply in general to problems which depend upon the theory of elasticity. The property of the solutions which we derive from these principles is to render the numerical applications easy, and to offer distinct and intelligible results, which really determine the object of the problem, without making that knowledge depend upon integrations or eliminations which cannot be effected. We regard as superfluous every transformation of the results of analysis which does not satisfy this primary condition.

429. 1st. We shall now make some remarks on the differential equations of the movement of heat.

If two molecules of the same body are extremely near, and are at unequal temperatures, that which is the most heated communicates directly to the other during one instant a certain quantity of heat; which quantity is proportional to the extremely small difference of the temperatures: that is to say, if that difference became double, triple, quadruple, and all other conditions remained the same, the heat communicated would be double, triple, quadruple.

This proposition expresses a general and constant fact, which is sufficient to serve as the foundation of the mathematical theory. The mode of transmission is then known with certainty, independently of every hypothesis on the nature of the cause, and cannot be looked at from two different points of view. It is evident that the direct transfer is effected in all directions, and that it has no existence in fluids or liquids which are not diathermanous, except between extremely near molecules.

The general equations of the movement of heat, in the interior of solids of any dimensions, and at the surface of these bodies, are necessary consequences of the foregoing proposition. They are rigorously derived from it, as we have proved in our first Memoirs in 1807, and we easily obtain these equations by means of lemmas, whose proof is not less exact than that of the elementary propositions of mechanics.

These equations are again derived from the same proposition, by determining by means of integrations the whole quantity of heat which one molecule receives from those which surround it. This investigation is subject to no difficulty. The lemmas in question take the place of the integrations, since they give directly the expression of the flow, that is to say of the quantity of heat, which crosses any section. Both calculations ought evidently to lead to the same result; and since there is no difference in the principle, there cannot be any difference in the consequences.

2nd. We gave in 1811 the general equation relative to the surface. It has not been deduced from particular cases, as has been supposed without any foundation, and it could not be; the proposition which it expresses is not of a nature to be discovered by way of induction; we cannot ascertain it for certain bodies and ignore it for others; it is necessary for all, in order that *the state of the surface may not suffer in a definite time an infinite change.* In our Memoir we have omitted the details of the proof, since they consist solely in the application of known propositions. It was sufficient in this work to give the principle and the result, as we have done in Article 15 of the Memoir cited. From the same condition also the general equation in question is derived by determining the whole quantity of heat which each molecule situated at the surface receives and communicates. These very complex calculations make no change in the nature of the proof.

In the investigation of the differential equation of the movement of heat, the mass may be supposed to be not homogeneous, and it is very easy to derive the equation from the analytical expression of the flow; it is sufficient to leave the coefficient which measures the conducibility under the sign of differentiation.

3rd. Newton was the first to consider the law of cooling of bodies in air; that which he has adopted for the case in which the air is carried away with constant velocity accords more closely with observation as the difference of temperatures becomes less; it would exactly hold if that difference were infinitely small.

Amontons has made a remarkable experiment on the establishment of heat in a prism whose extremity is submitted to a definite temperature. The logarithmic law of the decrease of the temperatures in the prism was given for the first time by Lambert, of the Academy of Berlin. Biot and Rumford have confirmed this law by experiment[1].

[1] Newton, at the end of his *Scala graduum caloris et frigoris*, *Philosophical Transactions*, April 1701, or *Opuscula* ed. Castillioneus, Vol. II. implies that when a plate of iron cools in a current of air flowing uniformly at constant temperature, equal quantities of air come in contact with the metal in equal times and carry off quantities of heat proportional to the excess of the temperature of the iron over that of the air; whence it may be inferred that the excess temperatures of the iron form a geometrical progression at times which are in arithmetic progression, as he has stated. By placing various substances on the heated iron, he obtained their melting points as the metal cooled.

Amontons, *Mémoires de l'Académie* [1703], Paris, 1705, pp. 205—6, in his *Rémarques sur la Table de degrés de Chaleur extraite des Transactions Philosophiques* 1701, states that he obtained the melting points of the substances experimented on by Newton by placing them at appropriate points along an iron bar, heated to whiteness at one end; but he has made an erroneous assumption as to the law of decrease of temperature along the bar.

Lambert, *Pyrometrie*, Berlin, 1779, pp. 185—6, combining Newton's calculated temperatures with Amontons' measured distances, detected the exponential law

To discover the differential equations of the variable movement of heat, even in the most elementary case, as that of a cylindrical prism of very small radius, it was necessary to know the mathematical expression of the quantity of heat which traverses an extremely short part of the prism. This quantity is not simply proportional to the difference of the temperatures of the two sections which bound the layer. It is proved in the most rigorous manner that it is also in the inverse ratio of the thickness of the layer, that is to say, that *if two layers of the same prism were unequally thick, and if in the first the difference of the temperatures of the two bases was the same as in the second, the quantities of heat traversing the layers during the same instant would be in the inverse ratio of the thicknesses.* The preceding lemma applies not only to layers whose thickness is infinitely small; it applies to prisms of any length. This notion of the flow is fundamental; in so far as we have not acquired it, we cannot form an exact idea of the phenomenon and of the equation which expresses it.

It is evident that the instantaneous increase of the tempera-

of temperatures in a long bar heated at one end. Lambert's work contains a most complete account of the progress of thermal measurement up to that time.

Biot, *Journal des Mines*, Paris, 1804, XVII. pp. 203—224. Rumford, *Mémoires de l'Institut*, Sciences Math. et Phys. Tome VI. Paris, 1805, pp. 106—122.

Ericsson, *Nature*, Vol. VI. pp. 106—8, describes some experiments on cooling in vacuo which for a limited range of excess temperature, 10° to 100° Fah. shew a very close approach to Newton's law of cooling in a current of air. These experiments are insufficient to discredit the law of cooling in vacuo derived by M. M. Dulong and Petit (*Journal Polytechnique*, Tome XI. or *Ann. de Ch. et de Ph.* 1817, Tome VII.) from their carefully devised and more extensive range of experiments. But other experiments made by Ericsson with an ingeniously contrived calorimeter (*Nature*, Vol. V. pp. 505—7) on the emissive power of molten iron, seem to shew that the law of Dulong and Petit, for cooling in vacuo, is very far from being applicable to masses at exceedingly high temperatures giving off heat in free air, though their law for such conditions is reducible to the former law.

Fourier has published some remarks on Newton's law of cooling in his *Questions sur la théorie physique de la Chaleur rayonnante, Ann. de Chimie et de Physique*, 1817, Tome VI. p. 298. He distinguishes between the surface conduction and radiation to free air.

Newton's original statement in the *Scala graduum* is "Calor quem ferrum calefactum corporibus frigidis sibi contiguis dato tempore communicat, hoc est Calor, quem ferrum dato tempore amittit, est ut Calor totus ferri." This supposes the iron to be perfectly conducible, and the surrounding masses to be at zero temperature. It can only be interpreted by his subsequent explanation, as above.

[A. F.]

ture of a point is proportional to the excess of the quantity of heat which that point receives over the quantity which it has lost, and that a partial differential equation must express this result: but the problem does not consist in enunciating this proposition which is the mere fact; it consists in actually forming the differential equation, which requires that we should consider the fact in its elements. If instead of employing the exact expression of the flow of heat, we omit the denominator of this expression, we thereby introduce a difficulty which is nowise inherent in the problem; there is no mathematical theory which would not offer similar difficulties, if we began by altering the principle of the proofs. Not only are we thus unable to form a differential equation; but there is nothing more opposite to an equation than a proposition of this kind, in which we should be expressing the equality of quantities which could not be compared. To avoid this error, it is sufficient to give some attention to the demonstration and the consequences of the foregoing lemma (Art. 65, 66, 67, and Art. 75).

4th. With respect to the ideas from which we have deduced for the first time the differential equations, they are those which physicists have always admitted. We do not know that anyone has been able to imagine the movement of heat as being produced in the interior of bodies by the simple contact of the surfaces which separate the different parts. For ourselves such a proposition would appear to be void of all intelligible meaning. A surface of contact cannot be the subject of any physical quality; it is neither heated, nor coloured, nor heavy. It is evident that when one part of a body gives its heat to another there are an infinity of material points of the first which act on an infinity of points of the second. It need only be added that in the interior of opaque material, points whose distance is not very small cannot communicate their heat directly; that which they send out is intercepted by the intermediate molecules. The layers in contact are the only ones which communicate their heat directly, when the thickness of the layers equals or exceeds the distance which the heat sent from a point passes over before being entirely absorbed. There is no direct action except between material points extremely near, and it is for this reason that the expression for the flow has the form which we assign to it. The flow then results from an infinite multitude of actions whose effects are added; but it is not from this cause that its value during unit of time is a finite and measurable magnitude, even although it be determined only by an extremely small difference between the temperatures.

When a heated body loses its heat in an elastic medium, or in a space free from air bounded by a solid envelope, the value of the outward flow is assuredly an integral; it again is due to the action of an infinity of material points, very near to the surface, and we have proved formerly that this concourse determines the law of the external radiation[1]. But the quantity of heat emitted during the unit of time would be infinitely small, if the difference of the temperatures had not a finite value.

In the interior of masses the conductive power is incomparably greater than that which is exerted at the surface. This property, whatever be the cause of it, is most distinctly perceived by us, since, when the prism has arrived at its constant state, the quantity of heat which crosses a section during the unit of time exactly balances that which is lost through the whole part of the heated surface, situated beyond that section, whose temperatures exceed that of the medium by a finite magnitude. When we take no account of this primary fact, and omit the divisor in the expression for the flow, it is quite impossible to form the differential equation, even for the simplest case; a fortiori, we should be stopped in the investigation of the general equations.

5th. Further, it is necessary to know what is the influence of the dimensions of the section of the prism on the values of the acquired temperatures. Even although the problem is only that of the linear movement, and all points of a section are regarded as having the same temperature, it does not follow that we can disregard the dimensions of the section, and extend to other prisms the consequences which belong to one prism only. The exact equation cannot be formed without expressing the relation between the extent of the section and the effect produced at the extremity of the prism.

We shall not develope further the examination of the principles which have led us to the knowledge of the differential equations;

[1] *Mémoires de l'Académie des Sciences*, Tome v. pp. 204—8. Communicated in 1811. [A. F.]

we need only add that to obtain a profound conviction of the usefulness of these principles it is necessary to consider also various difficult problems; for example, that which we are about to indicate, and whose solution is wanting to our theory, as we have long since remarked. This problem consists in forming the differential equations, which express the distribution of heat in fluids in motion, when all the molecules are displaced by any forces, combined with the changes of temperature. The equations which we gave in the course of the year 1820 belong to general hydrodynamics; they complete this branch of analytical mechanics[1].

430. Different bodies enjoy very unequally the property which physicists have called *conductibility* or *conducibility*, that is to say, the faculty of admitting heat, or of propagating it in the interior of their masses. We have not changed these names, though they

[1] See *Mémoires de l'Académie des Sciences*, Tome XII. Paris, 1833, pp. 515—530.

In addition to the three ordinary equations of motion of an incompressible fluid, and the equation of continuity referred to rectangular axes in direction of which the velocities of a molecule passing the point x, y, z at time t are u, v, w, its temperature being θ, Fourier has obtained the equation

$$C\frac{d\theta}{dt} = K\left(\frac{d^2\theta}{dx^2}+\frac{d^2\theta}{dy^2}+\frac{d^2\theta}{dz^2}\right) - C\left(\frac{d}{dx}(u\theta)+\frac{d}{dy}(v\theta)+\frac{d}{dz}(w\theta)\right),$$

in which K is the conductivity and C the specific heat per unit volume, as follows.

Into the parallelopiped whose opposite corners are (x, y, z), $(x+\Delta x, y+\Delta y, z+\Delta z)$, the quantity of heat which would flow by conduction across the lower face $\Delta x\Delta y$, if the fluid were at rest, would be $-K\frac{d\theta}{dz}\Delta x\,\Delta y\,\Delta t$ in time Δt, and the gain by convection $+Cw\,\Delta x\,\Delta y\,\Delta t$; there is a corresponding loss at the upper face $\Delta x\,\Delta y$; hence the whole gain is, negatively, the variation of $(-K\frac{d\theta}{dz}+Cw\theta)\,\Delta x\,\Delta y\,\Delta t$ with respect to z, that is to say, the gain is equal to $\left(K\frac{d^2\theta}{dz^2}-C\frac{d}{dz}(w\theta)\right)\Delta x\,\Delta y\,\Delta z\,\Delta t$. Two similar expressions denote the gains in direction of y and z; the sum of the three is equal to $C\frac{d\theta}{dt}\Delta t\,\Delta x\,\Delta y\,\Delta z$, which is the gain in the volume $\Delta x\,\Delta y\,\Delta z$ in time Δt: whence the above equation.

The coefficients K and C vary with the temperature and pressure but are usually treated as constant. The density, even for fluids denominated incompressible, is subject to a small temperature variation.

It may be noticed that when the velocities u, v, w are nul, the equation reduces to the equation for flow of heat in a solid.

It may also be remarked that when K is so small as to be negligible, the equation has the same form as the equation of continuity. [A. F.]

do not appear to us to be exact. Each of them, the first especially, would rather express, according to all analogy, the faculty of being conducted than that of conducting.

Heat penetrates the surface of different substances with more or less facility, whether it be to enter or to escape, and bodies are unequally permeable to this element, that is to say, it is propagated in them with more or less facility, in passing from one interior molecule to another. We think these two distinct properties might be denoted by the names *penetrability* and *permeability*[1].

Above all it must not be lost sight of that the penetrability of a surface depends upon two different qualities: one relative to the external medium, which expresses the facility of communication by contact; the other consists in the property of emitting or admitting radiant heat. With regard to the specific permeability, it is proper to each substance and independent of the state of the surface. For the rest, precise definitions are the true foundation of theory, but names have not, in the matter of our subject, the same degree of importance.

431. The last remark cannot be applied to notations, which contribute very much to the progress of the science of the Calculus. These ought only to be proposed with reserve, and not admitted but after long examination. That which we have employed reduces itself to indicating the limits of the integral above and below the sign of integration $\int$; writing immediately after this sign the differential of the quantity which varies between these limits.

We have availed ourselves also of the sign Σ to express the sum of an indefinite number of terms derived from one general term in which the index i is made to vary. We attach this index if necessary to the sign, and write the first value of i below, and the last above. Habitual use of this notation convinces us of

[1] The coefficients of penetrability and permeability, or of exterior and interior conduction (h, K), were determined in the first instance by Fourier, for the case of cast iron, by experiments on the permanent temperatures of a ring and on the varying temperatures of a sphere. The value of $\frac{h}{K}$ by the method of Art. 110, and the value of h by that of Art. 297. *Mem. de l'Acad. d. Sc.* Tome v. pp. 165, 220, 228. [A. F.]

the usefulness of it, especially when the analysis consists of definite integrals, and the limits of the integrals are themselves the object of investigation.

432. The chief results of our theory are the differential equations of the movement of heat in solid or liquid bodies, and the general equation which relates to the surface. The truth of these equations is not founded on any physical explanation of the effects of heat. In whatever manner we please to imagine the nature of this element, whether we regard it as a distinct material thing which passes from one part of space to another, or whether we make heat consist simply in the transfer of motion, we shall always arrive at the same equations, since the hypothesis which we form must represent the general and simple facts from which the mathematical laws are derived.

The quantity of heat transmitted by two molecules whose temperatures are unequal, depends on the difference of these temperatures. If the difference is infinitely small it is certain that the heat communicated is proportional to that difference; all experiment concurs in rigorously proving this proposition. Now in order to establish the differential equations in question, we consider only the reciprocal action of molecules infinitely near. There is therefore no uncertainty about the form of the equations which relate to the interior of the mass.

The equation relative to the surface expresses, as we have said, that the flow of the heat, in the direction of the normal at the boundary of the solid, must have the same value, whether we calculate the mutual action of the molecules of the solid, or whether we consider the action which the medium exerts upon the envelope. The analytical expression of the former value is very simple and is exactly known; as to the latter value, it is sensibly proportional to the temperature of the surface, when the excess of this temperature over that of the medium is a sufficiently small quantity. In other cases the second value must be regarded as given by a series of observations; it depends on the surface, on the pressure and on the nature of the medium; this observed value ought to form the second member of the equation relative to the surface.

In several important problems, the equation last named is replaced by a given condition, which expresses the state of the surface, whether constant, variable or periodic.

433. The differential equations of the movement of heat are mathematical consequences analogous to the general equations of equilibrium and of motion, and are derived like them from the most constant natural facts.

The coefficients c, h, k, which enter into these equations, must be considered, in general, as variable magnitudes, which depend on the temperature or on the state of the body. But in the application to the natural problems which interest us most, we may assign to these coefficients values sensibly constant.

The first coefficient c varies very slowly, according as the temperature rises. These changes are almost insensible in an interval of about thirty degrees. A series of valuable observations, due to Professors Dulong and Petit, indicates that the value of the specific capacity increases very slowly with the temperature.

The coefficient h which measures the penetrability of the surface is most variable, and relates to a very composite state. It expresses the quantity of heat communicated to the medium, whether by radiation, or by contact. The rigorous calculation of this quantity would depend therefore on the problem of the movement of heat in liquid or aeriform media. But when the excess of temperature is a sufficiently small quantity, the observations prove that the value of the coefficient may be regarded as constant. In other cases, it is easy to derive from known experiments a correction which makes the result sufficiently exact.

It cannot be doubted that the coefficient k, the measure of the permeability, is subject to sensible variations; but on this important subject no series of experiments has yet been made suitable for informing us how the facility of conduction of heat changes with the temperature[1] and with the pressure. We see, from the observations, that this quality may be regarded as constant throughout a very great part of the thermometric scale. But the same observations would lead us to believe that the value of the coefficient in question, is very much more changed by increments of temperature than the value of the specific capacity.

Lastly, the dilatability of solids, or their tendency to increase

[1] Reference is given to Forbes' experiments in the note, p. 84. [A. F.]

in volume, is not the same at all temperatures: but in the problems which we have discussed, these changes cannot sensibly alter the precision of the results. In general, in the study of the grand natural phenomena which depend on the distribution of heat, we rely on regarding the values of the coefficients as constant. It is necessary, first, to consider the consequences of the theory from this point of view. Careful comparison of the results with those of very exact experiments will then shew what corrections must be employed, and to the theoretical researches will be given a further extension, according as the observations become more numerous and more exact. We shall then ascertain what are the causes which modify the movement of heat in the interior of bodies, and the theory will acquire a perfection which it would be impossible to give to it at present.

Luminous heat, or that which accompanies the rays of light emitted by incandescent bodies, penetrates transparent solids and liquids, and is gradually absorbed within them after traversing an interval of sensible magnitude. It could not therefore be supposed in the examination of these problems, that the direct impressions of heat are conveyed only to an extremely small distance. When this distance has a finite value, the differential equations take a different form; but this part of the theory would offer no useful applications unless it were based upon experimental knowledge which we have not yet acquired.

The experiments indicate that, at moderate temperatures, a very feeble portion of the obscure heat enjoys the same property as the luminous heat; it is very likely that the distance, to which is conveyed the impression of heat which penetrates solids, is not wholly insensible, and that it is only very small: but this occasions no appreciable difference in the results of theory; or at least the difference has hitherto escaped all observation.

CAMBRIDGE: PRINTED BY C. J. CLAY, M.A. AT THE UNIVERSITY PRESS.

www.ingramcontent.com/pod-product-compliance
Lightning Source LLC
Chambersburg PA
CBHW020858210326
41598CB00018B/1709
9783743308831